making math success happen

Edited, with commentary, by

Ivan W. Baugh and Anne Raymond

International Society for Technology in Education
Eugene, Oregon • Washington, DC

Making Math Success Happen

The Best of *Learning & Leading with Technology* on Mathematics

Edited by Ivan W. Baugh and Anne Raymond

Director of Publishing
Jean Marie Hall

Acquisitions Editors
Mathew Manweller
Scott Harter

Production Editor
Tracy Cozzens

Copy Editor
Nancy Olson

Cover Design
Kim McGovern

Book Design and Production
Tracy Cozzens

International Society for Technology in Education (ISTE)
Washington, DC Office:
1710 Rhode Island Ave. NW, Suite 900, Washington, DC 20036-3132

Eugene, Oregon Office:
175 West Broadway, Suite 300, Eugene, OR 97401-3003
Order Desk: 1.800.336.5191
Order Fax: 1.541.302.3778
Customer Service: orders@iste.org
Book Publishing: books@iste.org
Rights and Permissions: permissions@iste.org
Web: www.iste.org

First Edition
ISBN 978-1-56484-180-3

ABOUT ISTE

The International Society for Technology in Education (ISTE) is a nonprofit professional organization with a worldwide membership of leaders in educational technology. We are dedicated to promoting appropriate uses of technology to support and improve learning, teaching, and administration in PK–12 education and teacher education. As part of that mission, ISTE provides high-quality and timely information, services, and materials, such as this book.

ISTE Book Publishing works with experienced educators to develop and produce practical resources for classroom teachers, teacher educators, and technology leaders. Every manuscript and product we select for publication is peer reviewed and professionally edited. We look for content that emphasizes the effective use of technology where it can make a difference—increasing the productivity of teachers and administrators; helping students who have unique learning styles, abilities, or backgrounds; collecting and using data for decision making at the school and district levels; and and creating dynamic, project-based learning environments that engage 21st-century learners. We value your feedback on this book and other ISTE products. E-mail us at **books@iste.org**.

ISTE is home of the National Educational Technology Standards (NETS) Project, the National Educational Computing Conference (NECC), and the National Center for Preparing Tomorrow's Teachers to Use Technology (NCPT³). To find out more about these and other ISTE initiatives and to view our complete book list or request a print catalog, visit our Web site at **www.iste.org**. You'll find information about:

- ISTE, our mission and our members
- Membership opportunities and services
- Online communities and special interest groups (SIGs)
- Professional development services
- Research and evaluation services
- Educator resources
- ISTE's National Educational Technology Standards (NETS) for Students, Teachers, and Administrators
- *Learning & Leading with Technology* magazine
- *Journal of Research on Technology in Education*

ABOUT THE AUTHORS

Ivan W. Baugh retired July 1, 2002, after 48 years as a K–12 and collegiate-post graduate educator in Florida, North Carolina, Mississippi, Texas, and Kentucky. He began his career teaching music and concluded it training preservice and inservice educators in the use of technology as an instructional tool across the curriculum.

He holds a B.A. from Mississippi College, an M.S.M. from Southern Baptist Theological Seminary, Rank 1 endorsement (30 hours beyond the master's degree) from the University of Louisville, and an Ed.S. in administration of technology in instruction from Spalding University. He has authored for publication in a variety of professional journals 20 articles on teaching with technology across the curriculum.

Ivan worked as an educational technology consultant in Jamaica, Barbados, and Surinam. He presented at state, national, and international conferences on teaching with technology across the curriculum. He served as an invited participant for the Vision: T.E.S.T. (Technologically Enriched Schools of Tomorrow) Forum, co-sponsored by the U.S. Department of Education and the International Society for Technology in Education (ISTE). He was also invited to participate in the U.S. Department of Energy National Information Infrastructure Education Forum. He authored and directed a U.S. Department of Education grant titled Developing a Technology-Centered Environmental Science Curriculum while working in the Computer Education Support Unit of the Jefferson County Public Schools in Louisville, Kentucky.

Anne Raymond is an associate professor of mathematics at Bellarmine University in Louisville, Kentucky. Anne earned her B.A. in mathematics at Bellarmine, her M.S. in mathematics at the University of North Carolina at Chapel Hill, and her Ph.D. in mathematics education at Indiana University.

She co-authored, with Joanna Masingila and Frank Lester, the Prentice-Hall textbook *Mathematics for Elementary Teachers via Problem Solving* . She has also published a range of articles in notable mathematics education and general education journals including *The Journal for Research on Mathematics Education, Educational Studies in Mathematics, FOCUS on Learning Problems in Mathematics Education, The Journal of Teacher Education, The Elementary School Journal,* and *Contemporary Education.* Anne also served as co-editor for a monograph on collaborative action research published by Indiana State University.

Currently, Anne is researching connections between multiple intelligences and mathematics, with an emphasis on strengthening students' ability in mathematics. This research also investigates the role of technology in concert with multiple intelligences to improve mathematics education. Anne lives in Louisville with her husband, Frank, and her three children, Ben, Katie, and Chris.

CONTENTS

CHAPTER 2 • Connections

CHAPTER 3 • Problem Solving

Interlude

Content Standards

PREFACE

Technology in Mathematics Education

When we discuss technology with teachers of mathematics, we often get the response, "We *use* technology—the graphing calculator." The graphing calculator is a worthy piece of equipment, but it does not empower the learner to conduct many investigations beyond graphing data. When we suggest looking at a spreadsheet, educators respond, "Well, I haven't looked at one for a long time."

Making Math Success Happen closely examines the use of technology in the teaching of mathematics—*beyond* using a graphing calculator. We have organized this book around the 10 standards of the National Council of Teachers of Mathematics (NCTM). These standards fall into two categories: process and content. Each of the standards has been integrated with the National Educational Technology Standards for Students, also known as NETS•S, that have been published by the International Society for Technology in Education (ISTE).

Process Standards

Communication
Do mathematicians communicate?

Connections
Does mathematics relate to other curriculum content?

Problem Solving
Why is everything in mathematics a problem?

Reasoning and Proof
Are the results really evident?

Representation
Are pictures worth a thousand words?

Content Standards

Geometry
Have you considered this angle?

Measurement
What measurement unit are you using?

Algebra
Can technology open the gateway?

Number and Operations
And the answer is?

Data Analysis and Probability
What are the odds?

Each of the 10 chapters of the book begins with a "conversation" between the two of us, continues with commentary we have written titled Theory Into Practice and Insight, and concludes with field-tested articles that have been written by classroom teachers and are relevant to the particular standard featured in each of the chapters.

Our goal is to share ideas to help you expand your view of technology in the teaching of mathematics. To that end, the articles in the book are exemplary articles from past issues of *Learning & Leading with Technology* and its forerunner, *The Computing Teacher*. In the articles, classroom teachers relate how they use technology in their instructional programs in mathematics. We have heard it said, "If you always do what you've always done, you'll always get what you've always got." A further goal for this book, then, is to stimulate your thinking and help you find new ways to do familiar tasks.

Making Math Success Happen has a companion Web site for students to use while working with the concepts presented in the book. The Web site can be found at www.iste.org/publications/books/besmth/. It offers hands-on tools, many of them Excel and AppleWorks files. Files that accompany concepts will be noted in the text.

Ivan W. Baugh and Anne Raymond

INTRODUCTION, Part 1

The NCTM Standards and the ISTE NETS

By *Anne Raymond*

In the 1980s, a call rang forth for change in pedagogical practices in the teaching of mathematics, science, and technology. New shifts in thinking regarding mathematics, science, and technology education began with the publication of key documents by the National Commission on Excellence in Education (1983); the National Science Board Commission on Precollege Education in Mathematics, Science, and Technology (1983); and the National Research Council (1989). These reports make clear the imperative both to update the mathematics and science curricula to be more problem solving in nature and to make use of the growing technologies available to facilitate problem solving in mathematics and science (Raymond, Raymond, & McCrickard, 2002). Conversely, the increased use of technology within society necessitates a greater development of mathematical ideals and students' ability to think critically and solve problems.

In a summary of research on technology and mathematics education, Kaput (1992) discusses pedagogical benefits derived from the use of calculators and computers in the mathematics classroom. These benefits include *(a)* their power to support and enhance rich problem-solving environments, *(b)* the decrease in the amount of time required for skill development, thus allowing for more concentration on conceptual understanding, *(c)* the graphical advantage of computing utilities, and *(d)* the potential for deeper student understanding of algebraic ideas.

The National Council of Teachers of Mathematics (NCTM, 1989) responded to the calls for change in the 1980s by developing *Curriculum and Evaluation Standards for School Mathematics*. These standards were designed according to developmental levels for Grades K–4, 5–8, and 9–12. Ten years later, after much discussion and feedback from the larger mathematics education community, the NCTM (2000) developed the next stage of national standards, *Principles and Standards for School Mathematics*. This revised set of standards is divided into process standards and content standards that are meant to permeate the entire K–12 curriculum.

When developing the standards, the NCTM determined six principles, including the technology principle, upon which the reshaping of mathematics education would be built. The technology principle states that "technology is essential in teaching and learning mathematics; it influences the mathematics that it taught and enhances students' learning" (p. 24). As essential tools for teaching and learning, calculators and computers provide visual images of mathematical ideas, facilitate organizing and analyzing data, and compute efficiently and accurately. Further, when such tools are available, students can focus more on mathematical process, such as decision making, reasoning, and problem solving (p. 24).

In addition to curriculum standards, the NCTM (1991, 1995) developed *Professional Standards for Teaching Mathematics* and *Assessment Standards for School Mathematics* to complement the curriculum standards. Together, these standards offer broad guidelines for creating, implementing, and assessing mathematics for the 21st century.

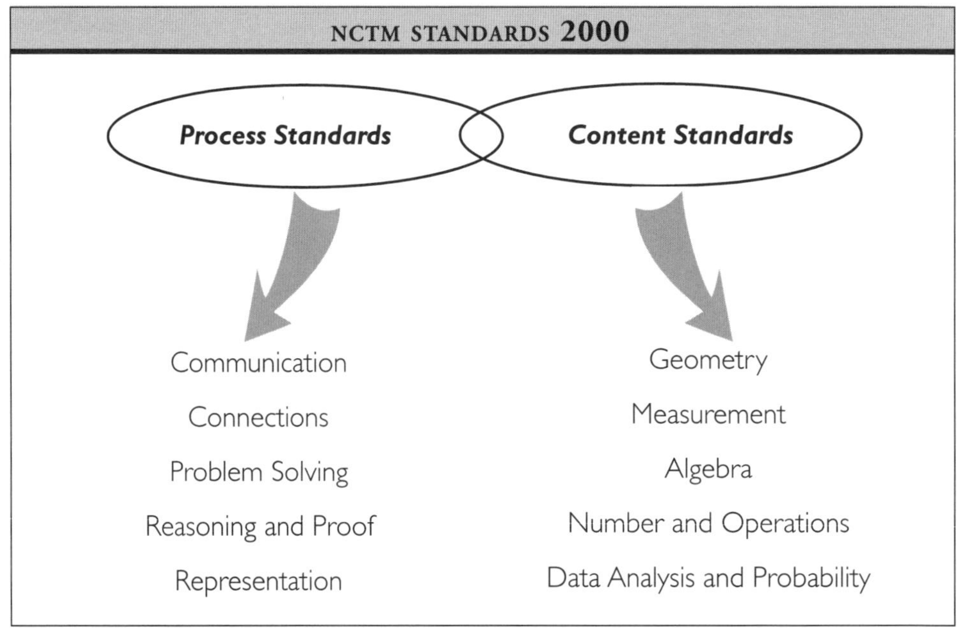

Figure 1.

Professional Standards for Teaching Mathematics (1991) describes key elements that ought to be considered when planning mathematics lessons in the spirit of the ideals reflected in the NCTM's (2000) mathematics content and process standards. Issues such as what types of discourse will take place during mathematics lessons or what type of environment will be established as we engage in our mathematics lessons are vital to take into account. It is particularly important to ponder these issues when planning lessons that integrate technology into the mathematics pedagogy because appropriate technology should be a tool that encourages discourse and establishes an engaging learning environment.

Professional Standards for Teaching Mathematics

- Worthwhile Mathematical Tasks
- Teacher's Role in Discourse
- Student's Role in Discourse
- Tools for Enhancing Discourse
- Learning Environment
- Analysis of Teaching and Learning

Part of planning for mathematics instruction is developing a means of assessing whether, and to what degree, lesson objectives have been met. The NCTM's *Assessment Standards for School Mathematics* suggests that when determining the assessment component of any lesson plan, the teacher ought to consider such issues as whether the assessment truly reflects the important mathematics concepts being taught and whether the assessment plan is equitable. That is, does it provide all students the opportunity to demonstrate what they know? Assessment plans that incorporate technological tools can greatly impact these issues, as long as the assessment process is consistent with the manner in which students learn the mathematics material.

Assessment Standards for School Mathematics

- The Mathematics Standard
- The Learning Standard
- The Equity Standard
- The Openness Standard
- The Inference Standard
- The Coherence Standard

The International Society for Technology in Education (ISTE) writes on its Web site: "In a unique partnership with teachers and teacher educators, curriculum and education associations, government, businesses, and private foundations, ISTE has responded to calls for educational technology standards, guidelines, and tools with its National Educational Technology Standards (NETS) Project." Out of this collaboration, ISTE (1998) published the *National Educational Technology Standards for Students*. These standards are centered around six categories:

National Educational Technology Standards for Students (NETS•S)

1. Basic operations and concepts
2. Social, ethical, and human issues
3. Technology productivity tools
4. Technology communications tools
5. Technology research tools
6. Technology problem-solving and decision-making tools

ISTE (2000) also developed technology standards for teachers which are set forth in its publication *National Educational Technology Standards for Teachers*. The intent is that all classroom teachers should be prepared to meet such standards and performance indicators through their teaching via technology.

National Educational Technology Standards for Teachers (NETS•T)

1. Technology operations and concepts
2. Planning and designing learning environments and experiences
3. Teaching, learning, and the curriculum
4. Assessment and evaluation
5. Productivity and professional practice
6. Social, ethical, legal, and human issues

With such mathematics and technology standards leading the way for education today, we decided to arrange this book around such standards. Consequently, we frame each of the 10 chapters around one of the NCTM standards. Within each mathematics standard, we discuss the technology standards that most closely connect to the process or content area.

To learn more details about the NCTM and ISTE and their standards, visit their Web sites:

www.nctm.org

cnets.iste.org

References

International Society for Technology in Education. (1998). *National educational technology standards for students.* Eugene, OR: Author.

International Society for Technology in Education. (2000). *National educational technology standards for teachers.* Eugene, OR: Author.

Kaput, J. J. (1992). Technology and mathematics education. In D. A. Grouws (Ed.), *Handbook of research on mathematics teaching and learning* (pp. 515–556). New York: Macmillan.

National Commission on Excellence in Education. (1983). *A nation at risk: The imperative for educational reform.* Washington, DC: U.S. Government Printing Office.

National Council of Teachers of Mathematics. (1989). *Curriculum and evaluation standards for school mathematics.* Reston, VA: Author.

National Council of Teachers of Mathematics. (1991). *Professional standards for teaching mathematics.* Reston, VA: Author.

National Council of Teachers of Mathematics. (1995). *Assessment standards for school mathematics.* Reston, VA: Author.

National Council of Teachers of Mathematics. (2000). *Principles and standards for school mathematics.* Reston, VA: Author.

National Research Council. (1989). *Everybody counts: A report to the nation on the future of mathematics education.* Washington, D.C.: National Academy Press.

National Science Board Commission on Precollege Education in Mathematics, Science, and Technology. (1983). *Educating Americans for the 21st century.* Washington, D.C.: National Science Foundation.

Raymond, F., Raymond, A. M., & McCrickard, M. (2002). *The role of technology in understanding mathematical concepts in intermediate economic theory: A rationale and a demonstration using MAPLE.* A paper presented at the annual meeting of the Kentucky Economics Association, Lexington, Kentucky.

Mathematics, Technology, and the Affective Domain

By Anne Raymond and Ivan W. Baugh

The affective domain and how it impacts the learning process has been studied. In the area of mathematics, the affective domain requires much attention because of long-standing attitudes and beliefs about mathematics that are not necessarily conducive to the learning process.

The affective domain in mathematics education includes beliefs about mathematics; about self; about mathematics teaching; and about the social context of learning, attitudes (such as likes, dislikes, and preferences), and emotions (such as joy or frustration in solving problems). Other related concepts from the affective domain in mathematics include confidence, self-concept, self-efficacy, and mathematics anxiety (McLeod, 1992). Research efforts in the area of mathematics anxiety have attempted to determine how beliefs, attitudes, and emotions impact mathematics anxiety.

In studies of beliefs about mathematics, it has been found that teachers' beliefs about mathematics can have an impact on both their teaching and their students' beliefs about mathematics. Teachers' beliefs range from "traditional beliefs" to "nontraditional

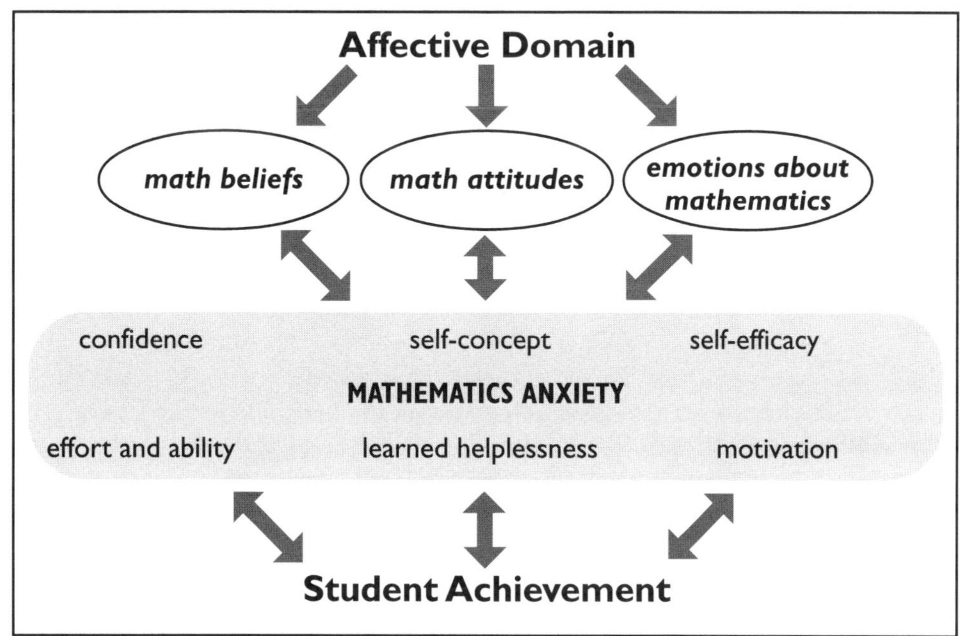

Figure 1. A Concept Map of the Affective Domain in Mathematics Education

beliefs" (Raymond, 1997), and their related teaching styles range similarly. Efforts to influence preservice teachers' beliefs have been attempted through the modeling of innovative teaching practices in the hopes of reducing their mathematics anxiety and improving their interest in mathematics (Emanaker, 1996; Raymond & Santos, 1995).

Students' beliefs about mathematics are often directly related to their teachers' beliefs about mathematics. Certainly, students' beliefs are connected to their mathematics experiences in the classroom (Kloosterman, Raymond, & Emanaker, 1996). If students experience mathematics as something they must memorize on their own or see through someone else's perspective, then they will have a disconnection to mathematics and perhaps view mathematics as irrelevant to their lives. On the other hand, if students engage in mathematical problem solving in a collaborative process, they come to believe that mathematics is something they can do and is something that can be a social endeavor.

As we will see in many of the articles within this book, when students engage in mathematical problem solving, particularly when using technology as a tool for solving problems, they become mathematically empowered and gain a confidence that students may never achieve in a traditional mathematics classroom.

A Personal Reflection on Students' Mathematics Anxiety

When I began teaching technology to inservice teachers in 1986, all I had to do was mention the word "spreadsheets" and the anxiety level significantly escalated. Over the 16 years that followed, I continued teaching inservice teachers as well as preservice teachers. You would think math anxiety would have declined over time. I found just the opposite to be true. Math anxiety became much more exacerbated. Increasingly, even among preservice teachers, I found the mere mention of spreadsheets sent the anxiety level through the ceiling.

One way I found to be effective for significantly reducing math anxiety was to introduce "creating charts" on the computer. I would start by creating a setting where we did a class survey of pets. After entering the word "Pets" in cell A1 and the word "Quantity" in cell B1, we entered the labels (pet types) starting in cell A2 and the number of pets in cell B2.

Concurrent with this activity, I taught them how to move from cell to cell using the tab key, shift plus tab, enter, and shift plus enter. When we had entered all the pets, we selected all cells containing labels (column A) and data (column B). We accessed the charting tools of the spreadsheet software. When they saw how easy it was to make charts using a spreadsheet, they were hooked. Math anxiety went out the window for most of the students; for others, it was significantly reduced.

Little by little we learned to enter formulas to calculate results. This allowed us to do simple arithmetic with which they had a good comfort level. Gradually, I introduced spreadsheet functions. As the learners (inservice and preservice teachers) acquired skills for using a spreadsheet, the anxiety level continued to diminish. For many, the power of the spreadsheet gave them new confidence in their ability to complete other mathematical activities.

What disturbed me so deeply was the large percentage of preservice and inservice participants with significantly high math anxiety levels. In fact, the more years I taught preservice teachers, the more I found disturbingly high math anxiety. Perhaps it was always there and I simply became more sensitive to it, but I don't think so. As I have conversed with math teachers throughout the years, I find they feel compelled to cover the textbook in a "damn the torpedoes—full speed ahead" style. While I had

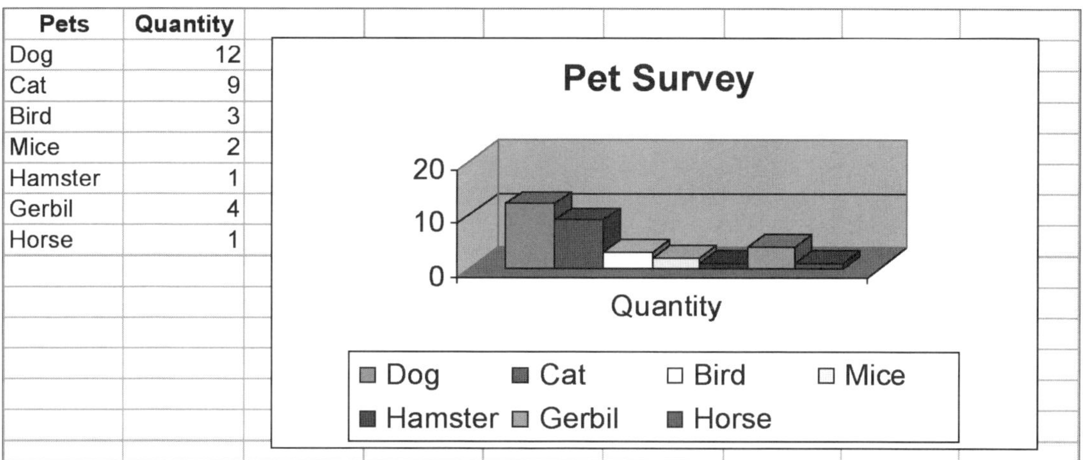

Figure 2. Creating graphs with spreadsheet programs can help relieve math anxiety.

no research to back up my hypothesis at the time, I believe this philosophy of teaching mathematics has created a sizable population that "hates" math! By disregarding the affective domain in the mathematics classroom, we create mathematical cripples for the rest of their lives.

How tragic! I believe effective mathematics instruction must include careful attention to the affective domain. Until mathematics educators make changes in their approach to classroom pace and instruction, I believe we will continue to have a disproportionately high level of math anxiety. How can we change that?

One strategy would be to cease pressing forward to teach the entire textbook and spend additional time helping students develop a higher comfort level with the skills as they acquire them. Another essential involves helping students grasp the abstract concepts through the use of manipulatives—yes, even in the secondary classroom. The gains that accrue from finding techniques and resources to bridge the gap between the abstract and the students' understanding make the effort very well worth the time it takes. In my more than 40 years of teaching, I found that the time spent using manipulatives for introducing abstract concepts enabled me to cover more material in the same amount of time. This will require a significant philosophical change for many mathematics educators. They must stop teaching as they were taught and start teaching in a manner that enables students to master the concepts. I believe this will greatly reduce mathematics anxiety, and, as a result, test scores will improve.

References

Emanaker, C. (1996). A problem-solving based mathematics course and elementary teachers' beliefs. *School Science and Mathematics, 96*(2), 75–84.

Kloosterman, P., Raymond, A. M., & Emanaker, C. (1996). Elementary students' beliefs about school mathematics: A longitudinal study. *Elementary School Journal, 97*(1), 39–56.

McLeod, D. B. (1992). Research on affect in mathematics education: A reconceptualization. In D. A. Grouws (Ed.), *Handbook of research on mathematics teaching and learning* (pp. 575–596). Old Tappen, NJ: Macmillan.

Raymond, A. M. (1997). Inconsistency between a beginning elementary school teacher's mathematics beliefs and teaching practice. *Journal for Research in Mathematics Education, 28*(5), 550–576.

Raymond, A.M., & Santos, V.M. (1995). Preservice elementary teachers and self-reflection: How innovation in mathematics teacher preparation can challenge mathematics beliefs. *Journal of Teacher Education 46*(1), 58–70.

CHAPTER 1

Communication

The Conversation Begins

Ivan: Many teachers believe children need only to know how to solve problems. When those same children enter the world of work, they find a strong need to communicate. Where will they develop those skills?

Anne: There are a variety of ways to communicate mathematical understanding. Not only do we communicate through writing and oral explanations and discussions, but we also communicate through representations such as models, charts, and drawings. In mathematics, pictures often express more than a thousand words when it comes to providing insights into concepts and theorems. Technology provides a means for students to create, adapt, and experiment with representation in order to develop and communicate their knowledge.

Ivan: William Glasser writes that we remember 95% of what we teach to others, but only 10% of what we read. Communication to others requires the speaker to have a high level of understanding. Figure 1.1 communicates the relationship of classroom practice to how much learners retain. However, as shown in Figure 1.2, that which we remember best, teachers use the least. The triangle below demonstrates that teachers use reading more often than more effective teaching strategies.

Student Retention Rates

teach to someone	95%
experience personally	80%
discuss with others	70%
see and hear	50%
see	30%
hear	20%
read	10%

Figure 1.1. The percentages represent how much a learner retains given various pedagogical approaches.

How Often Teachers Use It

teach to someone	95%
experience personally	80%
discuss with others	70%
see and hear	50%
see	30%
hear	20%
read	10%

Figure 1.2. The percentages represent how much a learner retains. The triangle symbolizes how frequently the method is used.

Where will our students learn to communicate mathematical concepts if we don't teach them how to do it in class?

Anne: The articles in this chapter illustrate the multiple means of communicating mathematics understanding through technology. Each article demonstrates using multiple means of communicating about mathematics. For example, in Elizabeth Holmes's article, "The Spreadsheet Absolutely Elementary!" students communicate via charts, graphs, and writing. Similarly, in Margaret Niess's article, "Bottles and Beyond: Analyzing and Interpreting Graphs in the Middle Grades," students develop charts to interpret pictures, and they provide written explanations.

THEORY INTO PRACTICE

Communication and Technology

Writing About Mathematics

By Anne Raymond

The National Council of Teachers of Mathematics proposes a process standard on communication in the mathematics classroom. In implementing this standard, teachers need to constantly ask students to communicate mathematical understanding orally and in written form.

NCTM Communication Standard

Communication: Instructional programs from prekindergarten through Grade 12 should enable all students to

- Organize and consolidate their mathematical thinking through communication
- Communicate their mathematical thinking coherently and clearly to peers, teachers, and others
- Analyze and evaluate the mathematical thinking and strategies of others
- Use the language of mathematics to express mathematical ideas precisely

The International Society for Technology in Education provided leadership in developing National Educational Technology Standards for Students (NETS•S). The group that developed the technology standards included representatives from learned societies, including NCTM.

NETS•S Communication Standard

4. Technology communications tools

- Students use telecommunications to collaborate, publish, and interact with peers, experts, and other audiences.
- Students use a variety of media and formats to communicate information and ideas effectively to multiple audiences.

Much has been written regarding the benefits of writing about mathematics (e.g., Carter & Carter, 1994; Countryman, 1992; Lambdin, Kehle, & Preston, 1996). There are many ways to include writing in the mathematics classroom. Consider the following possibilities:

- Problem solving—word answers for word problems
- Explanations and justifications of mathematical solutions

- Journal writing/reflections/writing prompts
- Mathematics autobiographies
- Portfolio assessment—student rationale
- Student-created word problems

As you consider the above list of opportunities for communication in mathematics, examine the resources available for students to share their learning. Word processing fits well with many of the items in the list. Mathematical autobiographies would work well as Web pages or a HyperStudio stack. Integrated applications software and office suites make it easy for students to include spreadsheets in a word processing document in activities such as explanations and justifications of mathematical solutions or portfolio assessment pieces. To support the NETS standards, ISTE published *National Educational Technology Standards for Students—Connecting Curriculum and Technology*. This book contains lesson plans integrating NCTM standards and NETS standards in the K–12 environment. ISTE also provides field-tested lesson plans that teachers have submitted since the book was released. These are on ISTE's Web page at http://cnets.iste.org.index2ns.html.

Of particular interest are writing prompts, which can encourage students to write about mathematics. Students can express their feelings about mathematics as well as provide explanations about mathematical ideas. The following list of writing prompts, in large part inspired by the writing of Stenmark (1991), provides examples of effective writing opportunities for students.

POSSIBLE WRITING PROMPTS IN THE MATHEMATICS CLASSROOM

1. Reflect on your participation in class today and then complete the following statement (teacher should select one):

   I learned that I (or we) . . .

   I was surprised that I . . .

   I discovered that I . . .

   I was pleased that I . . .
2. Write a letter to a classmate who could not attend class today so that she will understand what we did and learn as much as you did. Be as complete as possible.
3. Describe places you became stuck and how you became unstuck when solving the problem.
4. What I like most (or least) about math is . . .
5. Design two mathematical bumper stickers, one funny and one serious.
6. Something I would really like to know about mathematics is . . .
7. If math could be a sound (or a shape, or an animal), it would be . . . because . . .
8. What was the most difficult (or easiest) part of ______________________ (teacher fills in the blank).

When considering software to purchase for the mathematics classroom, there are selection criteria worth examining, particularly those features that affect writing possibilities. Van de Walle (2001) suggests that teachers ask vital questions when comparing software options:

- What does this do better than can be done without the computer?
- How are students likely to be engaged with the content?
- How easy is the program to use?
- What sort of conceptual information is provided?
- What controls are provided to the teacher?
- Is there freedom of manipulation?
- Is there a connection between the symbolism and the models?
- Are there "print and save" capabilities?
- Are there "journaling" capabilities?

It is important to consider the journaling capabilities, as illustrated in Deborah Gettys' article, "Journaling With a Database." This author describes using a database management system, which is not often thought of as a means of journal management, to do just that. She argues that "a journal, in its simplest form, is a collection of similar pieces of recorded thoughts or information," which naturally fits within a software system that can sort and organize data.

References

Carter, J.A., & Carter, D.E. (1994). *The write equation: Writing in the mathematics classroom.* Palo Alto, CA: Dale Seymour.

Countryman, J. (1992). *Writing to learn mathematics: Strategies that work.* Portsmouth, NH: Heinemann Press.

Lambdin, D.V., Kehle, P.E., & Preston, R.E. (Eds.). (1996). *Emphasis on assessment: Readings from NCTM's school-based journals.* Reston, VA: NCTM.

Stenmark, J. K. (Ed.). (1991). *Mathematics assessment: Myths, models, good questions, and practical suggestions.* Reston, VA: NCTM.

Van de Walle, J. A. (2001). *Teaching mathematics developmentally (4th ed.).* New York: Addison Wesley Longman.

Resources

WebQuests, Think Quests, and Other Web Sites

Math Magazine WebQuest: www.bellmore-merrick.k12.ny.us/webquest/math/mag.html

WebQuests and Mathematics: http://coe.west.asu.edu/students/msyrkel/webquestusemath.htm

THEORY INTO PRACTICE

Engaging in Mathematical Discourse

By Anne Raymond

In conjunction with the development of the National Council of Teachers of Mathematics' curriculum standards, the NCTM developed companion standards for mathematics teaching and professional development as well as mathematics assessment. In the *Professional Teaching Standards*, two standards focus on discourse in the mathematics classroom: The Teacher's Role in Discourse and The Students' Role in Discourse.

Standard 2: The Teacher's Role in Discourse

The teacher of mathematics should orchestrate discourse by

- Posing questions and tasks that elicit, engage, and challenge each student's thinking
- Listening carefully to students' ideas
- Asking students to clarify and justify their ideas orally and in writing
- Deciding what to pursue in depth from among the ideas that students bring up during a discussion
- Deciding when and how to attach mathematical notation and language to students' ideas
- Deciding when to provide information, when to clarify an issue, when to model, when to lead, and when to let a student struggle with a difficulty
- Monitoring students' participation in discussion and deciding when and how to encourage each student to participate

Standard 3: The Students' Role in Discourse

The teacher of mathematics should promote classroom discourse in which students

- Listen to, respond to, and question the teacher and one another
- Use a variety of tools to reason, make connections, solve problems, and communicate
- Initiate problems and questions
- Make conjectures and present solutions
- Explore examples and counterexamples to investigate a conjecture
- Try to convince themselves and one another of the validity of particular representations, solutions, conjectures, and answers
- Rely on mathematical evidence and argument to determine validity

The articles in this chapter address mathematical discourse in a variety of ways. In Deborah Gettys' article, "Journaling With a Database," students develop a database that houses their ongoing journal entries about their mathematics experiences. Students record what they learned each day and reflect on how they felt about the topic and what questions they have about the topic. This type of journaling provides data that students can bring to classroom discussions. The database presents data to teachers so that they can look for patterns of learning based on student reflections. The database also provides a means of allowing every student to express knowledge and feelings in a manner that is unthreatening.

Marge Cappo and Gail Osterman address discourse differently in their article, "Teach Students to Communicate Mathematically." These authors discuss how you can appeal to many students by encouraging them to use the computer to communicate mathematically in a wide range of ways. For example, students can engage in mathematics publishing through problem creation and problem discussion. Students can also learn to talk to each other about mathematics by engaging in math games on the computer. Also, students communicate mathematics through a variety of mathematical representations including graphs, charts, and symbols.

These articles, along with the others, provide examples of how mathematical discourse can be enhanced with technology because students become more focused. The technology helps students put their thoughts and ideas together prior to class discussion and allows for multiple means of communicating about mathematics.

INSIGHT

Bringing Words to Mathematics

By Anne Raymond

Consider the following problem:

> *Seventy-eight second graders are going on a field trip. Each bus holds 24 students. How many buses will have to be hired for the trip?*

Often students will ponder this problem, trying to figure out the most appropriate operation to use to solve it. Ultimately, many students decide that division will be most successful. They divide 78 by 24 and obtain the answer 3.25. This is likely where students will stop unless we teach them to think about actually answering problems rather than merely translating problems into an arithmetic situation. If we encourage students to communicate solutions beyond providing numeric responses—to, in other words, answer questions using complete sentences and thorough explanations—they will learn to combine numbers and words to communicate clear, reasonable solutions.

In the case of the field trip dilemma above, we know the answer to the question is not 3.25, but, "They will have to hire four buses for the field trip to fit all the students on a bus."

It can be said "the only way to truly solve word problems is with word answers." This is especially true if we want students to learn to effectively communicate mathematical solutions.

Clarity in mathematical communication is vital to mathematical understanding. This is particularly true given the growing perspective that mathematics is an "action word." That is, engaging in mathematics today involves such actions as: conjecturing, predicting, manipulating, explaining, experimenting, measuring, comparing, verifying, connecting, and creating. These types of actions call for increased interaction and increased (and new types of) communication.

What means of communication are called for when students are involved in the action of mathematics? Students who are actively engaged in mathematics can communicate through physical models, drawings, writing, and talking.

Physical models can be made by using manipulatives or other objects, or by students physically acting out a concept or idea. Drawings, diagrams, or sketches can be made by hand or with the computer. Written communications can be journal writing, student explanations of problem solutions, or other types of writing described in the Theory Into Practice commentaries in this chapter.

Oral communication can be one-on-one, small group, or whole class and can involve student-to-student communication or student-to-teacher communication. Oral communication can take the form of conversations, discussions, debates, or presentations. When students (and teachers) talk with each other to make sense of mathematics, they have to listen. Students can improve their listening skills by looking at the speaker and by learning to take notes about what they hear. The mathematics teacher can aid the development of listening skills by providing students with organizing tips or handouts about what types of information they should be noting.

For example, when students are engaged in a conversation during problem solving, the teacher might provide some sort of self-, peer, or small-group assessment about the interactions during the team's work. In the example that follows, students are prompted to reflect on their partner's level of communication during mathematical problem solving and assess how their own communication added to the problem-solving experience.

STUDENT SELF-ASSESSMENT ABOUT PARTNER COMMUNICATION

1. When problem solving with my partner, I (check one):
   ____ did most of the talking
   ____ did about half the talking
   ____ did less than half the talking
2. The most helpful thing my partner said in our discussion was:
3. The thing my partner said that was most confusing to me was:
4. The most interesting thing my partner said about our mathematics solution was:
5. One way I helped clarify things for my partner was:
6. One question I asked my partner was:
7. I would rate our communication about our problem as (check one):
   ____ very good
   ____ OK
   ____ needs improvement

INSIGHT

How Did Your Student Get *That* Answer?

By *Ivan W. Baugh*

Have you ever looked at an answer on a student's paper and wondered how on earth the student got *that* answer? If many students have difficulty with an assignment, have you thought about how you could have taught the concept differently? Consider having your students complete a writing activity such as the one that follows:

Scenario

> *One of your classmates was absent from class yesterday when we studied a new concept. Please write instructions which will help your friend learn the material that we studied in class and that you practiced in the homework assignment last night. Remember to include appropriate examples to clarify your writing.*

Classroom Activity

As the students write, move about the room pre-reading their papers. This will help you identify papers you believe will best help any student having difficulty with the skill studied. Collect the papers and read them. This will enable you to understand what the students understood and where gaps in their knowledge appear. It will also help decrease the chance they will repeat the same mistakes so many times that they will have difficulty mastering the correct procedures.

After completing the writing activity, work with the class to check their homework. When a disparity of answers occurs, select students to explain how they solved the problem to get the various answers. This will help the students as they refine the understanding they expressed in the writing activity.

After Class

Read their papers. Select two or three you believe are worthy of sharing with the rest of the class. If the students have completed the writing assignment on a word processor, you can easily post them on a class Web page that afternoon after school so that other students can use them. The papers can also be placed on a classroom bulletin board for others to read as they have the need. Why take time with a writing activity when students need to master the skills involved in using the studied concept? Consider these points:

- This will give you an opportunity to discover gaps in their understanding before the misinformation becomes fixed in their learning.

- Students will teach others as they write, thereby functioning at the 95% level on the learning retention scale developed by William Glasser (discussed in the conversation opening this chapter).
- At many of the Writing Across the Curriculum workshops I have attended, one of the reasons frequently stated in support of this practice is that it clarifies understanding.
- If you had any absentees on the day you initially taught the lesson, this provides an avenue to help the students learn the material independently.

The article in this chapter by Marge Cappo and Gail Osterman focuses on mathematical communication. Their article and the article by Margaret Niess will help you develop ways for your students to practice communication in mathematics.

Bottles and Beyond

Analyzing and Interpreting Graphs in the Middle Grades

By Margaret L. Niess

One of the most common means of communicating numerical data is through graphing. In fact, to reword an old phrase, "A graph is worth a thousand words." However, it is important to note that this is true *only if* the reader can interpret the message. The skill of analyzing and interpreting graphical messages has received increased attention in the National Council of Teachers of Mathematics (NCTM, 1989) curriculum standards for Grades 5–8 in its statement that "Communication with and about mathematics and mathematical reasoning should permeate the 5–8 curriculum" (p. 66). The following sequence of activities were designed to encourage students to do more than simply draw graphs; the emphasis is on analyzing and interpreting graphical communications.

The Bottle Activity

Begin with a hands-on experiment in which students store the data in a spreadsheet. ClarisWorks is used in this example, although you can use any integrated package for Macintosh or DOS-based machines. For this activity you will need:

- Clear glass bottles of different sizes and shapes.
- Graduated cylinder (for measuring volume in ml).

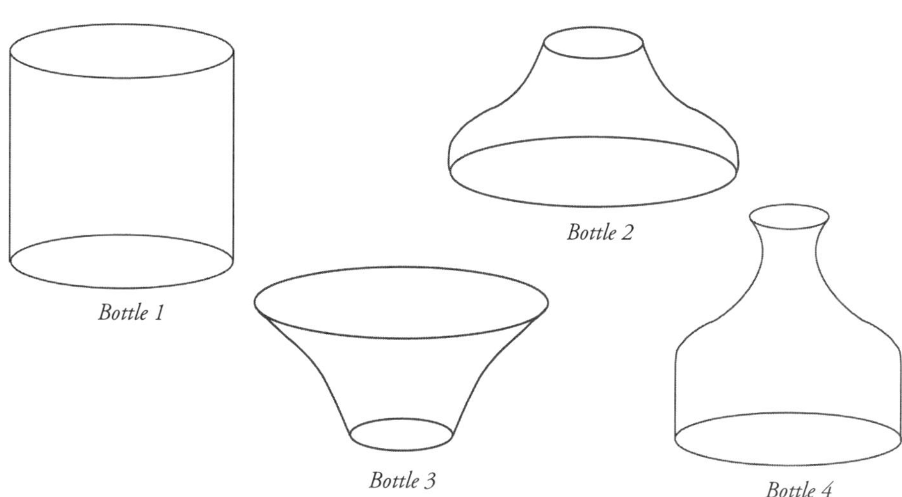

File Edit Format Calculate Options View

bottle problem (SS)

A1

	A	B	C	D	E	F	G	H	I	J	K	L	M	N	O
1			Filling a Bottle of Water												
2															
3	Volume		Height 1			Volume		Height 2							
4	10		8			10		1		Volume is measured in ML					
5	20		16			20		2		Height is measured in MM					
6	30		24			30		3							
7	40		32			40		6							
8	50		40			50		9							
9	60		48			60		14							
10	70		56			70		19							
11															
12	Volume		Height 3			Volume		Height 4							
13	10		5			10		1							
14	20		10			20		2							
15	30		15			30		8							
16	40		18			40		32							
17	50		21			50		64							
18	60		22			60		128							
19	70		23			70		160							
20															

Figure 1. Pictures and data for bottles.

- Ruler (for measuring height in mm).
- Water.

Have students draw pictures of their bottles in the Draw mode of the software. Instruct them to pour 10 ml of water into an empty bottle and measure the height of the water in the bottle. The data, a volume of 10 ml that results in a height of 8 mm, is recorded in the spreadsheet as in Figure 1. Continue adding water in 10 ml increments until the bottle is full, measuring the height of the water in the bottle and recording the data in the spreadsheet. Students must then collect similar data for each of the bottles that they have drawn. Figure 1 shows a completed table for different bottles along with pictures of each of the bottles.

Next, have students graph the data using the "charting" mode of the spreadsheet. Copy each graph, its picture, and the data to a word processing file. With this evidence, students are able to discuss the shape of the bottle and the shape of the graph, making connections between the actual picture of the bottle and the graphical representation of the bottle. For example, the first bottle has straight sides and the graph is a straight line, with no changes in the slope between any two points. Yet, the second bottle has a wide circumference at the bottom, with the circumference narrowing toward the top. The graph in Figure 2 displays this message with only a slight change in the height for the first 30 ml of water, yet the change increases for 40 ml and 50 ml, and finally the change is even steeper for a volume of 60 ml and 70 ml.

After students have discussed their graphs, have them summarize what they have learned about the relationship of the shape of the bottle to the graph. Give them other sets of data already stored in the spreadsheet, as in Figure 3. Their job is to graph the data and determine the shape of the bottle from the evidence in the graph. They can copy graphs to the draw package where they can draw predictions about the bottles used to create the graphs. Be sure to discuss the notion of the "slope" between each set of points to encourage students to begin using this terminology in their discussions.

"A graph is worth a thousand words." However, it is important to note that this is true *only if* the reader can interpret the message.

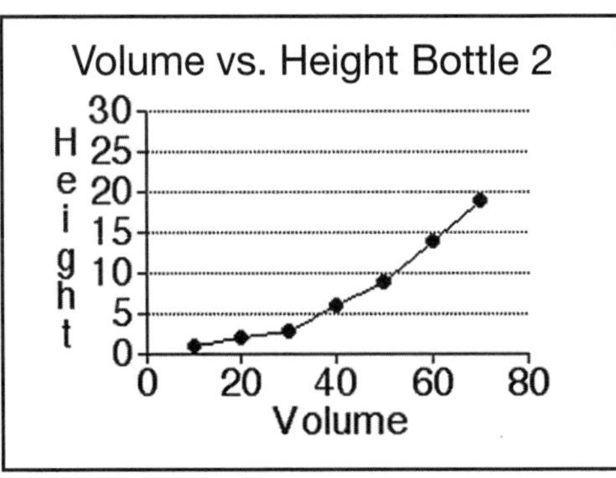

Figure 2. Graph for bottle 2.

File Edit Format Calculate Options View

bottle problem (SS)

H20

	A	B	C	D	E	F	G	H	I	J	K	L	M	N	O
1			Filling a Bottle of Water												
2															
3	Volume		Height 1			Volume		Height 2							
4	10		5			10		10		Volume is measured in ML					
5	20		12			20		25		Height is measured in MM					
6	30		18			30		37							
7	40		28			40		47							
8	50		43			50		69							
9						60		84							
10															
11	Volume		Height 3			Volume		Height 4							
12	10		10			10		15							
13	20		20			20		25							
14	30		35			30		30							
15	40		55			40		35							
16	50		70			50		45							
17	60		85			60		60							
18	70		92			70		70							
19	80		100			80		75							
20	90		105												

Figure 3. Mystery bottle data.

Figure 4. Interesting bottle to graph.

Finally, challenge students to sketch a graph for different shapes of bottles that they have not filled with water. Figure 4 provides a picture of an "interesting" bottle to graph. If you have access to HyperCard, you can have them draw a picture of the bottle, sketch the graph, and provide a defense for their assertion that the graph is correct for the particular bottle. Some important follow-up questions at this point include: Are you sure the inside of the bottle has the same shape as the outside of the bottle; is it possible that the ornamental sides are filled glass attached to the bottle?

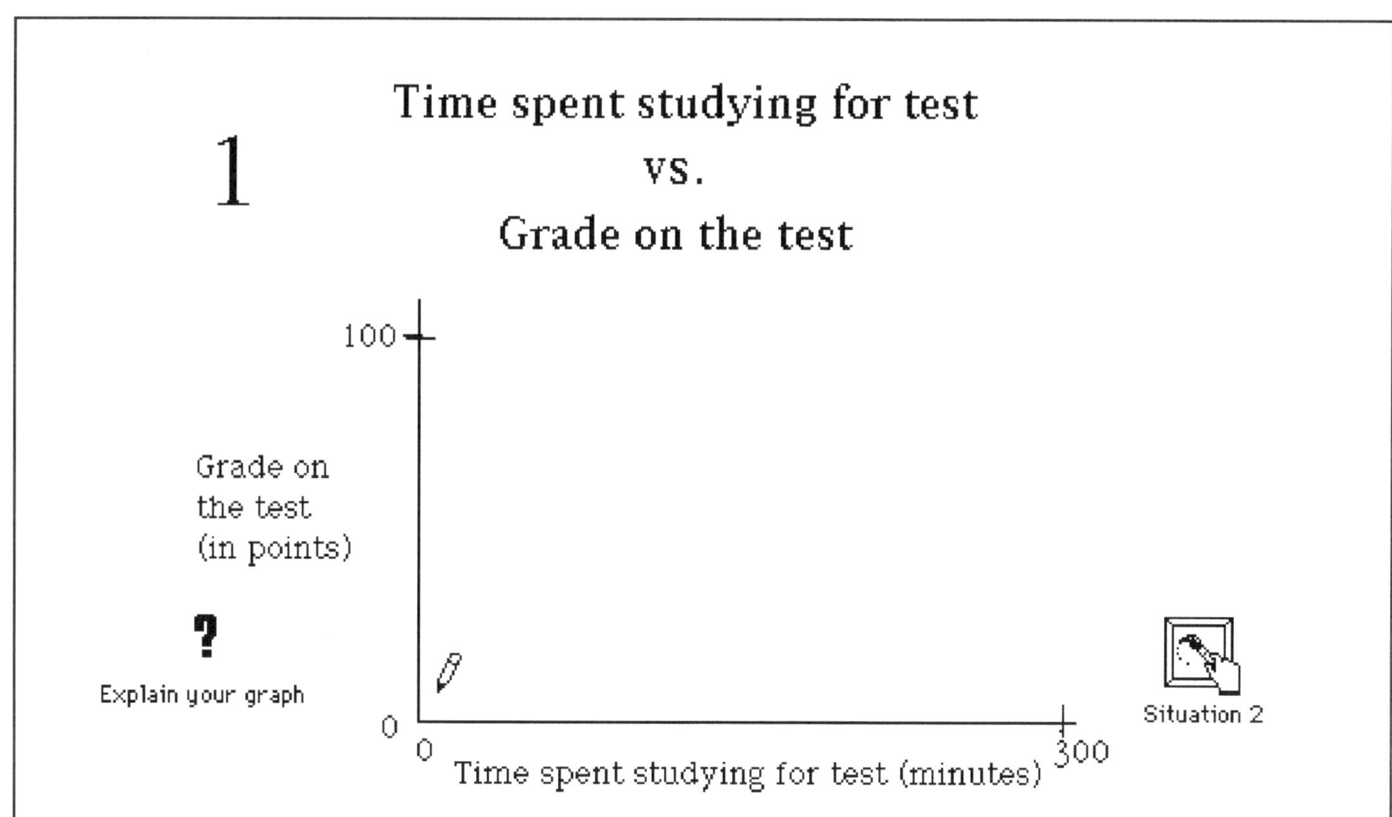

Figure 5. Situation 1 for Hypercard stack.

Beyond Bottles

To encourage students to continue analyzing and interpreting graphs, I created a HyperCard stack where the students must describe particular relationships in graphical form. (Copies of this stack may be obtained by sending a disk and appropriate return postage to the author.) The first situation in this stack (see Figure 5) instructs students to sketch the relationship between the time spent studying for a test (from 0 to 300 minutes) and the grade on the test (from 0 to 100 points). After sketching the graph, the student clicks on a button "Explain your graph." Clicking on this button moves the student to a card with a field for recording an explanation for the shape of the graph. A button on this card returns the student to the graph. The graph card also has a button to move to the next situation. The final situation in this stack challenges students to describe a graphical relationship of their own.

One of the situations in this HyperCard stack asks the students to predict the relationship between a person's height and shoe size. As a follow-up to the discussion of this situation, have students actually collect these data (as described [Froelich, 1993] in NCTM's *Addenda Series Grades 9–12: Connecting Mathematics*). Because men's and women's shoes sizes do not use the same scale, it is important to have them collect three pieces of data for each person: Gender, Height, Shoe Size. When recording the data in the spreadsheet, you must put men's and women's shoe sizes on the same scale, either by adding 1.5 to the men's sizes or by subtracting 1.5 from the women's sizes. This activity leads nicely to a discussion of scatterplots and lines of best fit. The line of best fit may be estimated by drawing an ellipse around the data, using the axis of the ellipse as the line of best fit. With this line, students can estimate shoe size given a particular height and discuss whether height is a good predictor of shoe size.

For all of these activities, technology fosters environments where students are encouraged to analyze and interpret graphical messages.

For all of these activities, technology fosters environments where students are encouraged to analyze and interpret graphical messages. The "freedom to explore, conjecture, validate and to convince others is critical to the development of mathematical reasoning in the middle grades ... (and) reasoning is fundamental to the knowing and doing of mathematics" (NCTM, 1989, p. 81).

References

Froelich G. W. (1993). *Curriculum and evaluation standards for school mathematics addenda series, Grades 9–12: Connecting mathematics.* Reston, VA: National Council of Teacher of Mathematics.

National Council of Teachers of Mathematics, Commission on Standards for School Mathematics. (1989). *Curriculum and evaluation standards for school mathematics.* Reston, VA: Author.

The Spreadsheet: Absolutely Elementary!

Elementary teachers and teacher trainers can use these activities to teach themselves and their students (ages 5–12) about spreadsheets and how they can illustrate complex math concepts. The spreadsheet templates are then placed in electronic "posters" to create wonderful interactive activities to help students visualize and communicate mathematical concepts.

By Elizabeth Dudley Holmes

Spreadsheets. The word conjures images of scientists, accountants, bankers, and business people busily manipulating mouse-driven stacks of figures with electronic accuracy. The image we have of the spreadsheet as a complex tool that requires sophisticated skills frequently discourages elementary teachers from attempting to use it with their students. The truth is, however, that spreadsheets can often illustrate difficult mathematical concepts better than many other tools, and because they are part of almost any integrated software package, such as Microsoft Works or ClarisWorks, they are readily accessible to almost any classroom teacher.

What Is a Spreadsheet, Anyway?

A spreadsheet is a computer application included in most integrated software packages alongside a word processor, database, and some basic drawing or painting tools. The spreadsheet software simply provides an electronic grid of columns and rows that allows the user to unfold, stretch, or "spread" out data for manipulation, analysis, and reflection.

Applications such as the spreadsheet allow students to shift from the rote drill-and-practice software commonly used in mathematics classrooms into an environment that allows students to gather, display, and change data while seeking reasonable solutions to a problem. Once relevant information is displayed on the grid, the spreadsheet offers tools that allow you to perform calculations ranging from very simple to extremely complex. The spreadsheet allows elementary students to follow simple computing steps that build operations and algorithms sequentially. This unique feature—the spreadsheet's ability to adjust to differing degrees of simplicity or difficulty—makes it a useful tool for elementary students conceptualizing the big ideas in mathematics.

Another compelling feature of the spreadsheet is its capacity to build graphics that visually represent the data entered. The opportunity to see data presented in a variety of ways enables students to make meaningful and important mathematical connections. The spreadsheets featured in this article were created in ClarisWorks, but they could easily be reproduced in other popular productivity packages such as Microsoft Works or Microsoft Excel.

Introductory Spreadsheet Activities for Teachers and Students

The activities presented are samples of spreadsheet activities I prepared for technology training sessions with elementary teachers. Focusing on the Curriculum and Evaluation Standards for School Mathematics (National Council of Teachers of Mathematics [NCTM], 1989) and the Professional Standards for Teaching Mathematics (NCTM, 1991), I created each activity to emphasize reasoning, problem solving, making connections, and communicating mathematical ideas through the use of technology.

I originally developed these lessons to introduce teachers to the basic concepts involved in using spreadsheets in the classroom. After receiving feedback from teachers who adapted these activities for their own elementary-age students, however, I realized they are also effective for helping students learn problem solving. These simple activities can help take some of the tedium out of using spreadsheets in mathematics and, at the same time, illustrate some fairly complex mathematical concepts.

Simple Spreadsheets

The first activity is designed to familiarize teachers with the spreadsheet and provide a vision for the power of its tools. The activity is grounded in an age-old problem posed by Mother Goose in the rhyme "As I Was Going to St. Ives."

As I Was Going to St. Ives
As I was going to St. Ives,
I met a man with seven wives;
Every wife had seven sacks,
Every sack had seven cats,
Every cat had seven kits;
Kits, cats, sacks, and wives,
How many were going to St. Ives?

The first step in problem solving with a spreadsheet is to spread out the problem within the columns and rows of the spreadsheet. Teachers (or students) should brainstorm in small groups, break down the elements of the problem, and enter the parts of the problem into the spreadsheet as illustrated in Figure 1. Teachers experience the power of the spreadsheet as they enter data from the problem, create formulas for subtotals and totals, and easily discover the rhyme's solution.

After constructing the spreadsheet, teachers learn that Mother Goose envisioned 2,802 travelers on that road to St. Ives—a far cry from the illustrations that generally accompany the rhyme when it appears in nursery rhyme books! A pie graph of the traveling troupe is added to depict the solution graphically, an extension that helps construct visual meaning for the facts and figures in the spreadsheet. Detailed step-by-step instructions for creating the St. Ives spreadsheet and poster template in ClarisWorks are available in the L&L section of the ISTE Web site at http://www.iste.org/publish/learning/learning.html. Click on the "Online Article Supplements" button to access the two files.

I Can Do This!

Success with a single spreadsheet is often enough to impress teachers with the value of spreadsheets in the classroom. One positive experience with the spreadsheet can help enthusiastic educators consider the possibilities of new interdisciplinary projects that link mathematics to science, social studies, health, and P.E.! They will begin to visualize using spreadsheets to solve money problems and measurement problems, to conduct surveys, to chart data, and more. Because most teachers will want to take their "St. Ives" spreadsheets back to their classrooms to share with students, we added some additional activities, including a method of preserving the "problem" so it can be used repeatedly by many students.

Electronic Posters

Electronic posters are created by combining the drawing feature of ClarisWorks with the spreadsheet and graphing features that we have already used. The Draw component is analogous to a piece of poster board with useful drawing tools. An *electronic* poster, however, allows users to display text, graphics, interactive spreadsheets, graphs, and even sound in one place. You can create such special projects as the St. Ives

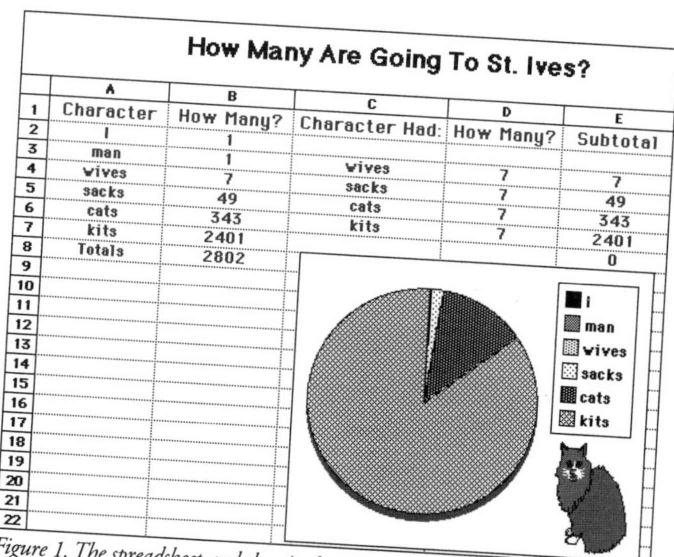
How Many Are Going To St. Ives?

	A	B	C	D	E
1	Character	How Many?	Character Had:	How Many?	Subtotal
2	I	1			
3	man	1			
4	wives	7	wives	7	7
5	sacks	49	sacks	7	49
6	cats	343	cats	7	343
7	kits	2401	kits	7	2401
8	Totals	2802			0

Figure 1. The spreadsheet and the pie chart for the St. Ives problem.

jodie's st.ives (DR)

As I Was Going to St. Ives

Study the Mother Goose rhyme. Add data to the spreadsheet to solve the problem.

As I was going
to St. Ives,
I met a man
with seven wives;
Every wife
had seven sacks,
Every sack
had seven cats,
Every cat
had seven kits;
Kits, cats, sacks,
and wives,
How many were going
to St. Ives?

Not enough chart data.

1	Characters	How Many?	Character Had:	How Many?	Subtotals
2	I		wives		
3	man		sacks		
4	wives		cats		
5	sacks		kits		
6	cats				
7	kits				
8					
9	Total				

Using a calculator, add the total percentages provided in the chart.
What is the total? Type your answer here:

Explain:

How many were going to St. Ives? Retell the story using accurate numbers.

Figure 2. The St. Ives electronic poster for students includes several study questions in addition to the interactive spreadsheet and pie chart.

spreadsheet and preserve them on a Stationery template created in the Drawing program. The Draw program allows a seamless presentation of the spreadsheet and graph on a thoughtfully constructed background that can include any text instructions, additional information, or student study questions. Most importantly, the template allows the students to watch the problem unfold as they enter, change, and add data. Finally, the skillfully constructed template provides opportunities for communicating new knowledge and articulating important connections, which can be assessed by the teacher after the activity. Figure 2 presents the original template for solving the "St. Ives" problem. The blank template offers a limited structure for solving the problem in a spreadsheet and an empty chart awaiting data.

When the student adds the traveler, the graph shows that he represents 100% of the troupe. When the student adds one man to the spreadsheet, the graph clearly illustrates that each traveler accounts for 50% of the whole. Figure 3 shows the chart as soon as the seven wives are added and the graph that corresponds to it.

When the kits are added to the problem, as in Figure 4, a curious thing happens to the graph and to the percentages. The percentages for "I" and the "man" are reduced to 0.0%! Did they disappear statistically as they did on the pie graph? Where are they? Can they be counted as travelers to St. Ives?

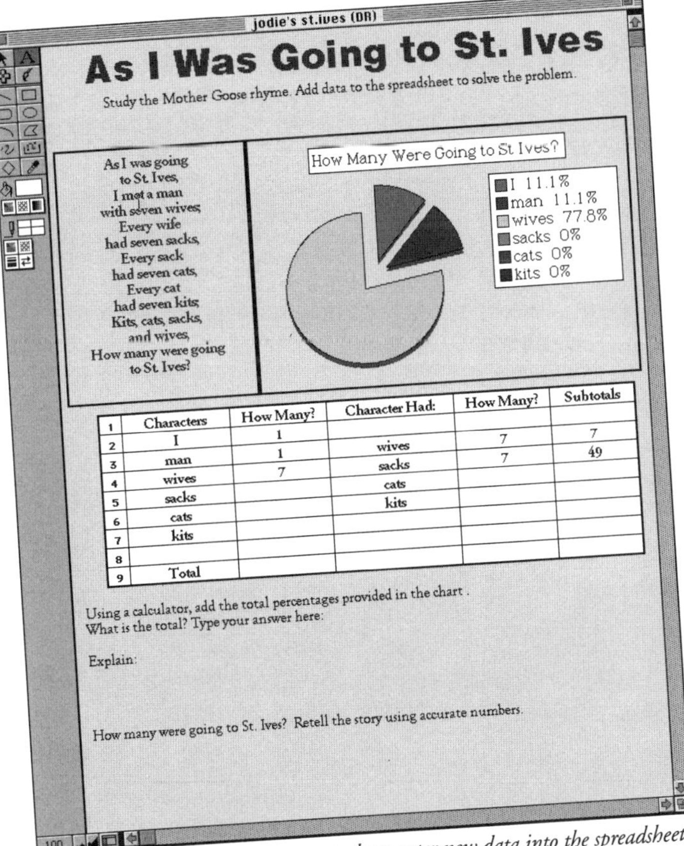

Figure 3. The pie chart changes as students enter new data into the spreadsheet.

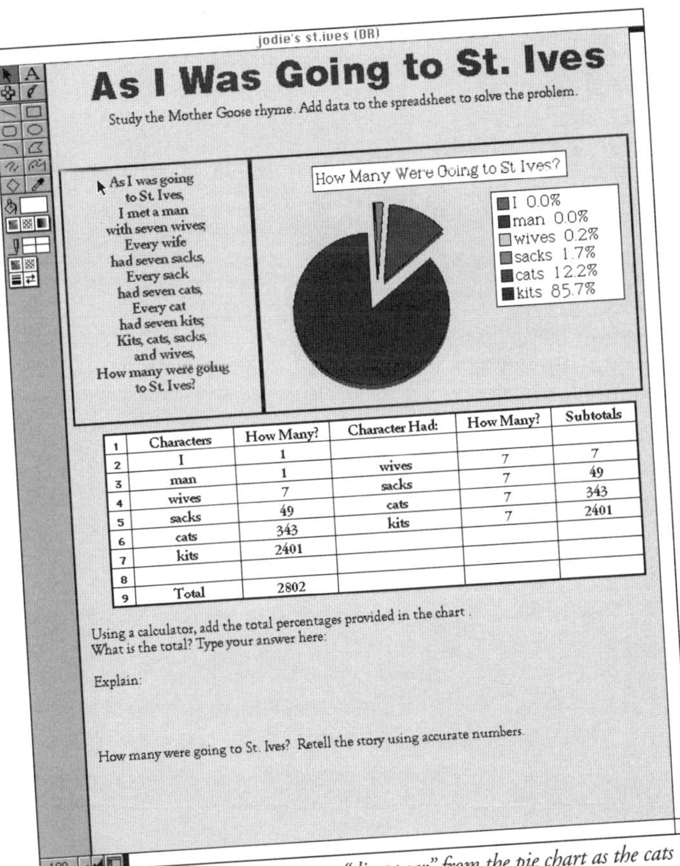

Figure 4. The traveler and the man "disappear" from the pie chart as the cats and kits are added to the spreadsheet.

The electronic poster encourages students to formulate a reasonable explanation by suggesting the use of a calculator to total the percentages displayed in the chart. What is the total? The calculator confirms that 99.8% of the travelers are displayed on the graph. The missing 0.2% are the two travelers who simply became too small to be visible among all those cats and kits! Yet they are accounted for mathematically—and that makes sense!

Finally, the electronic poster helps students communicate the mathematical ideas discovered and the connections made during the project by directing them to retell the story using numbers. This written reflection will offer insight into the depth and accuracy of the student's understanding and into the connections made between parts of a whole, ratios, proportion, and computation.

Other Examples for Development

After teachers work with the St. Ives poster, they always get so excited and ask me what's next and how they can make more. The following examples showcase other spreadsheet templates designed to meet curricular needs across the elementary grade levels.

Conclusion

Teacher and student reactions to the spreadsheet templates presented in this article suggest that the spreadsheet is a powerful tool for the development of mathematical habits of mind in the elementary school. The emphasis on data collection and a sensible, strategic migration toward solving a problem increases the authenticity and effectiveness of matching school tasks to the competencies needed in the world of work. The place to begin building these skills is in the elementary school. Why not begin with these easy-to-use spreadsheet templates?

References

National Council of Teachers of Mathematics. (1989). *Curriculum and evaluation standards for school mathematics.* Reston, VA: Author.

National Council of Teachers of Mathematics. (1991). *Professional standards for teaching mathematics.* Reston, VA: Author.

Resources

BodyScope 1.0. (1994). MECC, 6160 Summit Drive North, Minneapolis, MN 55430-4003; 800/685-6322 or 612/569-1500; fax 612/569-1551; www.mecc.com.

ClarisWorks 4.0. (1995). Claris Educational Software, 5201 Patrick Henry Drive, Santa Clara, CA 95952; 800/747-7483 or 408/987-7000; fax 408/987-7563; www.claris.com.

Excel 5.0. (1995). Microsoft Corporation, One Microsoft Way, Redmond, WA. Contact local software reseller for ordering information.

Microsoft Works 4.0. (1995). Microsoft Corporation, One Microsoft Way, Redmond, WA. Contact local software reseller for ordering information.

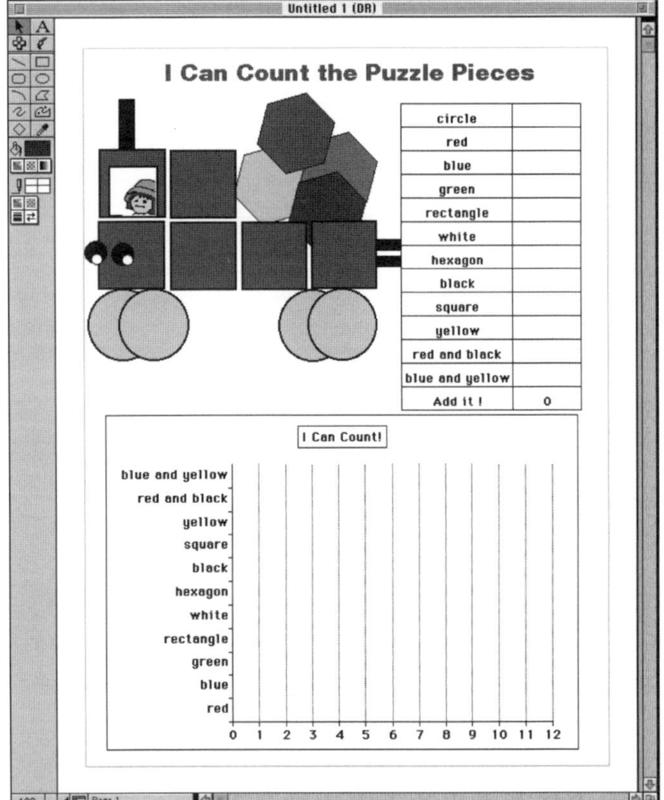

I Can Count the Puzzle Pieces

"I Can Count the Puzzle Pieces" was designed for kindergarten and first-grade students. The colorful puzzle, chart of data, and the interactive bar graph make the template a fun activity for students learning the concepts of shape, color, and numeration. Just as the "St. Ives" graph changed with each entry, the bar graph for "I Can Count the Puzzle Pieces" grows each time a student enters a numeric symbol into the data chart.

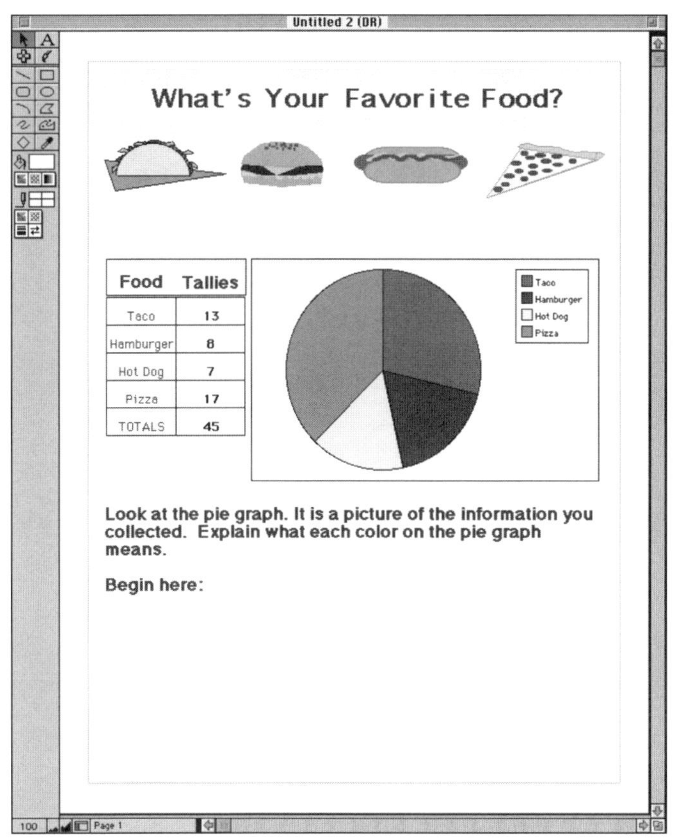

Food	Tallies
Taco	13
Hamburger	8
Hot Dog	7
Pizza	17
TOTALS	45

What's Your Favorite Food?

The template "What's Your Favorite Food" is designed for second graders to collect and analyze survey data. Students poll classmates about their favorite foods. They enter the data into a chart and interpret the graphic results displayed on the pie graph. This fun-filled activity, which introduces students to the basics of statistical data collection, is extended by using the survey results as material for writing word problems for mathematics study. Problems might include scenarios like the following teacher-generated idea: "All of the boys and girls in the survey were going to have a pizza party. The pizza lovers were very happy! Now find out how many children would not get to eat their favorite food at the party."

The natural curiosity of second graders makes elaboration on this project natural. Eight-year-old learners are fascinated when encouraged to speculate about the potential results of the same survey in other second-grade classrooms. Further research data is easily collected by placing the electronic poster on a computer in another classroom or on the network to make it available to the entire school!

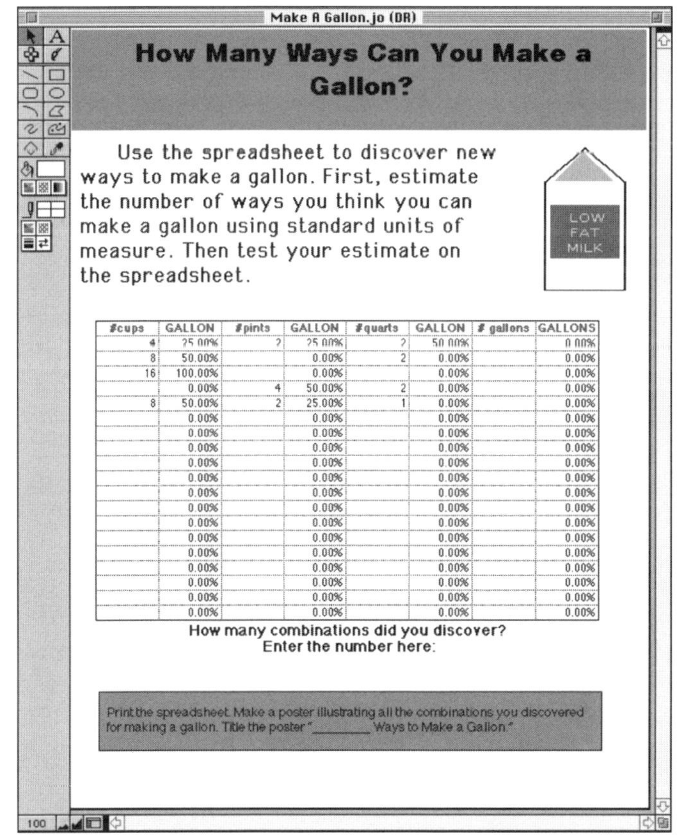

#cups	GALLON	#pints	GALLON	#quarts	GALLON	# gallons	GALLONS
4	25.00%	2	25.00%	2	50.00%		0.00%
8	50.00%		0.00%	2	0.00%		0.00%
16	100.00%		0.00%		0.00%		0.00%
	0.00%	4	50.00%	2	0.00%		0.00%
8	50.00%	2	25.00%	1	0.00%		0.00%
	0.00%		0.00%		0.00%		0.00%
	0.00%		0.00%		0.00%		0.00%
	0.00%		0.00%		0.00%		0.00%
	0.00%		0.00%		0.00%		0.00%
	0.00%		0.00%		0.00%		0.00%
	0.00%		0.00%		0.00%		0.00%
	0.00%		0.00%		0.00%		0.00%
	0.00%		0.00%		0.00%		0.00%
	0.00%		0.00%		0.00%		0.00%
	0.00%		0.00%		0.00%		0.00%
	0.00%		0.00%		0.00%		0.00%
	0.00%		0.00%		0.00%		0.00%
	0.00%		0.00%		0.00%		0.00%
	0.00%		0.00%		0.00%		0.00%

How Many Ways Can You Make a Gallon?

Third- and fourth-grade students explore the problem "How Many Ways Can You Make a Gallon?" Students make valuable connections between units of liquid measurement and the percentage of a gallon that each unit represents. The trial-and-error quality of the activity allows students to "play" with combinations that total 100%, or one gallon. The strategies employed by users of this template are varied and deeply rooted in the individual's previous knowledge and ability to make mathematical connections between the units of measurements of liquids and their percentages of the whole. Students with strong backgrounds in computation can "make gallons" by adding the percentages, and other students may choose other methods based on their own base knowledge of the subject. In this way, students use their previous mathematical knowledge to make valuable connections between units of liquid measurement and the actual size of a gallon. Once the template is completed, students can print their results and create a poster that illustrates all of the combinations they used to make a gallon.

How High Will the Ping-Pong Ball Bounce?

This template can be used to collect data, analyze the results, and make predictions in response to the problem "How High Will the Ping-Pong Ball Bounce?" The problem itself is an example of the many interesting questions that can be posed to students using spreadsheet tools. The calculating power of the spreadsheet allows students to discover complex scientific phenomena while applying mathematical skills.

The problem parameters require students to suspend a ping-pong ball at heights of increasing increments and collect data on the height of the first bounce of the ball. Text questions included on the template encourage students to hypothesize about the anticipated results and state the anticipated outcome. As they continue to collect data, students see a pattern in the line graph that may challenge their initial assumptions. The first few drops appear to support the notion that the height of the ball will always result in a proportional increase in the height of the bounce. Higher and higher drops demonstrate that the bounce (responding variable) eventually responds less to the height of the drop (manipulated variable). Why?

This is the "teachable moment" teachers dream of. The decline in proportionate "bounce" can be attributed to the principle of terminal velocity—a concept that is challenging to teach, but fairly easy to grasp when conducting an experiment of this kind. The free-falling ping-pong ball reaches terminal velocity as the upward force of air resistance becomes stronger and the ball's downward speed increases. This upward force eventually stops the ball from accelerating downward. The point when the ball stops accelerating is its terminal velocity, or maximum speed. Once terminal velocity affects the drop, the resulting bounce heights are relatively consistent. The effects of terminal velocity can be seen on the line graph and in the data. When you present the concept of terminal velocity in this fashion, it makes more sense to students.

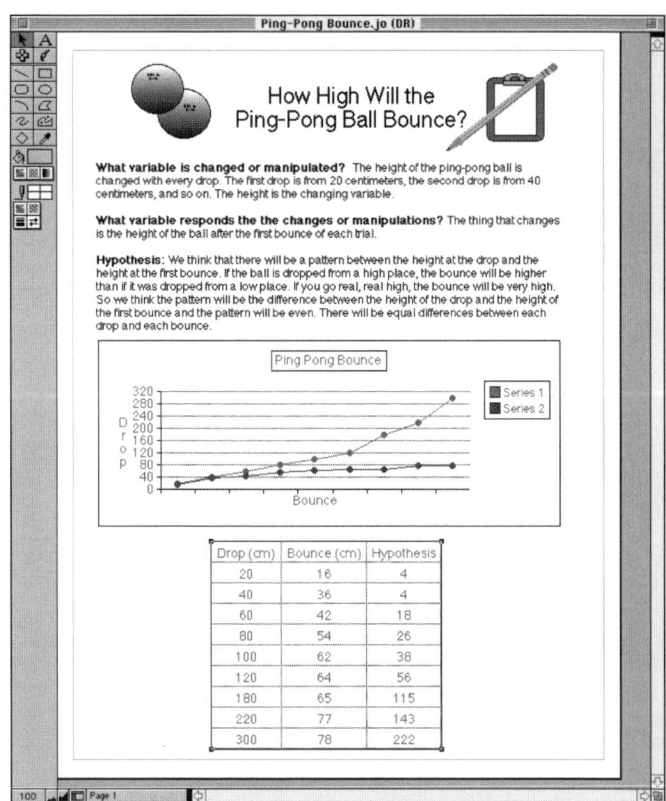
Ping-Pong Bounce.jo (DR)

How High Will the Ping-Pong Ball Bounce?

What variable is changed or manipulated? The height of the ping-pong ball is changed with every drop. The first drop is from 20 centimeters, the second drop is from 40 centimeters, and so on. The height is the changing variable.

What variable responds the the changes or manipulations? The thing that changes is the height of the ball after the first bounce of each trial.

Hypothesis: We think that there will be a pattern between the height at the drop and the height at the first bounce. If the ball is dropped from a high place, the bounce will be higher than if it was dropped from a low place. If you go real, real high, the bounce will be very high. So we think the pattern will be the difference between the height of the drop and the height of the first bounce and the pattern will be even. There will be equal differences between each drop and each bounce.

Drop (cm)	Bounce (cm)	Hypothesis
20	16	4
40	36	4
60	42	18
80	54	26
100	62	38
120	64	56
180	65	115
220	77	143
300	78	222

Body Maps-Body Math

This activity is designed to allow fourth- and fifth-grade students to construct knowledge about the proportions of the human body using data collected through measurement. For example, height is proportionately equal to a person's arm span. The addition of the bar graph to the template visually reinforces the proportional relationships between the sizes of certain parts of the human body. Although the data shown in our template focuses on a small number of body measurements taken from one student, the spreadsheet is capable of generating any number of graphic displays on any selected portion of the spreadsheet. Once the spreadsheet has been completed by all members of the class, students can generalize based on the graphic and numeric data drawn from the graph. The interdisciplinary focus of the activity makes this template a winner for health and science classes and provides plenty of data to explore during mathematics lessons.

Note: The skeleton graphic on the template was taken from MECC's BodyScope program.

Journaling with a Database

By Deborah Gettys

Journaling is frequently used to collect information about students and their learning; databases are an effective tool for managing information. Why not combine the two?

Journaling in mathematics—what a terrific idea! Its virtues and values have been written and extolled, workshops have been presented, and most of us have added the idea—if not the action—to our vast repertoire of classroom strategies. But as with any "new" idea, it will not make the gigantic leap from paper to practice in many classrooms without a means of management. A database management program, by its simplest definition, is a system that accesses and organizes any collection of similar pieces of information. And, a journal, in its simplest form, is a collection of similar pieces of recorded thoughts or information. It's a natural fit.

Creating a database for math journaling

Consider the type of information students can record in a mathematics journal: date, topic, summary of what happened in class, and a practical application of the topic, to include a few possibilities. The format might resemble the following:

Date	Name
Topic	Summary
My feeling about today's class	Feeling [numeric]
Material or activity used	Today I learned

Questions

Think about collecting a tall stack of spiral notebooks from your class of 30 students and reading each one every day. It becomes more and more difficult to keep track of what students have written up to the current entry. Tracking trends or patterns of likes, dislikes, questions, concerns, and successes is likely not to happen. There are too many pieces of information for one person to manage.

However, when this information is collected in a database, whether it's daily or several times each week, it can be accessed, sorted, manipulated and made usable! Figure 1 shows a few entries from a database created by a fourth grade class. By selecting only relevant data, the information becomes accessible and useful in assessing what's happened in the classroom and how it has affected individual students.

Date	Name	Today I learned	My feeling	Topic	#	Material	Questions
5/11	Jeremy	We did this last year.	Liked it.	Reading decimals.	10	Out of the book, at desk, board.	None.
5/11	Pinky	How to use decimals with whole numbers.	It was fun.	We're starting decimals.	09	At the desk.	Are these like fractions?
5/13	Jeremy	I sort of knew this before.	Just like yesterday.	More decimals.	10	Ruler, overhead.	
5/13	Pinky	I learned where to put the decimal with dimes and pennies and how to measure lines with decimals.	OK and sort of easy.	We were starting measuring with decimals.	05	Rulers, book.	

Figure 1. Database from a fourth-grade math class.

Student applications

• *What have you learned about this topic?*

All of us have a way of sometimes forgetting the details of past events. When Jamie was asked to describe what she had learned during the unit on decimals, her first response was to shrug and say, "I don't know." But when she had a chance to look at her records in the database (Figure 2), she saw that she had learned quite a bit. Since the information was recorded in her own words, it made sense to her and she could feel proud of what she had learned.

• *What have you enjoyed doing in class this past quarter?*

Have students sort their entries by Topic and then sort again by a Feeling (numeric) indicator. Students will have a summary of all the topics discussed in class, listed in order from their favorite to least favorite (or vice versa). Dealing with specifics rather than blanket generalities is important in developing critical thinking skills.

• *How does what we're doing today connect with what we've done already in this class?*

Most database management programs allow you to select records that contain key words in a given field. As a class, use this feature to search for key words in the Summary field and select those records for days on which the current topic or related concepts were addressed. This can be especially helpful if journal entries are made for several subject areas. For example, if students are beginning to study measurement, do a search on MEASURE (or any form of the word). They may be surprised when it shows up in science or social studies entries as well. What better way to help them begin to connect individual bits of information?

Teacher applications

• *How did today's lesson go?*

Select the information contained in the Summary field, copy it, and then paste it in a word processing document. You'll immediately have a collection of how students thought the lesson went (Figure 3). In the past, I would ask students at the end of each semester to write down what their favorite lesson or activity had been. Many times, their memories of what we had done didn't reach back more than a month or so. By gathering individual reactions immediately, it's much easier to keep track of what worked and what needs to be revised or thrown out.

• *What topics have we addressed this past quarter?*

Select records for the desired time using the Date field. Sort by Topic and print. This is much easier than going through pages of lesson plans to review what objectives have been worked on, what needs additional attention, and what remains to be addressed.

• *How can I communicate with parents what a student has learned, in the student's words?*

Select the Summary or any other desired field for all records and sort the previous grading period's entries by Date. For each student, print this collection for an easy-to-read mathematics summary to share with parents at conference time. One parent of a particularly "closed-mouth" student really

Name	Date	Today I learned
Jamie	5/11	I learned lots of things about measuring small lengths.
Jamie	5/13	How to do money with decimals and measuring. It was all new.
Jamie	5/14	How to compare decimals.
Jamie	5/17	First we compared again because some kids didn't understand that, then we added decimals.
Jamie	5/21	How to do story problems with decimals.

Figure 2. What have you learned about this topic?

Summary of what happened today

- We started on the overhead. Mrs. T. said decimals out loud. She chose kids to go up to the overhead and write them down. We went up to the chalkboard and she read more out of the book. Kids at desk wrote down, then we did a worksheet.
- I paid attention. People went up to the overhead and wrote down the answer to the problems she gave you. Looked up to the overhead and then the chalkboard at the answers of the other kids.
- We got our math books out. We looked at pictures of base 10 blocks in our book and then some people did some on the overhead. Eight people came up to the board and did some of the problems. Then she handed out a worksheet.
- We're learning about decimals. We'd already done the tenths and now we're going in to the hundredths. It's pretty easy.
- Mrs. T. told us numbers and we had to write them down on the board or at our desk.

Figure 3. How did today's lesson go?

appreciated having information available in this format. A word of caution for this application, however. If you intend to share any part of a student's journal, let the student know up front to maintain his/her trust.

• *What have students enjoyed? What should I keep doing? What should I change?*

A number of questions dealing with students' feelings about the content and class can be asked. While the range of student opinion will most likely vary greatly, it can be helpful to look for general trends among the complete class or within an individual student. For example, begin by sorting student records by Feeling (numeric) and printing a list of Topics or Activities in the resulting order (Figure 4). Some general conclusions may be drawn, such as feelings seemed to be most positive at the beginning of the unit! As a teacher, I can ask myself, what needs to happen differently so that enthusiasm remains high?

Now, if I look at individual student feelings (Figure 5), I can see that Jered's feelings about the unit never changed from mediocre and Jeremy thought everything was great. I can use my knowledge of the individual student to determine if Jered doesn't feel very positive about decimals because he doesn't understand them, and if Jeremy thinks it was all great because it was too easy. This can provide one more way of helping the teacher assess individual needs.

Date	Name	Feeling	Topic
5/11	Jeremy	10	Reading decimals.
5/13	Jeremy	10	More decimals.
5/14	Jeremy	10	Decimals.
5/17	Jeremy	10	Decimals.
5/11	Pinky	09	Starting decimals.
5/13	Jamie	09	Money with decimals and measuring.
5/17	Joel	09	Adding decimals.

Figure 4. What have students enjoyed?

Date	Name	Feeling	Topic
5/11	Jeremy	10	Reading decimals.
5/13	Jeremy	10	More decimals.
5/14	Jeremy	10	Decimals.
5/17	Jeremy	10	Decimals.
5/21	Jeremy	09	Decimals.
5/11	Jered	05	Decimals in the hundredths.
5/13	Jered	05	Still the hundredths.
5/14	Jered	05	Decimals.
5/17	Jered	05	Same thing as before.
5/21	Jered	05	Reviewing decimals and doing story problems.

Figure 5. Individual students' feelings.

• *What materials have I used? Have the materials made a difference in student response?*

Sort records on the Materials used field. Within each material grouping, sort records by the Feeling (numeric) field. If a particular material or activity is especially popular and effective with students, it will appear at the top of each grouping on a fairly consistent basis. Likewise, if a material is not well received, it will appear towards the bottom. This particular question was an eye-opener for me; I had always assumed that students didn't particularly enjoy working with base 10 blocks, but found some of the more exciting manipulatives such as geoboards and pattern blocks to be extremely motivating. I was surprised to discover through this question that because of the increased understanding gained with base 10 blocks, manipulative wound up at the top of the list!

• *What areas need reteaching or reviewing?*

Select the Questions I have field. Print a list of all entries in this field (Figure 6). Use the resulting list as a basis for regular review and reteaching.

Table 1 summarizes these applications; note that some are for student use and allow students to review and reflect upon what they learned and how they felt about the lesson. Others are intended primarily for teacher use and serve as an instructional or management tool.

Logistics for integrating database journaling into the classroom

The specifics for integrating the use of databases for journaling in the mathematics class depends on the accessibility to computers. If several computers are available in a classroom, each student can maintain an individual journal, making entries daily or several times each week. If only one computer and some type of projection unit such as an

LCD are available, the journal can be kept for the class as a whole. If a projection unit is not available, consider gathering students, half of the class at a time if necessary, around the computer, enlarging the font, and acting as recorder for students' input. For some classes, depending on their familiarity with expository writing, you may decide to begin as a total class and move gradually into individual journals.

There are many database management programs available, all operating on the same basic principles. Once you have worked with one, you'll feel comfortable with any. If you have an Apple II in your classroom, AppleWorks by Claris is suitable. Microsoft Works or Claris FileMaker Pro are available for both Macintosh and MS-DOS computers. Many database management programs do not allow multiple lines within a field, so writing extensive Summary or Comments fields may seem impossible. Simply create several Summary fields (e.g., Summary 1, Summary 2, etc.) and when doing a search or select, apply the procedure to all related fields. It may take a little experimenting or playing with your selected program, but most likely, there's a way to do what you want to do. The important part is realizing the potential of this tool and then deciding how it can support what happens in your classroom.

Date	Name	Questions
5/11	Jeremy	None.
5/14	Jeremy	Sometimes it's hard to know why a number that looks bigger is really smaller.
5/17	Jeremy	This is just like adding other numbers.
5/11	Pinky	Are these like fractions?
5/14	Pinky	Why is 2.3 bigger than 2.12 when 12 is bigger than 3?
5/11	Jered	Is this like money?

Figure 6. What areas need reteaching, clarifying, or reviewing?

Applications	Fields Required
STUDENT USES	
Ranking topics by student enjoyment.	Topic, numeric feeling.
Making connections between topics in mathematics.	Search for key words in summary.
Making connections between math and the real world.	Topic, search for key words in how the lesson applies.
TEACHER USES	
Summarizing content presented in a given time period.	Date, topic.
Student summary of the quarter for parent conferences.	Date, summary (could include feeling fields as well).
Relationship between student's feeling about class and different math topics.	Sort by topic and then sort by student numeric feeling.
Listing the use of specific materials, relating to effectiveness in class.	Date, topic, materials, numeric feeling.
Areas for reteaching or review.	Search for key words in questions.

Table 1. Summary of database applications.

Summary

Keeping database journals in mathematics class provides benefits for everyone involved. For students, it aids in individual construction of knowledge, facilitates connections within math and between math and the real world, and encourages communication—between teacher and student, and student and student. For teachers, it provides a means of managing valuable student input, monitoring an instructional "route," and serving as a vehicle of communication between teacher and student, and teacher and parent. For parents, it helps summarize and communicate what their child is learning and thinking. Using a database management program for journaling is a beginning to making effective use of the computer in the classroom. Explore, experiment, and expand—take advantage of this powerful tool!

Teach Students to Communicate MATHEMATICALLY

By Marge Cappo and Gail Osterman

Luis and Maria are sitting back to back with geoboards in hand. Luis is stretching rubberbands around the 9 x 9 grid to sketch a design—a sailboat. When he's finished, without telling Maria just what he's drawn, he gives her directions on how to recreate the same design. No peeking allowed!

Of course, Maria can ask questions. In what direction does the straight line go up and down or across? Is the triangle a right, isosceles, or equilateral triangle?

In a little while Maria and Luis progress to *Elastic Lines,* an electronic geoboard that allows them to use different sizes and types of geoboards as well as rotate and flip shapes. Their collaboration continues at the computer as they talk mathematics and record their discoveries in their math journals.

Appealing to many students

Luis's and Maria's experiences are a far cry from math classes in which students are required to memorize formulas, practice computations, and answer questions with "yes," "no," or a number. This "new" type of instruction is giving them an opportunity to communicate, and, in the process, develop a deeper understanding of mathematics.

The instructional impetus comes, in part, from the National Council of Teachers of Mathematics, who last year issued "The Standards." One standard, mathematics as communication, is encouraging teachers to employ various methods—writing, talking, representing, reading, and listening—to help students explore the language of numbers and their relationships. "As students communicate their ideas," states The Standards, "they learn to clarify, refine, and consolidate their thinking."

This standard is supported by Howard Gardner's theory of multiple intelligences, for employing a variety of teaching techniques increases the probability of reaching students. Luis' and Maria's exercises, for example, offered

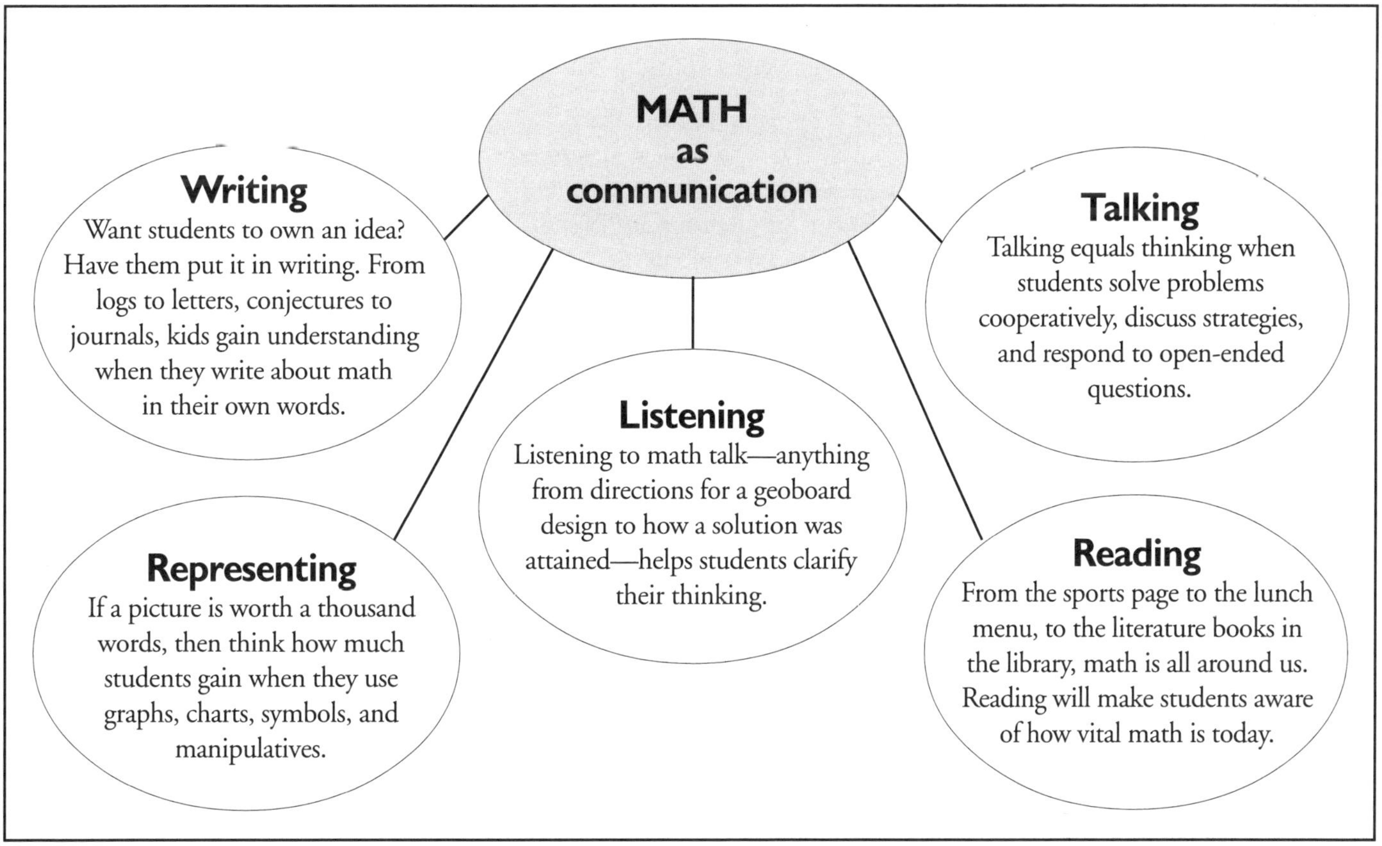

MATH as communication

Writing
Want students to own an idea? Have them put it in writing. From logs to letters, conjectures to journals, kids gain understanding when they write about math in their own words.

Talking
Talking equals thinking when students solve problems cooperatively, discuss strategies, and respond to open-ended questions.

Listening
Listening to math talk—anything from directions for a geoboard design to how a solution was attained—helps students clarify their thinking.

Representing
If a picture is worth a thousand words, then think how much students gain when they use graphs, charts, symbols, and manipulatives.

Reading
From the sports page to the lunch menu, to the literature books in the library, math is all around us. Reading will make students aware of how vital math is today.

Figure 1.

something for the kinesthetic learner (manipulating the rubberbands and the computer keyboard), the linguistic learner (writing in the journal), the interpersonal learner (listening and talking to a partner), and the logical/mathematical learner (the actual subject under exploration).

Today's math teachers do not limit their instruction to just one method, nor do they direct their teaching to only one intelligence. They "pull out all the stops" to help students become mathematically literate.

MATH PUBLISHING

As adults, we often write to help us think. We write to raise questions, to clarify what we don't understand, and to communicate our ideas to others. So it is not surprising that writing in math class helps students grow mathematically.

A major benefit of writing about math is that it affords students time to think. Too often students look for the quick answer and are not allowed sufficient time to ponder a problem.

Math publishing enables students to take ideas from the specific to the general. Through their writing and editing, students can move from the skills needed to solve a particular problem to the strategies needed to solve a class of problems.

Writing also provides students with a means to share their thinking. Teachers can gain insights into students' understanding and students can learn from one another. By reading aloud or displaying their writing around the classroom, students are able to exchange ideas and strategies.

Equally important is the fact that students witness growth of their mathematical thinking. Through the editing and rewriting process, students can observe the changes in their thoughts and see their growth in conceptualizing. In fact, students who keep math diaries during the year will have a detailed record of their growth.

1st day→ Today I tryed Building Perspectives. It was very hard. I don't know how you could possibly figure it out. I'm very frustrated!

a week later→ I have discovered 2 stratages to get the answer. number 1, is by using a piece of paper and making sqares to fill in the names of the colors like this→ 12345 I also remove the sqare so I can see what's behind then or beside them.

another way is to find out little places of information and then go to predict and do a little of problem solving there. you keep doing over + over and till all the sqares are filled in.

Figure 2. "Once the kids started writing, a new rhythm to the class developed. Their attention span increased, and they seemed more interested in developing strategies."—Marilyn Husser, Howard Elementary, Eugene, Oregon.

Ideas for writing in math class

There are many ways to bring writing into math class. Teacher Joan Countryman from the Germantown Friends Academy in Philadelphia uses learning logs in which students write a daily account of what they learned.

Marilyn Husser of the Howard Elementary School in Eugene, Oregon, combines journal writing with the use of problem-solving software. Frequently, after detailing their strategies in writing, students exchange journals with a peer. Students then act as robots and follow the strategies exactly as written—an enlightening exercise for most!

Sue Dolezal of Sentinel High School in Missoula, Montana, asks her geometry students to finish phrases, such as "The difference between an isosceles and an equilateral triangle is . . . " or "What I understand by a tangent line

What is the first thing you do when, you get the spinner's numbers?
I look at the screen and see what number I need to reach a bump, shortcut, or town. Then I try to get that number with the spinner's number

Figure 3. Students write questions about their strategies for playing How the West Was 1+3x4.

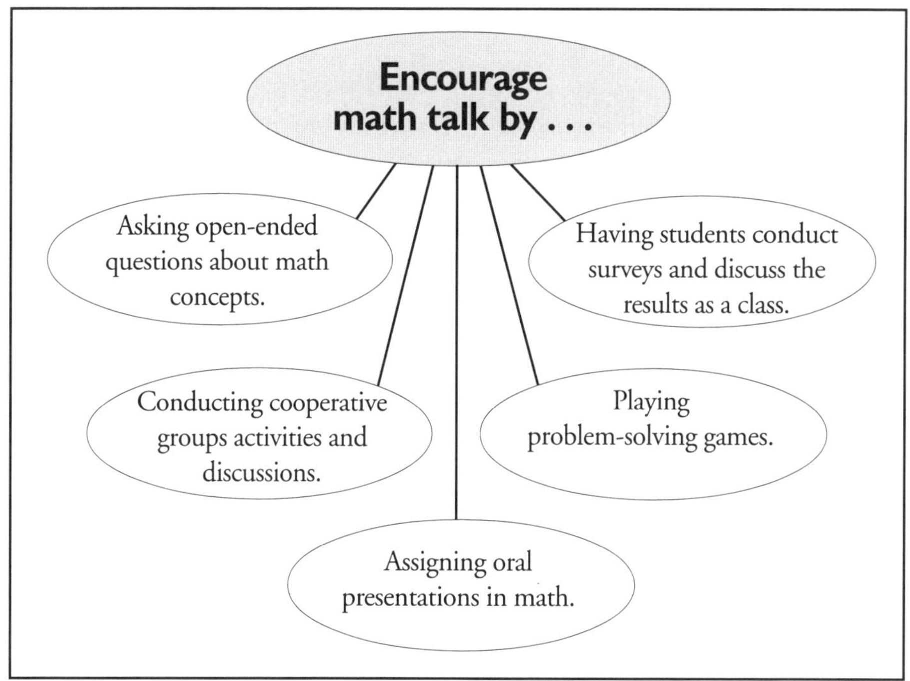

Figure 4.

is . . . " These exercises help students clarify what they do and don't understand.

Lynne Alper, a computer mathematics specialist for Project EQUALS, often concludes a math software lesson by asking students to write questions about strategies that can be used to play the software. She collates the questions and then has groups of students select two questions to answer.

Problem creating

Writing math problems can be as valuable as writing about math concepts. Andrew Kramer of the Jewish Community Day School in West Palm Beach, Florida, believes creating word problems helps students understand mathematics.

His students' favorite computer lab activity is Carnival Math. Using Magic Slate II, students develop word problems at the computer and then leave the answer on a sheet of paper face down next to the computer.

Students move from computer to computer to solve each other's problem. Beforehand they receive play money. It costs $10 to play a problem, which students place in a designated spot next to the computer. If their answer is wrong, they have the opportunity to play again for another $10.

If the contestant discovers that the author's answer is wrong, the contestant must first prove it to the teacher and then correct it. As a reward, the contestant may take all the money at the computer and transfer it to the computer that displays his or her problem.

MATH TALK

Do your students say they never use math except in math class? Do they think that math is unimportant in their lives? A simple game will illustrate otherwise.

Tell your students that for one day your classroom is the Kingdom of Nomath and in this kingdom...you guessed it...no one is allowed to talk math. Nomath citizens who believe someone violates this rule may refer names to the Council of Sages, students who determine what is and what isn't mathematics.

Your class will probably think it's easy to not talk math. But, just wait until they need the *liter* container for their science experiment, or it is *time* for a snack, or they have to *count* the lunch money. And what happens in music class when they try to sing in 3/4 *time?* Or at recess when the soccer *score* is 3 to 2? Or in social studies when you use *dates* to place the American Revolution in context? Taking attendance, instructing them to turn to page 106, mixing three drops of red and two drops of yellow paint in art . . . The list goes on and on and proves the point that we all talk math every day.

Games promote math talk

Games are an easy way to promote math talk, especially if they are played in cooperative learning groups. They come in many forms—from computer software to intriguing mind-teasers used as a "Problem of the Day."(See note below about resource materials.)

Lynne Alper, when using How the West Was 1 + 3 x 4 software, combines software *with* a problem of the day. After playing the game (which focuses on order of operations), she assigns her stu-

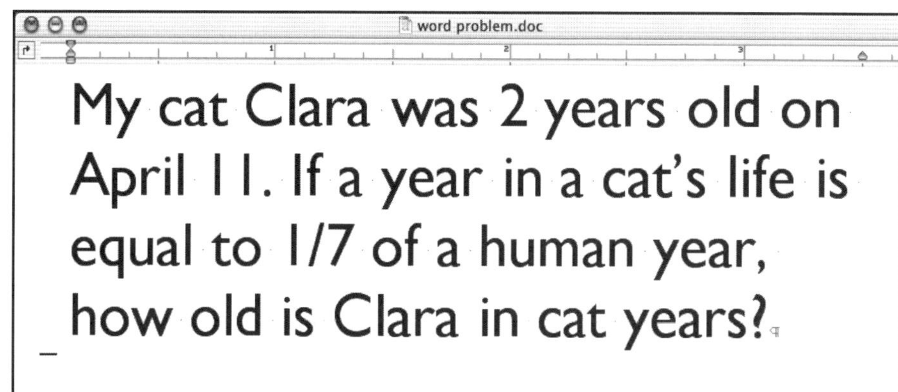

Figure 5. Students write word problems at the computer for others to solve.

MATH TALK

How many students had something for breakfast?

Only one student had everything that was offered for breakfast. She had juice, cereal, and eggs.	Nine students had juice for breakfast, but only two had nothing else.
The same number of students had juice and cereal (without eggs) as had cereal by itself.	Five students had eggs, but only one had them without anything else. Eight students had no eggs.
Four students had juice and eggs for breakfast.	

1. Give each group member a card with a clue on it to share with the group.
2. Each group must cooperatively determine the number of objects in every possible region on a Venn diagram.
3. When the groups are finished, compare the results. How different were the diagrams the groups created?
4. Promote additional math talk with the following questions:
   - Is there a better way to solve this problem? Can you think of another kind of diagram?
   - Must all Venn diagrams be circles? Can you make other shapes to represent the sets?
   - How else might the information be organized so that it leads to discovering the solution?

From Get It Together, Math Problems for Groups, Grades 4–12, *by EQUALS. This is an excellent source of fun problems. Teachers of younger students can get ideas from* Mental Math in the Primary Grades, *by Jack Hope, available from Dale Seymour Publications, or* The Problem Solver, *by Shirley Hoogeboom, available from Creative Publications.*

dents to select any three numbers. Lynne then challenges them, using all the math operations, to generate as many different answers as possible. Debating, arguing, challenging, collaborating, sharing, exchanging . . . Lynne encourages it all in an effort to get her students to talk mathematics.

Keep on talking

Another technique to promote math talk is the use of open-ended questions. Requiring students to explain something forces them to articulate their ideas. Sample questions could include: "How would you change this recipe for six to a recipe that feeds ten?" or "What is an easy way to determine the lowest common denominator?" or "How would you measure the classroom if you did not have a measuring stick?"

A related technique is the use of oral presentations. They are readily employed in other subjects, so why not math? Assignments could be as simple as telling the class what a parallelogram is, to more demanding tasks like discussing the strategy used to solve a Puzzle Tanks challenge, to involved assignments such as explaining how to design a tessellation.

Surveys have always been popular with students. They are also rich with potential for math talk. To start, focus on topics of interest to kids, such as how much money fellow students spend in one week or how many hours of television they watch in a week. As they become hooked on surveys and discussing the results, students can explore various ways to show the data... which leads into another method of communicating mathematics—representing.

READING IN MATH

Reading and interpreting mathematical information is another way of learning math. Math abounds in the printed word, from everyday reading material (menus, baseball cards, department store ads) to books in all the content areas.

One of the best teaching tools around is the newspaper. Take a look at any newspaper and scan the articles. You'll notice references to patterns, comparisons, relationships, statistics, graphs, charts, wages, temperatures, prime rates, and much, much more.

Don't be surprised if every article has a mathematical detail . . . or astounded by the large number of details. A recent review of the front page of The New York Times turned up more than 100 mathematical references, including a story about "Political Math."

Read all about it!

Use topics in the newspaper to encourage math communication. For example, why do some people fear the census? It may come down to math. If welfare authorities discover an elderly man sharing his one bedroom apartment with his son and daughter-in-law, he could lose his rent subsidy.

Or, let the frequent lotteries springboard you into a discussion on probability. For instance, a recent article explored the question of winning twice in one year. Such articles can lead to talks about "lucky numbers," how students would spend the money, or the amount they would keep after taxes.

How about projecting the impact of the postal rate on a mail order company? Or using the FDA's call for new labeling of packaged food to sensitize kids to their dietary habits? Can students calculate the percentage of the U.S. Required Daily Allowance their food provides them?

Or, have students use the newspaper to create business-related problems that include the costs of hiring employees (information in the Help Wanted section), interest rates for borrowing money, the rents for warehouse space.

Whole language . . . whole math

Don't stop with the newspaper. Once students start reading about math, they will recognize (just as they did with math talk) that math is present in much of the written word.

Jean Merrill's *Pushcart War* is a wonderful example of this fact. The book, appropriate for upper-elementary and middle school students, details in a satirical manner the methods pushcart vendors used to keep the trucks from taking over the city. Aside from being rich in language arts (character development, how the setting influences the changes that transpire, the message the author is trying to convey) and social studies (how people live and work together, government's role in solving problems, comparison of the Pea Blockade to the Boston Tea Party), mathematics permeates the novel!

The possibilities for math projects are unlimited. Students can make scale models of the city, draw maps to record the battles, and explore percentages needed to enlarge photos (as Morris the florist did) to life size. They can explain and evaluate the Flower Formula, create scenarios to illustrate the concept of public opinion and how it increases, and experiment with volume when the issue of pea storage arises. They can evaluate the Large Object Theory of History, calculate the traveling time of transporting goods across the city at various hours of the day . . . the list goes on and on.

Other books, such as the popular Anno books or Jonathan Swift's *Gulliver's Travels,* which is a component of the "My Travels with Gulliver" Math Module, can be used to round out math lessons and give students a whole math experience.

MATH IN LITERATURE

Book	Concept	Related Software
Anno's Counting House, by Mitsumasa Anno	One-to-one correspondence	Muppets on Stage Counting Critters Number Farm
How Much is a Million?, by David Schwartz	Number sense	Dazzle Draw (create a simple shape that can be replicated one million times)
Farmer Mack Measures His Pig, by Tony Johnston	Measurement	Learn About: Animals
The Tooth Paste Millionaire, by Jean Merrill	Business math	Whatsit Corporation Survival Math Algebra Shop
Anno's Hat Tricks, by Mitsumasa Anno	Logic	Recycling Logic High Wire Logic What Shape is That Color?

LISTENING IN MATH

Implicit in all math communication is listening. Talking, writing, representing, and reading all require listening—whether it is to other students or to a teacher.

There are many fun ways teachers can build students' listening skills in math class. Recite a series of mental math problems—humorous problems, estimation problems, fantasy problems—that must be solved without paper and pencil.

Or, give oral instructions to math tricks. For example, think of a number; add 5 to it; multiply the result by 2; subtract 4; divide by 2; subtract the number you first thought of. Surprise students by telling them that the answer is 3. After students try the trick several times, listen to them explain why they think the answer is always 3.

Another method is to assign lessons that require following directions, such as the geoboard activity in which Luis and Maria were involved. Or, if you use cooperative learning groups, require each group member to be able to explain the group's reasoning; this demand will certainly promote careful listening!

Teacher-student interviews

Listening is not limited to students. Teachers need to listen also. One of the best techniques of determining the level of mathematical understanding is to interview, or briefly question, students.

The questions could be about the methods for solving a problem or even why students think the answer they gave to a problem is right. Don't be surprised to find students who have correct answers, but lack a fundamental understanding of the concept. Or conversely, get the answer wrong but grasp the concept.

It all overlaps

It is not the specific communication mode that is important. Listening, writing, talking, representing, and reading all overlap. What is important is that these modes of communication become a part of students' math instruction. The more they are used, the easier it will be for students to communicate their mathematical ideas. They will learn to clarify, refine, and consolidate their thinking.

Achieving this goal will benefit all of us, for it will move us one step closer to becoming a mathematically literate population.

The authors wish to thank Marge Scherer, editor of *Sunburst Strategies,* for her contribution to this article.

CHAPTER 2

Connections

The Conversation Continues

Ivan: If you asked a group of mathematicians,"How does your discipline relate to other disciplines?" would many of them give you a ready answer? Why do so many of them have a limited view of the connections with other content?

Anne: There are many examples of people who studied mathematics and found ways to apply mathematics to other fields of interest. Two such examples include Lewis Carroll and Art Garfunkel. Lewis Carroll studied mathematics in college. He went on to become a writer, most noted for his tale *Alice in Wonderland,* in which his writing style involves sequential logical and illogical paths. Thus, through Carroll, we have an example of how mathematical thinking influenced creative writing. Carroll also went on to write a book of logical puzzle problems, again applying mathematical thinking to writing. In the case of Art Garfunkel, we see a trained mathematician who went on to write and perform music. There are many known connections between mathematics and music, and Art Garfunkel was able to apply his mathematics background in developing his unique musical style.

Ivan: Although we tend to be aware of mathematicians who applied math in other areas, we often overlook those people who were not trained in mathematics and yet made significant discoveries in mathematics through their endeavors in other fields. For example, M.C. Escher is a noted artist who, through his art, became interested in mathematics and developed mathematical ideals that many mathematicians continue to pursue. Another example of such an individual is Benjamin Bennaker. He was an African American descendant of slaves who became fascinated with the workings of a clock. From his personal exploration of a clock, he began to school himself in related mathematics concepts. He developed into an inventor and architect, creating mathematics connected to these interests. He was ultimately invited by Thomas Jefferson to help design and plan the layout of Washington, D.C. Bennaker is among the first African Americans to become a noted mathematician, and his studies of mathematics began through his interest in mechanical workings and inventions.

Anne: The articles that follow give you ideas for making the connection to other content. They provide examples that will make it more viable for many teachers. Marcie Zisow describes how her fourth-graders generate authentic questions while completing a fundraising activity incorporating technology and mathematics. Bobette M. Morgan and Joe Jernigan build on the work of Leonardo da Vinci to help students analyze data. Leon Roland uses a spreadsheet in the study of how our representative government developed.

THEORY INTO PRACTICE

Railroad Math

By Ivan W. Baugh

Consider this analogy. In railroading, a unit train consists solely of one kind of car carrying a single commodity to a single destination. A mixed train includes a variety of cars going to a yard for distribution to various locations at that destination. Sometimes, it stops to drop a car at intermediate points.

Mathematics education often follows the unit train model. It serves the purpose of moving a single commodity to a destination. It only stops for crew changes. By contrast, connected learning parallels the mixed freight train. It consists of a variety of cars (tank cars, open hoppers, closed hoppers, flat cars, gondolas, different types of freight cars representing the various areas of the curriculum) carrying an assortment of goods (learning opportunities).

In both instances, connections must be made. Couplers link the cars together and to the engine. Hoses connect each car's brake system to the engine, thus giving the train distributed stopping ability.

Covering the textbook in the time allocated often deters looking at the larger picture. In student-centered classes, the student becomes the engine that moves learning forward. The assortment of cars represents the various topics in the curriculum. As students make connections, they internalize the material and will more likely remember it beyond the next test. This approach creates a need to know through the fostering of real-world associations.

NCTM Connections Standard

Connections: Instructional programs from prekindergarten through Grade 12 should enable all students to

- Recognize and use connections among mathematical ideas
- Understand how mathematical ideas interconnect and build on one another to produce a coherent whole
- Recognize and apply mathematics in contexts outside of mathematics

Mathematics as one of the cars in the mixed freight train (one of the subjects in the curriculum) serves an essential function. An effective instructional program will make connections with prior mathematics instruction, using that as a basis that accrues toward a comprehensive understanding of the content. Bullets 1 and 2 in the standard frequently occur. Bullet 3 is the area most often neglected. Teachers frequently leave this for the student to do. Consequently, it seldom occurs.

Students learning statistical analysis in mathematics could connect with data analysis in a science class or a social studies class. In the science class the students would analyze data gathered from a laboratory project, such as reaction time. They would prepare their report including the conclusions drawn from the data analysis. In social studies, they could examine census data or population totals for various areas. Following the data study, the students would formulate plans for addressing an existing problem in a problem-solving context. Another connection could be made with physical education as they monitor their progress meeting the standards of the President's Council on Physical Fitness and Sports. This could include looking at the progress of males and females and comparing the results.

NETS•S Research Standard

5. Technology Research Tools
    - Students use technology to locate, evaluate, and collect information from a variety of sources.
    - Students use technology tools to process data and report results.
    - Students evaluate and select new information resources and technological innovations based on the appropriateness to specific tasks.

Locating information related to the studied content, whether mathematics or a connected content, often presents a challenge for students. Technology, whether used to locate print materials in a card catalog or used in an online environment to search available resources, makes conducting research much more efficient when an individual takes the time to develop a thorough working knowledge of how to use the available tools. One common short-coming I find among students of elementary, secondary, and postsecondary levels is a lack of skills to take advantage of advanced search features. They load a search engine, type in a word and spend hours looking through "gazillions" of links to find those relevant to the topic under investigation. With a small investment of time developing a working knowledge of search techniques using a Web site such as Search Engine Watch (www.searchenginewatch.com), research becomes significantly more powerful. My students have found the material on the Search Engine Watch Web site under Power Searching for Anyone, Search Features Chart, and Search Links, all of which are particularly beneficial to enhancing or developing effective search skills. The user learns to structure the search instructions in a manner that will produce relevant results.

With the search yielding relevant connections, the researcher devotes time and energy to making important connections. Through the expanded insights and the newly constructed understandings developed from synthesizing the discovered information, students increase the knowledge base of all involved as they communicate the results of their research.

The articles in this chapter provide examples of ways to use connections in the teaching of mathematics content. Nowhere does the mathematics content become secondary. Through modeling the making of connections while learning the skills, students associate them with daily life and better internalize them for future use.

Note: ISTE has published a resource for teachers who want to help students develop their Web searching skills. It is titled *Web Searching Strategies: An Introductory Curriculum for Students and Teachers,* by Sam Miller.

THEORY INTO PRACTICE

Connecting Mathematics and Literature

By Anne Raymond

Educators have recognized the value in connecting mathematics to children's literature for quite some time. Many "how to" books have been written to illustrate lessons that connect mathematics and literature (e.g., Bresser, 1995; Sheffield, 1995; Welchman-Tischler, 1992; Whitin & Wilde, 1992). Further, numerous articles have been published that demonstrate the integration of mathematics and problem solving (e.g., Lietze, 1997).

Some purposes for using children's literature to teach mathematics include

- Providing a context or model for an activity involving mathematics
- Reviewing, introducing, developing, or preparing for a mathematics concept or skill
- Posing an interesting problem or a context in which students can pose their own problems

Example 1. In the book *How Many Feet in the Bed?* (Hamm, 1991), we hear the story about one family waking up in the morning. As family members wake up and climb in and out of the parents' bed, the book keeps track of how many feet there are in the bed. The context of this first-grade level book connects nicely to concepts of counting and adding. After reading the book, appropriate problem-solving questions to pose are:

1. If everyone in your family jumped on the bed on Saturday morning to watch cartoons, how many feet would be in the bed?
2. Show three ways with people and pets that there can be 10 feet in the bed.

Example 2. In *The Doorbell Rang* (Hutchins, 1986), two children are sitting down to eat a plate of 12 cookies Mom just baked. The doorbell rings multiple times, and each time, friends arrive in groups that form a total number of children who can evenly share the cookies. The story suggests a context for division. Even very young students can relate to the notion of sharing cookies, and through acting it out or making a chart, they can engage in the act of preparing for division to solve these problems:

1. When Sam and Victoria were first sitting down with the 12 cookies, how many would each of them get if they split the cookies evenly?
2. With 15 cookies, find how many people could share the 15 cookies evenly with no cookies left over.

Example 3. In Silverstein's poem "Smart" from his book *Where the Sidewalk Ends* (1974), we hear the lyrical story of a boy who keeps making trades with his money for other denominations of coins. Unfortunately, in the poem, the boy's trades are not always the best deal for him. This provides the context for the following problem:

Suppose the boy has $1 to trade for coins and he wants to make a fair trade. Using only nickels, dimes, quarters, and half-dollars, list all the ways he could trade his dollar for the same value in coins.

Example 4. In Flourney's *The Patchwork Quilt* (1985), we hear a touching tale surrounding a special quilt. The story of the quilt sets the stage for many geometric possibilities that are associated with quilts. One possible geometric problem-solving scenario is:

> *In the quilt design in Figure 2.1, how many different rectangles can you find?*

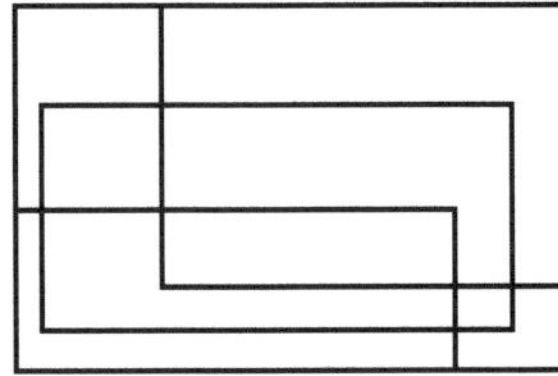
Figure 2.1

Example 5. In the book *If You Made a Million,* Schwartz (1989) poses many questions dealing with large numbers. He stirs up the context of thinking beyond what we can necessarily see, thus encouraging the reader to imagine and estimate. Connected to this book, one might pose the following questions:

> How many paper clips does it take to make a foot? How many paper clips would it take to make a mile? How many paper clips would it take to measure the distance from your home to school? How many paper clips would it take to measure from Louisville, Kentucky, to Lexington, Kentucky?

No matter what mathematical concept students are studying, it is helpful to provide them with a context in which they can imagine or see the concept. Literature provides a setting and a set of characters from which we can develop a wide range of scenarios within which we can build the mathematics. With such characters and images, students can also imagine their own mathematical problems and pose their own creative solutions.

References

Bresser, R. (1995). *Math and literature: Grades 4–6.* Sausalito, CA: Math Solutions.

Flournoy, V. (1985). *The patchwork quilt.* New York: Dial Books.

Hamm, D. J. (1991). *How many feet in the bed?* New York: Simon & Schuster.

Hutchins, P. (1986). *The doorbell rang.* New York: Greenwillow Books.

Lietze, A. (1997). Connecting process problem solving to children's literature. *Teaching Children Mathematics, 3*(7), 398-403.

Schwartz, D. M. (1989). *If You Made a Million?* New York: Lothrop, Lee, & Shepard Books.

Sheffield, S. (1995). *Math and literature: Grades K–3.* Sausalito, CA: Math Solutions.

Silverstein, S. (1974). *Where the sidewalk ends.* New York: Harper & Row.

Welchman-Tischler, R. (1992). *How to use children's literature to teach mathematics.* Reston, VA: National Council of Teachers of Mathematics.

Whitin, D. J., & Wilde, S. (1992). *Read any good math lately? Children's books for mathematical learning, K–6.* Portsmouth, NH: Heinemann.

THEORY INTO PRACTICE

Ethnomathematics

By Anne Raymond

Ethnomathematics can be defined as "the relationship between culture and mathematics". Broken down into its roots, *ethno* describes "all the ingredients that make up the cultural identity of a group: language, codes, values, jargon, beliefs, food and dress, habits, and physical traits" and *mathematics* includes ciphering, arithmetic, classifying, ordering, inferring, and modeling. The study of ethnomathematics challenges us to understand how "mathematics continues to be culturally adapted and used by people around the planet and throughout time" (D'Ambrosio, 2001, p. 309).

Traditionally in mathematics classrooms, the connections between mathematics and culture have not been emphasized in the content and instruction. However, a growing interest in helping students see such connections has emerged in the mathematics education community. Linking mathematics to culture provides both motivation for students and a sense of the usefulness of mathematics.

Some examples of how mathematics has been taught through an ethnomathematics approach include: studying geometry through Native American beadwork (Barkley & Cruz, 2001), learning about our number system by examining characteristics of number systems from other cultures (Zaslavsky, 2001), and building thinking skills by playing games from other cultures, such as the African game Oware (Powell & Temple, 2001). Students benefit not only from engaging in interesting mathematics, but also by learning about other cultures by examining societies through their histories, relationships, industries, and forms of entertainment (Voolich , 2001). Students particularly relate to other cultures and the power of mathematics through games (Zaslavsky, 1998).

Much about other cultures and the mathematics that are a part of the cultures can be investigated using the Internet. Further, many of the games and activities used in other cultures can be simulated on the computer, offering a technological connection to many ancient cultures.

References

Barkley, C. A., & Cruz, S. (2001). Geometry through beadwork designs. *Teaching Children Mathematics, 7*(6), 362-367.

D'Ambrosio, U. (2001). What is ethnomathematics, and how can it help children in schools? *Teaching Children Mathematics, 7*(6), 308-310.

Powell, A. B., & Temple, O. L. (2001). Seeding ethnomathematics with Oware: Sankofa. *Teaching Children Mathematics, 7*(6), 369-375.

Voolich, E. D. (2001). *A peek into math of the past.* Parsippany, NJ: Dale Seymour.

Zaslavsky, C. (1998). *Math games and activities from around the world.* Chicago, IL: Chicago Review Press.

Zaslavsky, C. (2001). Developing number sense: What can other cultures tell us? *Teaching Children Mathematics*, 7(6), 312-319.

INSIGHT

Statistics in Real Life

By *Ivan W. Baugh*

While participating in the Vision T.E.S.T. Forum (Vision: Technologically Enriched Schools of Tomorrow Forum co-sponsored by the U.S. Department of Education and ISTE, 1992), a member of the group described how a high school mathematics class and a football team partnered to help the football team prepare for their next game. The mathematics class analyzed game films of the football team's upcoming opponent. The class measured the forward progress of every play and recorded which position achieved the progress.

After collecting and recording the data, the class provided a statistical analysis to the football team. The analysis indicated the strengths and weaknesses of the upcoming opponent. With this information, the football team devised a new game plan, and the win/loss record for the team improved dramatically. Consider partnering with your school football team to compile this type of data to help them develop their game plan. Examine the impact on the school's win-loss record.

How could this type of analysis be applied to other school sports? What type of data would you need to gather for basketball or soccer? In what manner will you present the compiled data to facilitate the use of the information in preparing a game plan? In order to answer these questions, you will want to talk with the coaches to see what type of information they would want to have. Then investigate how technology can help you generate the desired information.

For example, you could create a grid that represents one-half of a basketball court (Figure 2.2). Use this grid to chart the shots taken by each player. A small circle could represent each shot taken. An X inserted inside the circle would show when the player made the basket.

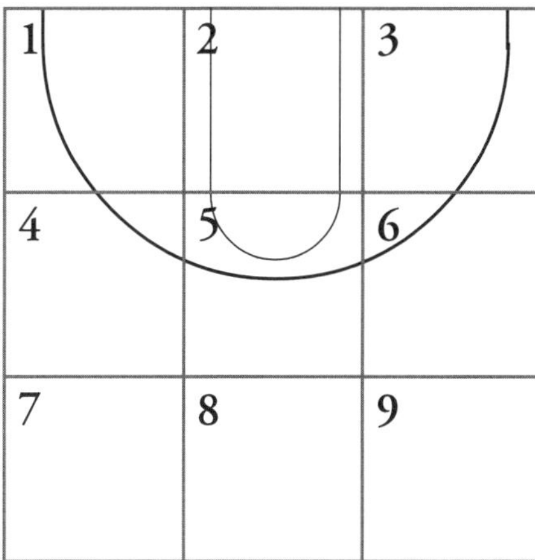

Figure 2.2

Enter the data in a spreadsheet and calculate the shooting percentage at each location on the grid. Use the spreadsheet to compute the players' and team's shooting percentages. Each week you could supply each team member with a progress report that would include a chart cumulatively showing the player's season totals. This would help the player find areas to work on.

Another student could chart the shooting percentages of opponents against each player. This would help players, in consultation with the coaches, decide how to improve their defensive play. Again, a chart showing the cumulative season totals would help the player monitor progress in this area.

Each of the above examples provide the student studying statistics with hands-on experience gathering data, analyzing the data, and providing information and interpretations to the players and coaches to help them improve the game and the team. As the statistics students complete the total process, they move from memorization to mastery learning as they work in a real-world environment. The activity gives them a real-world experience—not a classroom activity completed for the teacher's eyes only. It also gives them an opportunity to see that statistics have a valuable application in daily activities. This type of experience will look very impressive on a resumé.

THE TWELVE DAYS OF CHRISTMAS

A popular Christmas song can lead to exciting activities across the curriculum.

Have you ever wondered about the origin of "The Twelve Days of Christmas"? I remember thinking about it as a youth, wondering where and why they had such an observance. Then it occurred to me to consider how many gifts were received. Back then the math would have taken quite a bit of time. Today, students can use a spreadsheet to determine the number of gifts enumerated in the Christmas song. And they can search the Web for information about the background of the song.

By Ivan W. Baugh

Subject: Math, history, economics, science, language arts, multicultural studies, music

Grade Level: 4–12 (Ages 9–18)

Technology: Internet/Web, spreadsheet

Standards: *NETS* 3, 4, & 5. (Find out more about the NETS project at www.iste.org—select Standards Projects.) *NCTM* 1, 5, 6, 8, & 9. (View the updated math standards online at www.nctm.org.) *NCSS* 1, 5, & 9. (Read the social studies standards at www.ncss.org.) *NCTE/IRA* 1, 3, 7, 8, & 11. (Find the NCTE/IRA standards online at www.ncte.org.)

Online Supplement: www.iste.org/L&L

History. Challenge your learners to discover the origin and use of the song. I used the MetaCrawler Internet search engine and found that, according to some sources (e.g., **www.byrum.org/misc/christmas/origin.html**), the song originated in England during the rise of Protestantism. Beginning in the reign of Elizabeth I, members of the Catholic Church were forbidden to practice their religious faith. Encourage your students to figure out how long this ban lasted. Here are a few Web resources to help with these explorations.

http://www.geocities.com/alexstevenson.geo/christmas/12days.htm

This Web site discusses the origins of the song and provides insights into the period of English history during which it was supposedly written.

http://graceland.gentle.org/xmascel/12days.html

www.kencollins.com/Question-17.HTM

These pages describe the hidden meanings of the song. Have some of your students visit the first page and others visit the second. They can then compare the information they gather.

www.luminarium.org/renlit/eliza.htm

Visit this site to learn about other events during the reign of Queen Elizabeth I.

www.catholic-history.org.uk/

This site discusses the history of Catholics in England. It mostly contains bibliographic citations of primary source documents.

Science. Students could research the differences between doves and turtledoves. Information I found at **http://library.northernlight.com/BM19981221010119436.html?cb=0&sc=0#doc** indicates they appear on the endangered species list in England. How do French hens differ from other hens? And what are "colly birds"? What do they look like and what sounds did they make? Finding this information provides a challenge to students. According to Rupp (1996), "The four calling birds ... were probably 'colly' birds, from the Old English colly, meaning to blacken, as with soot or coal dust. ... Colly birds were thus European blackbirds" (p. 7).

What are some other winter holidays celebrated throughout the world?

Hanukkah (or Chanukah)

Kwanzaa

Ramadan

GRAPHIC COURTESY OF MELANET.COM.

Language Arts. Through the years, people have created parodies of the song. Challenge your students to create their personal versions of the song.

Music. An interesting extension of this historical insight involves looking at other uses of music to convey hidden meanings. Researchers will find that music played a significant role in African-American history. For example, African Americans used songs to help children learn to count ("Children Go Where I Send Thee") because they could not attend school. They also used song to send secret messages, for example, "There's a Meeting Here Tonight," which communicated information about a special meeting. Students could find different songs and describe their hidden meanings to the class.

Parents often think that current popular music contains hidden meanings. Have students examine the lyrics of their favorite songs to see if they contain any hidden messages.

Multicultural Studies. Which countries or groups of people also celebrate Christmas? How do their traditions differ from ours? To research Christmas traditions in other countries, visit **www.santaland.com/tradit.html**. Use the World Wide Web to locate schools around the globe that have e-mail. (Go to **www.mightymedia.com/keypals**, **http://web66.coled.umn.edu/schools.html**, or **www.epals.com** to find keypals for your students.) Write letters to exchange holiday customs. Conduct a survey to determine the number of different countries where the song is sung. How does the version sung in other countries vary from the one you traditionally sing?

What other winter holidays are celebrated throughout the world? Start by visiting the Diversity Calendar at **www3.kumc.edu/diversity/december.html** or Christmas.com's Winter holiday WorldView at **http://christmas.com/worldview**.

Hanukkah (or Chanukah). Have your students research the origin of Hanukkah at **www.ort.org/ort/hanukkah/**. Read personal accounts of Hanukkah celebrations around the world at **www.caryn.com/holiday/holiday-Hworld.html**. What numbers have a significant role in the observance of this holiday?

Kwanzaa. Students can research the history of Kwanzaa on these two Web pages: **www.cnn.com/EVENTS/1996/kwanzaa/history.html** and **www.melanet.com/kwanzaa/whatis.html**. Have them compare the information on these pages. Which numbers have special significance in Kwanzaa? Find out more about celebrating Kwanzaa at **www.dc.peachnet.edu/~mhall/mypage/holidays/winter/Kwanzaa/kwanzaa.htm**.

Ramadan. Get the background information you need as a teacher at **www.submission.org/teachers.html**. Send your students to **www.holidays.net/ramadan/**. They can read about the history and the celebration of the Fast of Ramadan. Do particular numbers have any special significance in Ramadan? Students can search the Web to find the starting and ending dates of Ramadan for the past few years. See what they can learn about the lunar calendar and have them search for patterns to predict the starting and ending dates this year.

Find other lesson plans for the winter holidays at **www.4teachers.org/premier/holidays.shtml**, **library.hilton.kzn.school.za/Interest/events.htm**, and **http://members.aol.com/Donnpages/Holidays.html**.

Conclusion

Through these experiences, learners gain new historical insights, develop an appreciation for the many different ways people can and do communicate, and construct personal knowledge as they extend their understandings. Happy holidays to all, and happy learning, too.

Reference

Rupp, R. (1996). 12 days and 184 birds. *Early American Homes, 27*(SPEISS), 4–9.

A TECHNOLOGY UPDATE: *Leonardo da Vinci and the Search for the Perfect Body*

Creating their own spreadsheets to generate information that verifies the work of a 15th-century artist might not intrigue many secondary school students. In the lesson provided here, though, they'll be actively involved in learning about Leonardo da Vinci's incredible work with proportions, ratios, and applied technology. The accuracy that he achieved without a computer will surprise many mathematics and art students, as the authors explain in this article.

This integrated lesson is designed for students in Grades 7 through 10 and can be used in technology, general math, algebra, and art classes. The charts were created using an AppleWorks spreadsheet, but the data can be adapted easily to any spreadsheet.

By Bobette M. Morgan and Joe Jernigan

Subject: General Mathematics, Algebra, Art, Technology

Grade Level: 7–10 (Ages 12–16)

Technology: spreadsheet software (e.g., AppleWorks or Microsoft Excel)

Standards: *NCTM*—Encourage students to become mathematical problem solvers. *NAEA*—Make direct connections between the visual arts and mathematics. *NETS*—Use technology tools to enhance learning, increase productivity, and promote creativity.

The Role of Standards

In its recently revised standards (National Council of Teachers of Mathematics [NCTM], 1998), the NCTM provides a broad, concept-driven curriculum that reflects the breadth of relevant mathematics and its relationships with technology. This vision includes five goals for each student:

1. To learn to value mathematics.
2. To become confident in his or her ability.
3. To become a mathematical problem solver.
4. To learn to communicate mathematically.
5. To learn to reason mathematically.

Furthermore, the National Art Educators Association (NAEA, 1995) has written that "there are many routes to competence in the visual arts." The NAEA considers such competency to be best achieved through "the ability to use an array of knowledge and skills." The association also suggests that, by the end of high school, a student should be able to do the following:

1. To understand and apply visual arts media, techniques, and processes.
2. To use knowledge of visual arts structures and functions.
3. To choose and evaluate a range of subject matter, symbols, and ideas.
4. To understand the visual arts in relation to history and cultures.
5. To reflect on and assess the characteristics and merits of his or her own work and that of others.
6. To make connections between the visual arts and other disciplines.

Both of these standards encourage active involvement through cross-disciplinary lessons. The lesson presented here encourages students to become mathematical problem solvers and to make direct connections to the visual arts.

This lesson, an example of constructivism, represents an eclectic view of learning that emphasizes four key components:

1. Learners construct their own understanding rather than have it deliv-

ered or transmitted to them.

2. New learning depends on prior understanding
3. Learning is enhanced by social interaction.
4. Authentic learning tasks promote meaningful learning (Kauchak & Eggen, 1998).

The Lesson

Objectives. Students will create a spreadsheet to calculate ratios and proportions using their classmates' measurements as a data base. Using Leonardo da Vinci's calculations—which are based on his famous drawing *The Proportions of the Human Figure* (1492)—students compare their results. Finally, they will draw conclusions about da Vinci's work, and modern technological applications will be identified and discussed.

Thinking Skills Required. Following Bloom's taxonomy, the students will use knowledge, comprehension, and application of information. To complete the assignment, they also will need to analyze and evaluate data.

Prerequisite Learning. Students should know how to create spreadsheets from data and work with formulas to calculate ratios and proportions. These formulas, of course, can be taught to students as they work through the assignment.

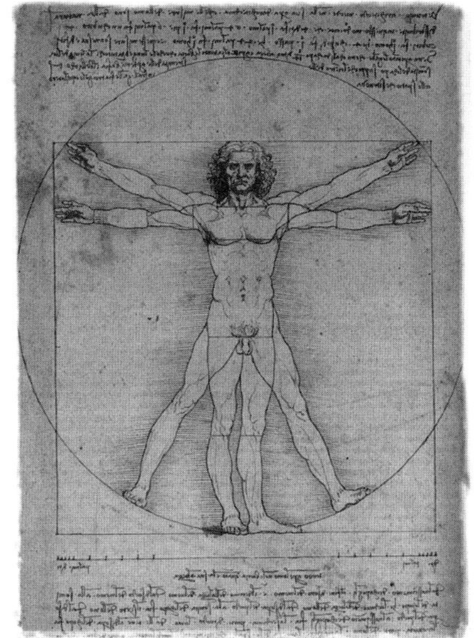

Leonardo da Vinci's The Proportions of the Human Figure.

Materials. The lesson requires the following:

- four measuring tapes,
- tacks,
- computers (one per student or one per classroom if presented as a group activity),
- student data-collection sheets,
- transparencies of da Vinci's *The Proportions of the Human Figure,*
- a transparency of Golden Section proportions (more on these proportions can be found in the extension activity at the end of the lesson), and
- a blank homework body proportions spreadsheet for each student.

Procedure. The procedure uses the following 10 steps.

1. Mount the measuring tapes on the wall in two different areas of the room before class. Mount two from the floor up and the other two crossing the first tapes, each making a letter "t" at four feet above the floor. The students should take measurements without wearing shoes. Each student should complete a body proportions spreadsheet. After they have determined their total height and height to their belly buttons in inches, ask them to determine their arm spans from fingertip to fingertip. Enter this information on the body proportions spreadsheet. (*Hint:* By holding a marker in each hand, students can mark their own arm span on the board and then measure the distance.)
2. Review the work students have done using spreadsheets. Describe how to calculate ratios and proportions. Show students the drawing that da Vinci used to illustrate the perfect body proportions for the human figure; do not share his results (those numbers are provided for you later in the lesson plan).

	A	B	C	D	E	F	G
1	**Body Proportions Spreadsheet**						
2							
3	**Student**	**Sex**	**Total Height**	**Height to Belly Button**	**Arm Span**	**Ratio of Belly Button Height to Total Height**	**Ratio of Arm Span to Total Height**
4	Suzy Q.	F	64	38.25	63.5	=D4/C4	=E4/C4
5	Martha P.	F	70	42.75	72	=D5/C5	=E5/C5
6	John B.	M	72	48	72	=D6/C6	=E6/C6
7	Sam S.	M	60	35.5	63	=D7/C7	=E7/C7
8							
9	Class Average		=AVERAGE(C4..C7)	=AVERAGE(D4..D7)	=AVERAGE(E4..E7)	=AVERAGE(F4..F7)	=AVERAGE(G4..G7)
10	Male Average		=AVERAGE(C6..C7)	=AVERAGE(D6..D7)	=AVERAGE(E6..E7)	=AVERAGE(F6..F7)	=AVERAGE(G6..G7)
11	Female Average		=AVERAGE(C4..C5)	=AVERAGE(D4..D5)	=AVERAGE(E4..E5)	=AVERAGE(F4..F5)	=AVERAGE(G4..G5)

This spreadsheet shows the formulas required to calculate class averages.

3. Direct students to use their collected data to create a spreadsheet complete with formulas to determine the accuracy of da Vinci's work.
4. Once the students' data has been gathered, you can enter the data on a transparency for students to use in creating their individual spreadsheets. In a one-computer classroom, you might have students enter their information directly with the appropriate projection equipment so that all students can see.
5. Enter the formula to calculate the ratio by dividing the total height of each individual into the belly button height. Round this number to three decimal places. Use a computer spreadsheet program to create a "Body Proportions Spreadsheet" template such as the one shown. Students could be instructed to determine their own formulas, although the template has formulas to help with data analysis if needed.
6. Determine the average proportions for each student and for the entire class. Students should note that even with large differences in their heights, the ratios of belly button heights to total heights produce a smaller range. At this point, you might have students sort their data by gender to determine the average proportions for each group.
7. Calculate the next ratio by dividing each arm span measurement in inches by each student's total height in inches. Again, even though the arm spans vary greatly, the actual ratios will be close.
8. Have students review their spreadsheet data and predict the perfect numbers according to da Vinci. Ask them to explain their reasoning.
9. Answer: The belly button height divided by the total height is the ratio of 1 to 1.618, or .618. Leonardo da Vinci used .618, but the original ratio was 5:8 or .625. The ratio of fingertip to fingertip and total height should be 1 to 1—that is, 1.000.

   Da Vinci wrote this in his notebook: "If you set your legs so far apart as to take a fourteenth part from your height, and you open and raise your arms until you touch the fingers, you must know that the center of the circle formed by the extremities of the outreached arms is equal to his height." Subsequent analysis has shown that parts of his figure share proportions that fall within the range of the Golden Section (Frayling, Frayling, & Van Der Meer, 1992). (See the extension activity at the end of this article for more information on the Golden Section.)
10. Did anyone in the class have the perfect proportions? How far is the class average from perfect proportions? (Hint: If you use this lesson in several classes, wait until the next day to give the answer and then determine which class was closest to having perfect proportions.)

Reflection

Encourage students to reexamine their thought processes. Here are suggested questions for teacher-led discussion.

- How did you know if your numbers were reasonable early in the data-collection stage?
- What reinforced your belief that you were on the right track?
- What other information can you find about da Vinci's Golden Section?
- What implications does this have in the art world? in fashion? architecture? science? math? sports?
- Are athletes more likely to be closer to perfection? How can you find out?
- Does age, weight, or sex affect the results?
- Are females closer to perfect proportion than males? How can you determine this?
- What did Leonardo da Vinci believe about perfect proportions?

Additional background information on ratios and human proportions can

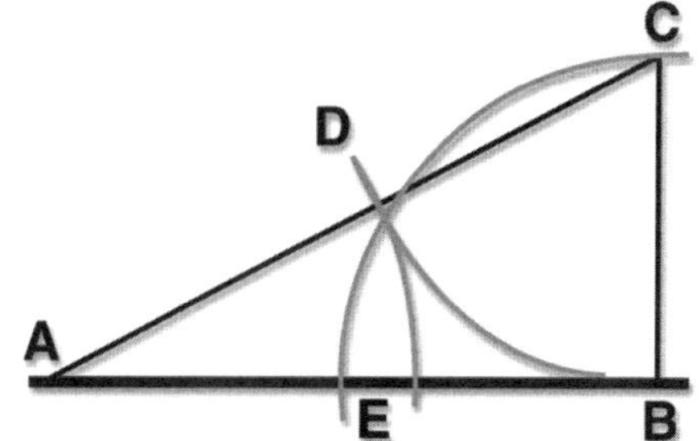

be found in *The Art Pack* (Frayling et al., 1992). The earliest book with technical advice for artists was Cennini's *Handbook for Artists,* which included a section on "the proportions which a perfectly formed man's body should possess" (Frayling et al., p. 11):

> There were great disagreements about what the correct proportions were and their relationship to beauty. In response Albrecht Dürer in 1528 wrote that "If you desire to compose a fine figure, you must take the head from one, the chest, arm, leg, hand, foot from others; I hold that the perfection of form and beauty is contained in the sum of all men" (p. 11).

Proportion remains the best method for judging relationship of parts to the whole. Artists still use it (often unconsciously), and so do art publics, whether the work in question is figurative or abstract.

Homework Assignment

Gather data from at least 10 people, perhaps parents, siblings, friends, or neighbors. Create the same sort of chart that you did in class, enter your data in spreadsheet format, and calculate the ratios. Compare these new results with your class results. If you do not have a computer, compile your results on paper and use a calculator or enter the data in the spreadsheet during the next class. Discuss your findings and conclusions.

Assessment

By the time students have completed this lesson, they will be able to create a spreadsheet, enter a formula to calculate ratios using their collected data, and explain their findings by drawing logical conclusions. Ask students to speculate about how Leonardo da Vinci might have arrived at his conclusions without using technology. Have them discuss the implications of his work in the modern world.

Extension Activity: Learn More About the Golden Section

The Golden Section (Plato). To divide a line using Golden Section proportions, divide the line AB into two sections of equal length. From B, draw an arc from the midpoint of AB to C at right angles to AB. Draw in CB and CA. Then from C, draw an arc with radius CB to cut AC at D. From A draw an arc with radius AD to cut AB at E. In proportion, EB is to AE as AE is to AB (Frayling et al., 1992).

References

Frayling, C., Frayling, H., & Van Der Meer, R. (1992). *The art pack.* New York: Alfred A. Knopf.

Kauchak, D., & Eggen, P. (1998). *Learning & teaching: Research-based methods.* Boston: Allyn & Bacon.

National Art Educators Association. (1995). *A vision for art education reform* [Online document]. Available: www.naea-reston.org/recentpapers.html.

National Council of Teachers of Mathematics. (1998, June 12). *Curriculum standards for Grades 5–8* [Online document]. Available: www.nctm.org.

Resource

AppleWorks for Macintosh, Apple Computer Inc., www.apple.com/appleworks/

Distributing Representatives

Using Spreadsheets to Study Apportionment

Are you wondering how to explain to students why states receive certain numbers of federal representatives without getting bogged down in mathematical details? Spreadsheets can help students perform the necessary mathematical calculations and allow them to analyze the results in depth.

By Leon Roland

Can mathematics ensure that each person is fairly represented when the number of representatives is assigned according to the population of a state? For students who are familiar with spreadsheets, the study of apportionment is an excellent interdisciplinary activity to teach mathematics, history, politics, and technology use. Spreadsheets are particularly helpful when using those methods of apportionment in which the proper divisor is found only by "guessing and checking," often requiring many different calculations. It also provides the opportunity to ask a number of "what-if" questions that go beyond the basic "What is the apportionment for State X?"

Any of the methods of apportionment provide an opportunity to use spreadsheets and see how changes in population change the distribution. For the political and historical study, however, it is necessary to look at the many methods used since the first representatives were elected in the United States. The first method used was the Jefferson Method. Jefferson's method was used after Washington vetoed Hamilton's method, which Congress had already selected. It just happens that Jefferson's method gave Virginia one more seat than the Hamilton Method. In 1876, both Hamilton's and Webster's methods were being used, but the fight to establish the final apportionment was done in such a way that neither of the methods was correctly calculated. Because the electoral college is dependent on each state's apportionment, this error caused Rutherford B. Hayes to be elected president rather than Samuel J. Tilden. These two examples provide the beginning of many interesting historical discussions. The resource list on the following page gives articles and books that discuss other problems which have occurred when apportioning the U.S. House of Representatives.

Getting Started

Before students determine how the number of representatives is chosen for each state, it is necessary to understand some of the basic terms. To determine the number of representatives, divide the total population of the nation by the number of seats to be selected. The result is called the standard divisor. You can then calculate each state's standard quota by dividing the state population by the standard divisor.

The results are rarely integers, making it impossible to ensure "fair" representation for every state. Because it is impossible for a state to have a fraction of a representative, some method of rounding the standard quota must be established. If a state's standard quota is 8.5, how many representatives will be given to that state? The real problem is simply a rounding problem, but any of the methods selected will make some state feel it has been cheated. Small states oppose the large states getting the "extra" representative. A state that just misses getting an "extra" representative feels slighted. All of the methods have both positive and negative arguments. In many of the methods, the standard divisor is modified in such a way that when all the states' populations are divided by this modified divisor and rounded, the correct number of seats is achieved. These two values are called the modified divisor and the modified quota. Students can use Statistical Abstracts or Internet sources such as www.census.gov to find actual populations to determine the representation.

Calculating Apportionment

There are many different methods for apportioning representatives. In my classes, I discuss six that have been used throughout U.S. history: the Jefferson Method, the Hamilton Method, the Lowndes Method, the Huntington-Hill Method, the Adams Method, and the Webster Method. Instructions for calculating apportionment using these six methods appear in the student worksheet on page 30 titled "How Many Representatives Does Your State Get?" Each of the different methods demonstrates different concepts. For example, the modified divisor is used in the Jefferson, Huntington-Hill, Adams, and Webster Methods. And, different solutions to the problem of rounding are provided in the Huntington-Hill, Adams, and Webster Methods.

Analyzing the Data

Some interesting points emerge when students calculate apportionment using the Jefferson Method. Results from the Jefferson Method and the formulas used to calculate them are shown in Figures 1 and 2. The modified divisor 2,350 is not unique; it can only be found by guessing and checking. This is an ideal situation in which to use spreadsheets. The standard quota is shown to allow comparison of the final apportionment and apportionment figured with the standard quota, which is the exact distribution of seats by population.

Note that State 3, which had only a small fraction more than 40 when calculated using the standard quota, received 41 seats in the final apportionment. Compare this to States 1 and 5. Both of these states had fractions greater than 0.5 more than the integer calculated with the standard quota, but did not receive extra representation. As other populations are studied, other results that may seem unfair occur.

Some other questions lend them-

Untitled 1 (SS)

	A	B	C	D	E
1	Jefferson Method				
2					
3	State	Population	Standard Quota	Modified Quota	Round Down
4	1	25,698	10.597	10.935	10
5	2	55,825	23.021	23.755	23
6	3	97,487	40.202	41.484	41
7	4	43,241	17.832	18.400	18
8	5	18,235	7.520	7.760	7
9	6	55,355	22.828	23.555	23
10					
11	Total	295,841			112
12					
13		Standard Divisor	2,424.926		
14		Modified Divisor	2,350.00		
15					

100

Figure 1. The apportionment of representatives calculated for six states using the Jefferson Method.

Untitled 1 (SS)

	A	B	C	D	E
1	Jefferson Method				
2					
3	State	Population	Standard Quota	Modified Quota	Round Down
4	1	25,698	=B4/$C$13	=B4/$C$14	=INT(D4)
5	2	55,825	=B5/$C$13	=B5/$C$14	=INT(D5)
6	3	97,487	=B6/$C$13	=B6/$C$14	=INT(D6)
7	4	43,241	=B7/$C$13	=B7/$C$14	=INT(D7)
8	5	18,235	=B8/$C$13	=B8/$C$14	=INT(D8)
9	6	55,355	=B9/$C$13	=B9/$C$14	=INT(D9)
10					
11	Total	295,841			=SUM(E5..E10)
12					
13		Standard Divisor	2,424.926		
14		Modified Divisor	2,350.00		
15					

100

Figure 2. The formulae for calculating apportionment using both the standard quota and a modified quota. When students change the modified quota, the new apportionment figures are automatically recalculated.

Resources

Balinski, M. L., & Young, H. P. (1975, August/September). The quota method of apportionment. *American Mathematical Monthly, 82,* 701–730.

Balinski, M. L., & Young, H. P. (1982). *Fair representation: Meeting the ideal of one man, one vote.* New Haven, CT: Yale University Press.

Saari, D. G. (1978). Apportionment methods and the House of Representatives. *American Mathematical Monthly, 85,* 792–802.

Steen, L. A. (1982, May 8). The arithmetic of apportionment. *Science News, 121,* 317–318.

Tannenbaum, P., & Arnold, R. (1992). *Excursions in modern mathematics.* Englewood Cliffs, NJ: Prentice Hall.

Young, H. P. (1985). *Fair allocation.* Providence, RI: American Mathematical Society.

selves to spreadsheet analysis. You will note in the modified quota for State 1 that the value is almost 11. There is always the question when a census is done that some of the homeless population is not counted. How much would the first state have to increase its population to gain the extra seat? Which state will lose? If every state has a population growth of 3%, will the apportionment change? A special interest group has a large following in both states 2 and 5. If they want to move a number of their members from one state to another to gain an extra seat in that state, should they move from 2 to 5 or 5 to 2? How many do they need to move?

Many of the mathematical activities that require spreadsheet use can be studied using one method, but for a study of apportionment from a historical and political perspective, some of the other methods should be presented. Each of the methods produce slightly different results. Some favor smaller states, and some larger. Students should compare the results they get with the different methods (see Table 1).

Conclusion

Using spreadsheets to calculate apportionment allows you to find the desired correct apportionment for methods that require multiple calculations to determine the correct modified divisor. The spreadsheet also allows you to ask questions such as the following:

- How much must the population of State 5 increase before it will gain a seat?
- Which state will lose a seat?
- Is the apportionment the same if each state's population increases by 10%?

Because the complex computations are done by the spreadsheet, students can focus on the questions being considered rather than correct computations. In a classroom situation, you may want to use only a few of the methods to discover how different methods affect final apportionment. Or you can use other data to determine the apportionment, such as land area or state budget.

Table 1. Summary of Apportionment for Six Methods

State	Population	Standard Quota	Jefferson	Hamilton	Lowndes	Huntington-Hill	Adams	Webster
1	25,698	10.597	10	11	11	11	11	11
2	55,825	23.021	23	23	23	23	23	23
3	97,487	40.202	41	40	40	40	39	40
4	43,241	17.832	18	18	18	18	18	18
5	18,235	7.520	7	7	8	7	8	7
6	55,355	22.827	23	23	22	23	23	23
Total	295,841	122	122	122	122	122	122	122
Standard Divisor		2,424.926	2,350			2,440	2,500	2,440

How Many Representatives Does Your State Get?

Use the following apportionment methods to allocate 122 representative seats to six states with different populations. Remember:

$$\text{Standard Divisor} = \frac{\text{Nation's Population}}{\text{Number of States}} \qquad \text{Standard Quota} = \frac{\text{State's Population}}{\text{Standard Divisor}}$$

Jefferson Method

Divide each state's population by some modified divisor that is slightly smaller than the standard divisor. Round the resulting modified quota down (e.g., both 5.1 and 5.9 become 5). Adjust the modified divisor until the sum of all the states' modified quotas equals the number of seats you wish to apportion.

Hamilton Method

Determine the standard divisor. Then divide each state's population by the standard divisor. Give each state the number of representatives equal to the integer value of its standard quota. Give the remaining seats, in order, to the states with the largest fractional parts.

Lowndes Method

First, determine the standard divisor. Then divide each state's population by the standard divisor. Give each state the number of representatives equal to the integer value of its standard quota. Then find the relative fractional part by dividing the fractional part by the integer value for each state. Give the remaining seats, in order, to the states with the largest relative fractional parts.

Huntington-Hill Method

Divide each state's population by some modified divisor, which will be very close to the standard divisor. Determine the geometric mean of the state's modified quota by taking the square root of the product of the integer value of the modified quota and the next greater integer. For example, if the modified quota is 13.67, the geometric mean is

$$\sqrt{(13)(14)}$$

or 13.49. If the modified quota is greater than the geometric mean, the value is rounded up. If the modified quota is less than the geometric mean, the state is given the integer value. Adjust the modified divisor until the sum of all the states' modified quotas equals the number of seats you wish to apportion.

The Huntington-Hill Method is the current method used to determine apportionment for the U.S. House of Representatives.

Adams Method

Divide each state's population by some modified divisor that is slightly larger than the standard divisor. Round the resulting modified quota up. Both 5.1 and 5.9 become 6. Adjust the modified divisor until the sum of all the states' modified quotas equals the number of seats you wish to apportion.

This method may result in some states receiving less representatives than the integer value figured using the standard quota.

Webster Method

Divide each state's population by some modified divisor, which will be very close to the standard divisor. If the fraction part is greater than or equal to 0.5, round the integer value up (e.g., 7.5 becomes 8). If the fraction part is less than 0.5, assign the integer value (e.g., 7.49 becomes 7). Adjust the modified divisor until the sum of all the states' modified quotas equals the number of seats you wish to apportion.

Fundraising with Technology

Fourth graders generate authentic questions and use spreadsheets to bring new life to a fundraising project that integrates math and technology skills.

Each year the fourth grade at Krieger Schechter Day School in Baltimore, Maryland, sponsors a Mother's Day Plant Sale. Proceeds from the sale fund an excursion on the Chesapeake Bay aboard two floating classrooms. This activity allows the children to learn firsthand about the importance of the world's largest estuary. This year we decided to enrich the fundraising program by using Microsoft® Excel spreadsheets.

By Marcie A. Zisow

Subject: Math, science, language arts, technology

Audience: Teachers, technology coordinators, teacher educators

Grade Level: 4–8 (Ages 9–13)

Technology: Spreadsheet and word processing software

Standards: *NETS•S* 3. *NETS•T* II–III. (Read more about the NETS Project at www.iste.org—select Standards Projects.) *NCTM* 5, 6, 8, & 9. (Read the math standards at http://standards.nctm.org.)

The Project

Fourth graders study Maryland in social studies for the entire school year. Units on Maryland's geography, history, and ecology are presented and explored thoroughly. The Chesapeake Bay, the world's largest estuary, is the focus of an integrated science, math, language arts, and art unit. Children read about the Chesapeake Bay and learn about the importance of protecting it. They study its plant and animal life and create collages to display what they have learned. As a culminating activity, the children plan and implement a Mother's Day Plant Sale.

Planning for the Sale

Before the actual sale, children work in small groups to plan how best to conduct the sale. First they brainstorm questions, then they use problem-solving skills to brainstorm methods to answer them.

- Where should we buy plants?
- How much will the plants cost?
- What types of plants should be purchased?
- How many of each type of plant should be purchased?
- Where will the money come from to purchase the plants?
- How should the sale be advertised and to whom?
- When should the sale be conducted?
- How much money is needed to reach the goal of renting two floating classrooms for the day?
- How much additional money is needed to purchase box lunches for the trip?

Presale Initiative

After groups have had time to discuss these questions, they share information and prepare for a Mother's Day Presale. Orders are taken in advance of the actual sale. Proceeds from the Presale are then reinvested so that plants can be purchased for the actual sale.

Children analyze data from past Mother's Day Plant Sales to decide which plants to order and in what quantities based on prior popularity. They also write letters to the administration requesting that Presale order forms be distributed in the school's weekly newsletter for three consecutive weeks. Children design the order form.

Integrating Math

Coinciding with the planning of the Mother's Day Plant Sale, children learn basic economic terms such as profit, loss, supply, and demand. Children practice addition, subtraction, multiplication, and division as they solve math story problems related to the upcoming sale.

For example, this problem may serve as the impetus for one such math lesson:

> The Mother's Day Presale is a success! Use Table 1 to answer the following questions. Working in your cooperative learning group, write a plan for answering the following questions. Do not solve the problems!
>
> 1. On which type of plant was the most profit made?
> 2. How much more profit was made on the most profitable plant than the least profitable plant?
> 3. How much profit was made on each type of plant?
> 4. What was the total amount of profit made from the Mother's Day Plant Presale?

Following group work, children share their plans for answering the above questions. They then actually *do* the math computations to answer the questions. Depending on the math ability of the group, some choose to do the calculations with paper and pencil while others choose to use calculators.

Table 1. Sample math practice problem.

Plant Type Ordered	*Wholesale Price (per plant)*	*Presale Price (per plant)*	*Total Number Sold*
Begonias	$.75	$4	123
Impatiens	$.75	$4	302
Geraniums (potted)	$1.75	$6	59
Hanging Baskets	$7.50	$15	78

Table 2. Mother's Day Plant Sale data.

Type of Plant	*Sale Price*	*Number Sold (Presale)*	*Number Sold (Actual Sale)*
Packs			
Ageratum	$4	9	18
Alyssum	$4	2	12
Begonias	$4	23	96
Coleus	$4	8	36
Dahlias	$4	11	12
Dusty Miller	$4	4	24
Impatiens	$4	106	224
Marigolds	$4	26	66
Petunias	$4	37	120
Salvia	$4	8	24
Snapdragons	$4	25	72
Vinca	$4	10	48
Tomato Plants	$4	7	36
Baskets			
Foliage	$15	4	10
Impatiens	$15	8	40
Ivy Geraniums	$15	4	30
Fuschia	$15	25	40
Pots			
Geraniums	$6	28	100
Shasta Daisy	$6	8	25
Black-Eyed Susan	$6	8	25
Jacob's Ladder	$6	5	0

How to Use Spreadsheets

As the Presale drew to a close, the computer teacher took students to the computer lab and instructed them on how to:

- access the spreadsheets
- enter information
- insert headers and footers
- resize cells
- enter formulas for calculations
- insert graphics
- create and insert graphs

Mother's Day Plant Sale: Worksheet A

Now that the Mother's Day Plant Sale is over, using the data supplied, what would you like to learn?

____

____

____

Write three questions you and your partner would like to use a computer spreadsheet to answer.

1. ______________________________
2. ______________________________
3. ______________________________

Write a paragraph with your partner describing what you will do with your spreadsheet to find your answers.

____

____

____

____

____

After you have completed your worksheet in the lab answer the following questions:

1. What is the answer to your #1 question?

____

____

2. What is the answer to your #2 question?

____

____

3. What is the answer to your #3 question?

____

____

4. What did you learn about using spreadsheets?

____

____

5. What would you do differently next time?

____

____

____

____

Children were given a template and directions for practice. They worked in pairs for approximately two math periods practicing spreadsheet skills.

Integrating Spreadsheets

The Mother's Day Plant Sale finally arrived. Proceeds from the Presale were used to purchase additional plants. As in the past, the project was well supported and—with the exception of a few plants—all were sold by day's end.

Children were given a Mother's Day Plant Sale Worksheet A and Mother's Day Plant Sale Data Sheet (Table 2). Pairs of children filled in three questions they would like answered about the Presale and actual sale.

Student-generated questions included:

1. How many plants did we sell for the Presale compared to the number sold at the actual sale?
2. Which plant was the most popular?
3. Did we sell more $4, $6, or $15 plants?
4. How much money did we make altogether?
5. What was the average number of plants sold for each type?
6. What percentage of plants were sold at the Presale?
7. About how many plants did each child at our school buy?
8. How much did we pay our supplier?
9. Did we make enough money to buy boxed lunches for our trip?

Children were then taken back to the computer lab and told they had approximately three lab sessions to work with their partners, use their new spreadsheet skills, and answer their own plant sale questions. As there were an odd number of children, the teacher served as partner to one child, placing herself strategically in the center of the lab to act as a model to the other groups.

What They Did

Interestingly, when faced with using spreadsheets to answer real questions, the majority of the students became confused over what to do and how to do it. Although spreadsheet skills had been taught before, students needed time to figure out how to connect what they had learned to what they needed to do now. Several groups began to type data information onto the spreadsheet without thought of how to better organize it for later computation. Other groups began to gather around the teacher and her partner to listen to their plans. The teacher noted several "Oh yeah's" as children listened in on the conversation and then quickly left to work on their own questions.

At the end of the first session all groups were busily engaged. One group claimed to be finished. However, when they printed out their spreadsheet and were asked their own answers, they could not use the spreadsheet information to answer.

The next two sessions proved invaluable as children refined the information on their spreadsheets. The meaning of the data in cells was clarified through the use of titles, and children added graphics and headers and footers to make the spreadsheets "look better." Before returning to the lab, children were asked to write a paragraph telling what they had done so far and what they still needed to do (Figure 1). This short writing assignment helped children to recall their purpose for returning to the lab and to stay on task.

After completing their work in the lab, children returned to the math classroom. The student pairs presented their spreadsheet and discussed the questions and answers they had discovered. They also reflected on what they would do next time and what they'd like to fix if time allowed. The computer lab teacher provided a final session for those who wanted to make last-minute changes to their spreadsheets. The lab teacher printed two copies of each child's spreadsheet, one for the student and one for the teacher (Figure 2).

Student Pair 1

Casey and I learned and accomplished many things! We set our goals and beat everyone! We found the price of all the hanging baskets we sold altogether. We did the same with pots and packs. We learned how to enter formulas and how to find out what they were. We figured out how to fix most of our mistakes. We learned how to make a pie chart and a bar graph of our data. We made our charts very decorative. We added pictures, colors, and changed the fonts. We pretty much covered everything! We learned a lot about spreadsheets!

Student Pair 2

David and I put in all of our data and figured out how many plants we sold for preorders, how much money we spent to get the plants, and which plant was the most popular. We need to go back and fix our formulas so we can get the right answers.

Student Pair 3

What Noah and I did so far was first we planned our questions. They were: What plant sold the least? How much money did we make? What plant did we sell the most of? We wrote down the names of all the plants first. Second, we put down all the prices. We're in the middle of putting in the quantities sold so we can enter our formulas to answer our questions.

Figure 1. Students write about their accomplishments with spreadsheets.

Lessons Learned

As research indicates, effective learning activities must be connected to real problems. Although children could follow the recipe on how to manipulate a spreadsheet, when asked to find answers to their own questions, they needed time to reflect and plan how to use technology skills to reach their goals. It was also important for the teacher to model the planning stage of the lesson by thinking aloud and being available to ask leading questions to help students formulate what it is they are trying to accomplish. Modeling proved a much more efficient and effective way to help students than visiting each group and telling them what to do.

This lesson represented a week taken from the regular math class. The time was well spent as evidenced by the final spreadsheets. Had the lessons been conducted weekly, the momentum for the project would have waned. Instead, by presenting it on five consecutive days, children did not lose focus.

The computer teacher noted how well the children worked, eager to accomplish their goal. The teacher attributed this to the fact that the children were involved in something they wanted to know, rather than on an artificial problem.

Conclusion

Integrating technology into the curriculum is best accomplished when it involves problem solving related to real-life situations of interest to children.

A Mother's Day Plant Sale presented an authentic situation where the children benefited directly from its success. Language arts, math, social studies, science, and computer technology were all combined into a meaningful learning experience. When technology is used as a tool to enrich instruction, it enables useful, meaningful, and transferable learning for the students who employ it.

Resource

Microsoft Excel is available at http://shop.microsoft.com/store/products.

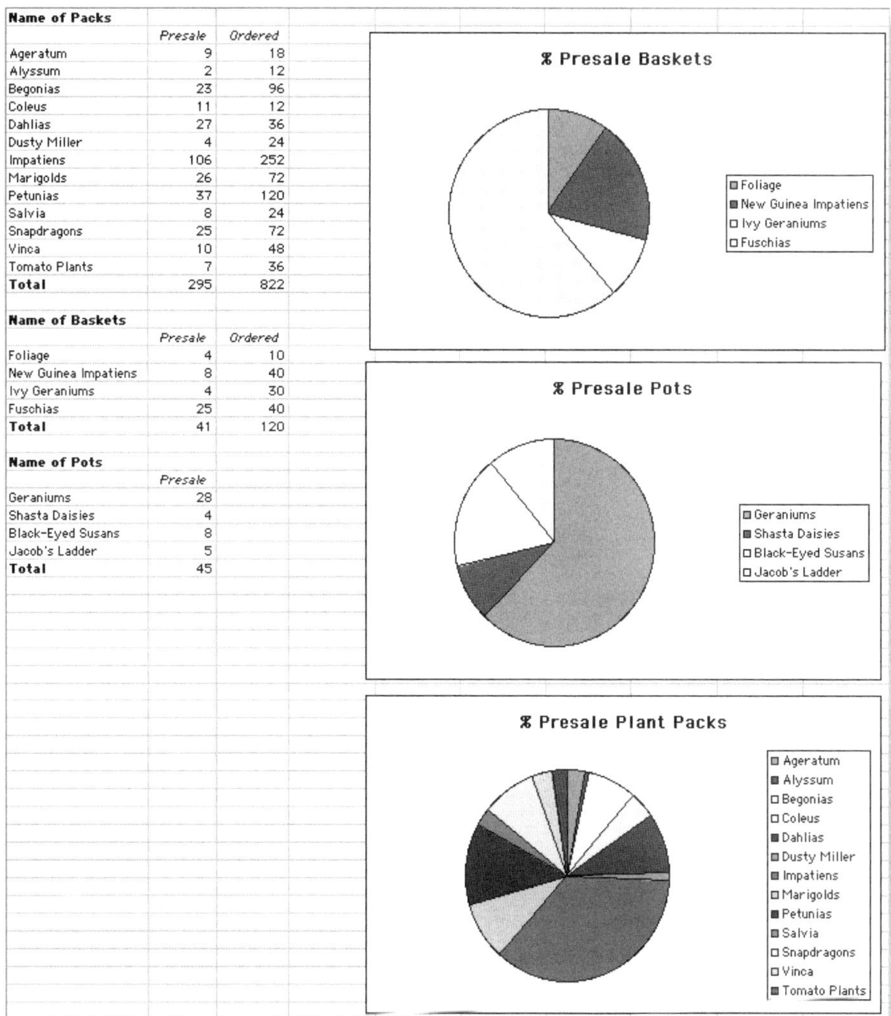

Name of Packs	Presale	Ordered
Ageratum	9	18
Alyssum	2	12
Begonias	23	96
Coleus	11	12
Dahlias	27	36
Dusty Miller	4	24
Impatiens	106	252
Marigolds	26	72
Petunias	37	120
Salvia	8	24
Snapdragons	25	72
Vinca	10	48
Tomato Plants	7	36
Total	295	822

Name of Baskets	Presale	Ordered
Foliage	4	10
New Guinea Impatiens	8	40
Ivy Geraniums	4	30
Fuschias	25	40
Total	41	120

Name of Pots	Presale
Geraniums	28
Shasta Daisies	4
Black-Eyed Susans	8
Jacob's Ladder	5
Total	45

Figure 2. Samples of student-generated spreadsheets and pie charts.

CHAPTER 3

Problem Solving

The Conversation Continues

Ivan: Consider flying in an airplane with either a pilot who has read everything there is to know about flying or a pilot who has actually flown a plane. Most of us would choose the one who has actually experienced flying. In reality, one cannot become a pilot without logging many hours of flying time. Thus, we don't consider someone a pilot unless the person has actually flown a plane.

Anne: The pilot scenario brings to mind a similar question: Are you really a problem solver if you haven't personally solved a problem? Students are often told the steps of problem solving or have watched or listened to someone else solve a problem. But, if they aren't engaged in problem solving themselves, they really can't know what problem solving is.

Ivan: Technology plays a key role in engaging a problem solver. When using technology, students are enabled and empowered to solve problems in ways that wholly engage them.

Anne: This talk about problem solving as an "active" pursuit reminds me of the Confucian proverb that says: I hear and I forget, I see and I remember, I do and I understand. This suggests how problem solving affects mathematical understanding. When you are "doing" mathematics through problem solving, you have opportunities to truly understand the mathematics.

Ivan: The articles included in this chapter on problem solving illustrate examples of opportunities that engage students in solving problems that are meaningful to the students. Whether they are using the technology to facilitate "guessing and checking" as in Lou Feicht's article, "Guess and Check: A Viable Problem-Solving Strategy," or pursuing an answer to a question about vampires in Hollylynne Drier's article, "Do Vampires Exist?: Using Spreadsheets to Investigate a Common Folktale," the students described in the articles are truly engaged in problem solving and thus empowered mathematically.

THEORY INTO PRACTICE

Problem Solving and Technology

By Anne Raymond

Beginning in the early 1980s, calls for reform in mathematics education have repeatedly included the need to improve the teaching and learning of mathematics by engaging students in more problem-solving tasks (National Commission on Excellence in Education, 1983; National Science Board Commission on Precollege Education in Mathematics, Science, and Technology, 1983; National Research Council, 1989). Consequently, in the formation of national standards for excellence in mathematics education, the National Council of Teachers of Mathematics developed a standard on problem solving that was to become an integral process throughout the K–12 curriculum.

NCTM Problem-Solving Standard

Problem Solving: Instructional programs from prekindergarten through Grade 12 should enable all students to

- Build new mathematical knowledge through problem solving
- Solve problems that arise in mathematics and other contexts
- Apply and adapt a variety of appropriate strategies to solve problems
- Monitor and reflect on the process of mathematical problem solving

Stemming from the work of George Polya (1957), educators strive to teach students about the art of problem solving, primarily by teaching them the four steps of problem solving. The four steps, as first designed by Polya, are:

Understand the Problem. At this stage one determines what the relevant pieces of information in the problem are and determines what the question is that needs to be answered. This is also where the problem solver determines what constraints might be present in the problem.

Devise a Plan. At this step, the problem solver determines or chooses a strategy by which to solve the problem. Such strategies might include making an organized list, drawing a picture, guessing and checking, etc.

Carry Out the Plan. At this step, the problem solver carries out the plan determined in the step above. The problem solver will determine, through implementing the chosen plan, if another approach may be needed and, also, whether he or she truly understands the problem. It is at this step that the problem solver generally comes to some solution to the problem.

Look Back. At this step, the problem solver provides a definite answer to the problem, checking to make sure that the answer is reasonable. Also, at this step, the problem solver may determine whether there may be additional answers or whether the solution may be generalized in any way.

Further development of tools by which we can teach students about problem solving include helping students see various types of mathematical problems that exist and teaching them how to apply steps of problem solving for each type of problem. Charles and Lester (1982) explain that there are two large categories of mathematical problems: routine and nonroutine. Within each category are three varieties of problems:

Routine	**Nonroutine**
Drill Exercises	Applied Problems
Simple Translation Problems	Puzzle Problems
Complex Translation Problems	Process Problems

When solving the nonroutine problems, students need to select from or develop their own strategies for solving the problem. In today's world, using appropriate technology is a viable heuristic tool of choice, particularly in solving applied and process problems.

NETS•S Problem-Solving Standard

6. Technology problem-solving and decision-making tools

- Students use technology resources for solving problems and making informed decisions.
- Students employ technology in the development of strategies for solving problems in the real world.

A number of software packages have been published that help students develop problem-solving skills across the curriculum. Many software companies now have preview copies of their software available online. In the resources at the end of this document, look at the online catalogs listed to find mathematics problem-solving software.

Beyond teaching students about problem solving and about steps for problem solving, educators need to think about teaching students via problem solving (Masingila, Lester, & Raymond, 2001). From this perspective, it is thought that a key means of teaching students meaningful mathematics is by engaging them in worthwhile mathematical problem solving.

In the age of increased technology, we can now engage students in problems beyond what we were once able, and at earlier ages (Van de Walle, 2001). For example, elementary students can now engage in probability experiments or simulations where technology allows them to run thousands of trials in seconds, thus generating more data from which they can develop graphs or draw conclusions. Middle school students can use graphing calculators or software programs to examine functions more complicated than typical linear or quadratic functions, which are normally the only types of graphs middle school students were challenged to study.

Whether we are using graphing calculators or computer software packages, problem solving can focus on more real-life issues and on the process of problem solving, even when problems present situations involving complicated mathematical computations. Since technology can do the computations quite simply, and even provide both arithmetic and visual information in the process, students can focus on posing

problems, suggesting possible solution paths, and interpreting outcomes from technological applications.

References

Charles, R., & Lester, F.K. (1982). *Teaching problem solving: What, why, and how.* Palo Alto, CA: Dale Seymour.

Masingila, J. O., Lester, F. K., & Raymond, A. M. (2001). *Mathematics for elementary teachers via problem solving.* Upper Saddle River, NJ: Prentice-Hall.

National Commission on Excellence in Education. (1983). *A nation at risk: The imperative for educational reform.* Washington, DC: U.S. Government Printing Office.

National Research Council. (1989). *Everybody counts: A report to the nation on the future of mathematics education.* Washington, DC: National Academy Press.

National Science Board Commission on Precollege Education in Mathematics, Science, and Technology. (1983). *Educating Americans for the 21st century.* Washington, DC: National Science Foundation.

Polya, G. (1957). *How to solve it* (3rd ed.). Garden City, N.J.: Doubleday.

Van de Walle, J. A. (2001). *Teaching mathematics developmentally* (4th ed.). New York: Addison Wesley Longman.

Resources

Book

Spencer, Karla H. (2002). *Best lesson plan websites for educators.* Westminster, CA: Teacher Created Materials.

Representative Online Catalogs

Edmark: www.riverdeep.net/edmark/ (Visit "Products" for a full list of titles. Click on "Play Online" for mathematics and critical-thinking games.)

Inspiration: www.inspiration.com (Choose "Product Information," and then use the pull-down menu to see "Inspiration Examples," which include a link to mathematics.)

Sunburst: www.sunburst.com (Choose "Products" from the first pull-down menu to access the catalog.)

Tom Snyder: www.tomsnyder.com (This company provided early leadership in capitalizing on the power of technology in problem solving.)

WebQuests, ThinkQuests, and Web Sites

WebQuest Math Lesson Plan 1: www.pavenet.org/FTP/Users/all_share/Cohort1/5319Project/murphy_rice_webquest/LessonPlans_PDF/MathLesson1.PDF

New York Times Learning Center: www.coolmath4kids.com/games/

THEORY INTO PRACTICE

Problem Solving and Worthwhile Mathematical Tasks

By Anne Raymond

If we want students to engage in mathematical problem solving, we ought to consider what problems students will be motivated to solve. We also need to consider what mathematical problems are worth solving. Too often, textbooks offer contrived problems that, though good experiences in problem solving, do not thoroughly connect students to real-world problems which mean something to the problem solver.

The National Council of Teachers of Mathematics has developed teaching standards wherein they describe what they believe is the standard for worthwhile mathematical tasks. This standard directly addresses the issue of what characteristics problems ought to have to fully engage students in valuable problem-solving moments.

Worthwhile Mathematical Tasks

The teacher of mathematics should pose tasks that are based on

- Sound and significant mathematics
- Knowledge of students' understandings, interests, and experiences
- Knowledge of the range of ways diverse students learn mathematics and that
    - engage students' intellect
    - develop students' mathematical understandings and skills
    - stimulate students to make connections and develop a coherent framework for mathematical ideas
    - call for problem formulation, problem solving, and mathematical reasoning
    - promote communication about mathematics
    - represent mathematics as an ongoing human activity
    - display sensitivity to, and draw on, students' diverse background experiences and dispositions
    - promote the development of all students' dispositions to do mathematics

In the article by Lou Feicht, "Guess and Check: A Viable Problem-Solving Strategy," we see the vehicle of spreadsheets as a means of engaging students in problem solving and pushing them to develop critical thinking in the process. The examples described within the article engage students in worthwhile mathematics tasks where the technology is appropriate and not contrived. In one example, high school students deal with the constraints involved in wanting to own their own DVD, including how many hours they would have to work to earn enough money to afford a DVD and how this affects the brand and quality of DVD they choose. Students use spreadsheets as a means of configuring various options.

In another worthwhile problem-solving task, these same students tackle the job of learning about investing money and earning interest in order to save for some long-term objective. Again, the spreadsheet facilitates the engagement in the meaningful problem-solving endeavor.

INSIGHT

Problem Solving Using Spreadsheets

By Ivan W. Baugh

Spreadsheets can provide an efficient tool for investigating solutions to a problem. Claris Corporation published *ClarisWorks in the Classroom* (1992) a collection of activities to use their integrated applications software package as an instructional tool. They included some worthy examples of how to use a spreadsheet as a problem-solving tool.

A spreadsheet can help in solving many different types of problems. Here is one useful example from the book. Samantha had $.33 in her pocket. She had nine coins. A user can find multiple ways to have $.33, but only one way that uses nine coins. The user is challenged to find the solution. The book provides a prepared spreadsheet that includes formulas to complete the calculations. The investigator needs only to plug in the number of each type of coin in column C. (I suggest that you protect columns A, B, and D, and cell C5 to avoid a student's accidentally changing or deleting the entry in one of those cells.) Since we know Samantha has $.33, we know she will have how many pennies? From there we use guess and check to arrive at the correct solution. The setup in Figure 3.1 shows the formulas we examine and discuss. In cells C5 and D5, we talk about functions.

◇	A	B	C	D	E
1		A	B	C	D
2	1	COIN	VALUE	NUMBER	WORTH
3	2	Pennies	0.01		=C3*D3
4	3	Nickels	0.05		=C4*D4
5	4	Dimcs	0.1		=C5*D5
6	5			=SUM(C3:C5)	=SUM(D3:D5)
7					

Figure 3.1

A second problem from the book states that a candy bar costs $.85. How many ways can you pay for the candy bar? The example provided allows the user to find one solution. I frequently ask the class to solve the problem one time. I ask them to share their solutions. As they share their solutions, I compile a list on the board. A sample list appears in Figure 3.2. Students enter their solutions in columns A, B, and C; the spreadsheet will verify that their choices produce the desired results. Again, protect column D to avoid accidental changes in that column.

We then use the list in Figure 3.2 to find a setup that will allow one user to find as many solutions as possible. The user enters the formula in cell D2 to calculate the value, and then uses Fill Down to replicate the formula to other cells in column D.

	A	B	C	D	E
1		A	B	C	D
2	1	$ 0.05	$ 0.10	$ 0.25	Value
3	2		1	3	$ 0.85
4	3	2		3	$ 0.85
5	4	17			$ 0.85
6	5	1	3	2	$ 0.85
7	6	7		2	$ 0.85
8	7	1	8		$ 0.85
9	8	8	2	1	$ 0.85
10	9	7	5		$ 0.85
11	10				$ -
12	11				
13					

Figure 3.2.

Immediately they find incorrect results because of the relative cell references. This provides an opportunity to teach about absolute cell references in formulas.

Placing $ before the column identifier in a formula anchors the cell references to that column ($A1). Placing $ before the row identifier in a formula anchors the cell references to that row (A$1). Placing $ before the column identifier and $ before the row identifier anchors the cell references to that cell ($A$1). Figure 3.3 shows the amended formula, incorporating absolute cell references. When they use Fill Down a second time, the correct values appear in column D. This occurs because rows 2 and below multiply the numbers in columns A, B, and C by the values in Row 1, not the row immediately above as happened before with relative cell references.

I then challenge the students to find other solutions. One strategy is to start with the smallest number of coins and gradually progress to the largest number of coins one can use to purchase the candy bar. You can ask questions such as, "What is the least number of coins one can use to purchase the candy bar?" "What is the largest

	A	B	C	D	E
1		A	B	C	D
2	1	0.05	0.1	0.25	Value
3	2		1	3	=$B$2*B3+$C$2*C3+$D$2*D3
4	3	2		3	=$B$2*B4+$C$2*C4+$D$2*D4
5	4	17			=$B$2*B5+$C$2*C5+$D$2*D5
6	5	1	3	2	=$B$2*B6+$C$2*C6+$D$2*D6
7	6	7		2	=$B$2*B7+$C$2*C7+$D$2*D7
8	7	1	8		=$B$2*B8+$C$2*C8+$D$2*D8
9	8	8	2	1	=$B$2*B9+$C$2*C9+$D$2*D9
10	9	7	5		=$B$2*B10+$C$2*C10+$D$2*D10
11	10				=$B$2*B11+$C$2*C11+$D$2*D11
12	11				
13					

Figure 3.3.

◇	A	B	C	D	E
1		A	B	C	D
2	1	Item	Cost	Quantity	Spent
3	2	Asparagus	1.19		=$B3*C3
4	3	Banana	0.07		=$B4*C4
5	4	Stawberry	0.05		=$B5*C5
6	5	Grapes	0.49		=$B6*C6
7	6	Watermelon	0.97		=$B7*C7
8	7	Eggplant	0.39		=$B8*C8
9	8	Carrots	0.69		=$B9*C9
10	9	Corn	0.88		=$B10*C10
11	10	Lemonade	0.75		=$B11*C11
12	11	Ice Cream	0.95		=$B12*C12
13	12			Spent	=SUM(D3:D12)
14	13			Balance	=9.99-D13
15					
16					

Figure 3.4.

number of coins one can use to purchase the candy bar?" "What is the maximum number of each coin that you can use while using all coins to purchase a candy bar?"

A third problem- solving challenge gives them a list of items to purchase at the grocery store. They are to spend $9.99 by purchasing different quantities of each item. We set up the spreadsheet as shown in Figure 3.4.

Students have experienced significant difficulty solving the problem without the formula in D13 that calculates the balance remaining. I have had many students simply give up, saying, "It can't be done." With the formula in D13, those same students much more easily solved the problem. In fact, they copied the formulas in column D, pasted them in column F and completed a second solution. They entered the new quantities in column E.

By using the spreadsheet, the students spent their time solving the problems rather than focusing on the mechanics of mathematics. I seriously doubt I could have gotten one solution from them without the spreadsheet. Before starting, many of them quickly stated, "It can't be done!" After doing it one time, they readily found multiple solutions.

Spreadsheets become a tool that facilitate problem solving, empowering the students to make discoveries in the same manner used by people in the world of work. They function as knowledge workers.

Reference

Claris Corporation (1992). *ClarisWorks in the classroom.* Santa Clara, CA: Author.

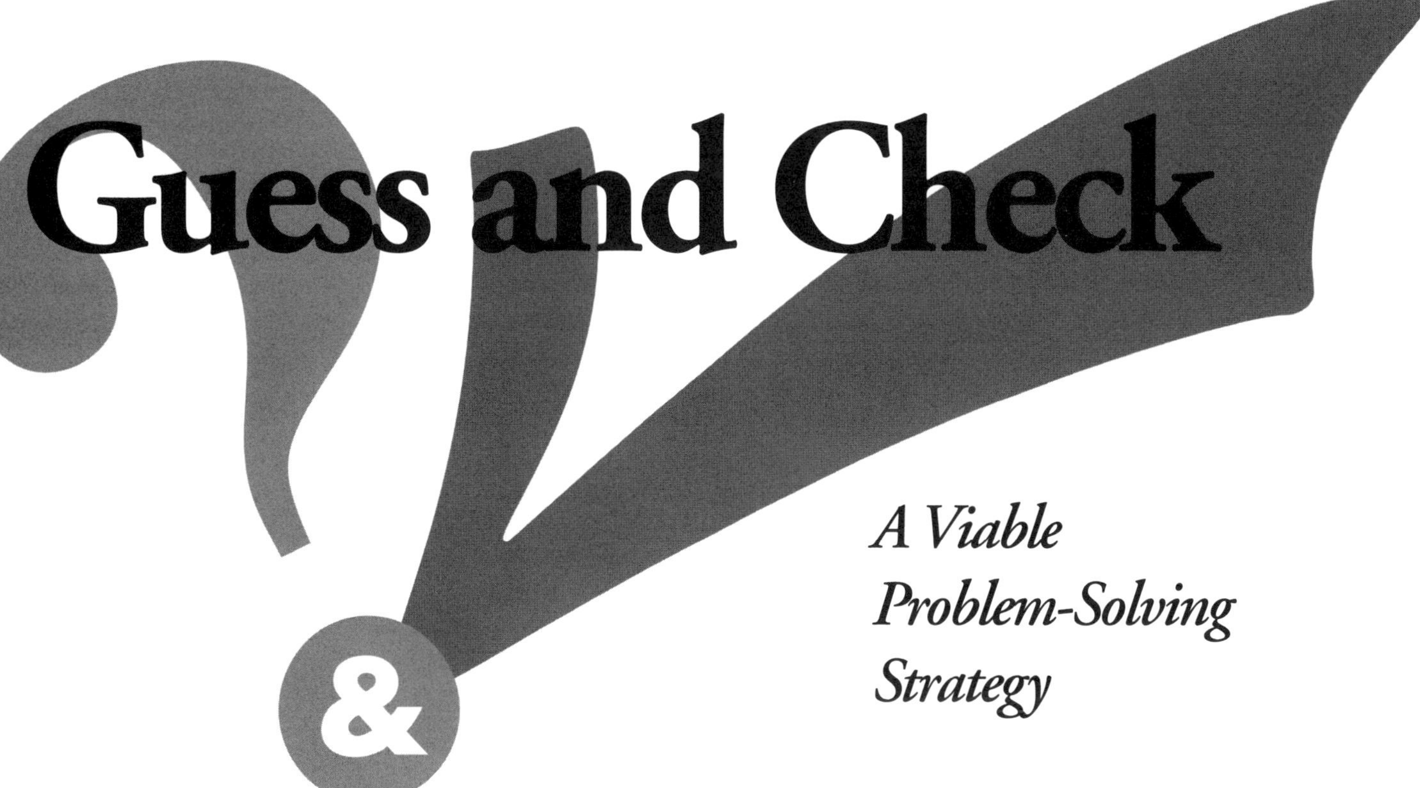

Guess and Check

A Viable Problem-Solving Strategy

Spreadsheets can help students follow their intuition to a better understanding of algebra.

By Louis Feicht

Subject: Pre-Algebra/Algebra

Grade Level: 6–9 (Ages 11–14)

Technology: Spreadsheet

Standards: NETS•S 3–6. (Read more about the NETS Project at www.iste.org—select Standards Projects.)

When students use a spreadsheet, they become immersed in its language. A spreadsheet can act as a bridge that students use to cross into the world of mathematics. They can use the more intuitive problem-solving method of making a table, because the software generates tables so easily. A spreadsheet proves to be powerful for students because it allows them to follow their intuition by using a guess-and-check strategy. My students have always solved problems such as those in this article by guess-and-check, despite my best efforts to introduce other methods. Problems that can be attacked by finding patterns, guessing and checking, and making tables naturally fit within the realm of computer spreadsheets. In fact, any problem that requires finding a numerical value where two quantities are equal lends itself well to the spreadsheet strategy of "don't spend as much time guessing and just check all possible values."

The great advantage of using a computer is that we can check every relevant possibility for an answer. Setting up the spreadsheet to solve the problems forces students to break problems into smaller pieces.

Modeling Problems Using Spreadsheets

The first example is a typical algebra word problem:

> There were 166 paid admissions to a game. The price was $2.10 for adults and $.75 for children. The amount taken in was $293.25. How many adults and children attended?

Figure 1 shows one way to set up this problem using a spreadsheet, while Figure 2 shows the spreadsheet and the underlying formulas. Most students will understand how to create a table. The number of adults at the game ranges from 0 to 166, which we know from a traditional algebraic approach. However, when we use the spreadsheet to solve this problem, subtle yet important differences emerge. In the column for the number of adults in attendance, we list every possibility, or in essence every possible domain value for the independent variable. The number of adults can be an integer from 0 to 166. However, we have not yet incorporated the level of abstraction that would result from introducing a variable such as *x* into the problem.

To set up the columns for the number of children, we reason that because there were 166 people at the game, the number of children in attendance must be 166 minus the number of adults. For example, if 3 adults attended, then 166–3 or 163 children must have attended. If your students are familiar with spreadsheets, they may already know the correct formula to enter for the number of children. Otherwise, this is an excellent example to develop the formula process. We use a notation such as "=166–A2" to tell the spreadsheet to subtract the quantity contained in, stored in, or represented by cell A2. In addition to developing the concept of variable, it allows students to see all specific instances, together with the generalization on the same page.

Next we set up the column for the money that would be taken in from adults paying $2.10 for a ticket. Use the spreadsheet to develop formulas such as "=A2*2.10". Similarly, the money from the number of children attending will be represented by a formula such as "=B2*.75". The last column for all the money collected would have the formula "=C2+D2".

We can now find a solution to the original question. Look in the column headed "Total Money" for a total of $293.25 to see how many adults and children attended. The answer is 125 adults and 41 children.

	A	B	C	D	E
1	Total Number of People = 166				
2					
3	Number of Adults	Number of Children	Money from Adults @ 2.10	Money from Children @ .75	Total Money
4	0	166	$0.00	$124.50	$124.50
5	1	165	$2.10	$123.75	$125.85
6	2	164	$4.20	$123.00	$127.20
7	3	163	$6.30	$122.25	$128.55
8	4	162	$8.40	$121.50	$129.90
9	5	161	$10.50	$120.75	$131.25
10	6	160	$12.60	$120.00	$132.60
11	7	159	$14.70	$119.25	$133.95
12	8	158	$16.80	$118.50	$135.30
13	9	157	$18.90	$117.75	$136.65
14	10	156	$21.00	$117.00	$138.00
15	11	155	$23.10	$116.25	$139.35
16	12	154	$25.20	$115.50	$140.70
17	13	153	$27.30	$114.75	$142.05
18	14	152	$29.40	$114.00	$143.40
19	15	151	$31.50	$113.25	$144.75
20	16	150	$33.60	$112.50	$146.1[illegible]
165					
166	162	4	$340.20	$3.00	$345.20
167	163	3	$342.30	$2.25	$344.55
168	164	2	$344.40	$1.50	$345.90
169	165	1	$346.50	$0.75	$347.25
170	166	0	$348.60	$0.00	$348.60
171					

The Coin Problem

Another typical example that arises in the same section of an algebra text is the coin problem.

> On a table are 20 coins consisting of quarters and dimes. Their combined value is $3.05. How many of each kind of coin are there?

This problem follows the same format using a spreadsheet in the same fashion as the ticket problem (Figure 3). It's reasonable to assume that more students could solve the coin problem than the ticket problem using only pencil and paper, because they are more likely to successfully guess and check, having a better understanding of money than tickets. Using a spreadsheet involves a variation of the guess-and-check strategy; we simply guess and check all possibilities and in the process begin to develop the language and abstract reasoning skills necessary to understand algebra.

The Temperature Problem

> Find a temperature where degrees Celsius and degrees Fahrenheit are equal.

Seventh-grade students were given the formulas to convert from degrees Celsius to Fahrenheit and back again. They were instructed to set up a spreadsheet as in Figure 4, using the formula for degrees Celsius—C=5/9(F–32) —to find where degrees Celsius was equal to degrees Fahrenheit. (It should be noted that these students had experience with spreadsheets and formulas.) When I decided on the domain of the problem and how to label the columns, I was essentially solving the problem. When students get enough experience, they will be able to choose their own inde-

	A	B	C	D	E
1	Total Number of People = 166				
2					
3	Number of Adults	Number of Children	Money from Adults @ 2.10	Money from Children @ .75	Total Money
4	0	166	=A4*2.1	=B4*0.75	=C4+D4
5	=A4+1	=166-A5	=A5*2.1	=B5*0.75	=C5+D5
6	=A5+1	=166-A6	=A6*2.1	=B6*0.75	=C6+D6
7	=A6+1	=166-A7	=A7*2.1	=B7*0.75	=C7+D7
8	=A7+1	=166-A8	=A8*2.1	=B8*0.75	=C8+D8
9	=A8+1	=166-A9	=A9*2.1	=B9*0.75	=C9+D9
10	=A9+1	=166-A10	=A10*2.1	=B10*0.75	=C10+D10
11	=A10+1	=166-A11	=A11*2.1	=B11*0.75	=C11+D11
12	=A11+1	=166-A12	=A12*2.1	=B12*0.75	=C12+D12
13	=A12+1	=166-A13	=A13*2.1	=B13*0.75	=C13+D13
14	=A13+1	=166-A14	=A14*2.1	=B14*0.75	=C14+D14
15	=A14+1	=166-A15	=A15*2.1	=B15*0.75	=C15+D15
16	=A15+1	=166-A16	=A16*2.1	=B16*0.75	=C16+D16
17	=A16+1	=166-A17	=A17*2.1	=B17*0.75	=C17+D17
18	=A17+1	=166-A18	=A18*2.1	=B18*0.75	=C18+D18
19	=A18+1	=166-A19	=A19*2.1	=B19*0.75	=C19+D19
20	=A19+1	=166-A20	=A20*2.1	=B20*0.75	=C20+D20
[illegible]	[illegible]	[illegible]	[illegible]	[illegible]	
167	=A166+1	=166-A167	[illegible]	[illegible]	[illegible]
168	=A167+1	=166-A168	=A168*2.1	=B168*0.75	=C168+D168
169	=A168+1	=166-A169	=A169*2.1	=B169*0.75	=C169+D169
170	=A169+1	=166-A170	=A170*2.1	=B170*0.75	=C170+D170
171					

Figure 2. Spreadsheet revealing one way formulas could be set up.

pendent and dependent variables, column labels, and formulas.

Even though I typed in the formula for degrees Celsius, it is not immediately apparent to all students what to do next. Most of my sixth- and seventh-grade students began by typing a formula in cell B4 such as "=5/9*(–256–32)" (Figure 5). Some students continued to type in this kind of formula for the rest of the class, and I allowed them to do so. They were checking each of the Celsius formulas by using the spreadsheet as a calculator and doing a lot of typing! Notice that solving the problem this way offers no advantages over using a calculator or pencil and paper, but it demonstrates the beginning of some level of abstraction, because students show the ability to apply the formula and replace the variable "F" in the formula with a number.

However, it won't take long for most students to tire of this and look for an easier solution. After typing in about 10 specific formulas, many of them began to look for ways to use the power of the spreadsheet. They were already familiar with the use of formulas and using the Fill Down function instead of typing in each cell. They were looking for a way to generalize and use a variable. Observing the pattern they created in Figure 5, they discovered that the only number that was changing (varying) in column B was the Fahrenheit temperature. They next recognized that each of these varying numbers was contained in column A. They quickly replaced their specific numbers with a more *general* formula, filled it down (Figure 6), and found the solution of –40° (Figure 7).

Bridging the Gap Between Specific and General

Using the first problem as an example,

	A	B	C	D	E
1	Total Number of Coins =		20		
2					
3	Number of Quarters	Number of Dimes	Number of Quarters *.25	Number of Dimes *.10	Total Money
4	0	20	$0.00	$2.00	$2.00
5	1	19	$0.25	$1.90	$2.15
6	2	18	$0.50	$1.80	$2.30
7	3	17	$0.75	$1.70	$2.45
8	4	16	$1.00	$1.60	$2.60
9	5	15	$1.25	$1.50	$2.75
10	6	14	$1.50	$1.40	$2.90
11	7	13	$1.75	$1.30	$3.05
12	8	12	$2.00	$1.20	$3.20
13	9	11	$2.25	$1.10	$3.35
14	10	10	$2.50	$1.00	$3.50
15	11	9	$2.75	$0.90	$3.65
16	12	8	$3.00	$0.80	$3.80
17	13	7	$3.25	$0.70	$3.95
18	14	6	$3.50	$0.60	$4.10
19	15	5	$3.75	$0.50	$4.25
20	16	4	$4.00	$0.40	$4.40
21	17	3	$4.25	$0.30	$4.55
22	18	2	$4.50	$0.20	$4.70
23	19	1	$4.75	$0.10	$4.85
24	20	0	$5.00	$0.00	$5.00
25					

Figure 3. The coin problem is set up similarly to the ticket problem.

the difference from a traditional approach is that all 167 cases, 0 adults through 166 adults, have their own unique equations. We are using equations, variables, and formulas, but we have not yet jumped immediately to using a single equation to represent the whole problem. Students without training in algebra can set up this problem and solve it. Students will be able to see every possible situation on the spreadsheet page, so the level of abstraction has initially been reduced from one single equation as in a traditional approach, laying the foundation for algebraic reasoning. Students who could set up the problem using a traditional algebra approach, but maybe not solve the equation correctly, now can get a correct solution.

Now we can remove the training wheels and give students a more general model of the problem. Although each case has its own distinct spreadsheet formula, the short leap to setting up one equation representing the whole problem uses the last three columns—"A18*2.10 + B18* .75=293.25". The substitution property of equality can be employed to replace B18, which is intuitively understood by students, who see that cell B18 contains 166–A18. Our equation would then become "A18*2.10 + (166–A18)*.75= 293.25", which is precisely where we may have started the problem without a spreadsheet, only using a different notation—"2.10x + .75(166–x)=293.25".

Using the spreadsheet can help teachers assist students to reach the final level of abstraction. Evidence suggests that students without algebra training will retain the ability to solve these types of problems even with pencil and paper, using the language of the spreadsheet (Sutherland & Rojano, 1993).

	A	B	C
1	Find the temperature that is the same in both Celsius and Farenheit Scales		
2		C=5/9*(F-32)	
3	Temperature in degrees Farenheit (Domain)	Temperature in degrees Celsius	
4			
5			

Figure 4. The temperature problem set up for younger students.

	A	B	
1	Find the temperature that is the same in both Celsius and Farenheit Scales		
2		C=5/9*(F-32)	
3	Temperature in degrees Farenheit (Domain)	Temperature in degrees Celsius	
4	-256	=5/9*(-256-32)	
5	-255	=5/9*(-255-32)	
6	-254	=5/9*(-254-32)	
7	-253	=5/9*(-253-32)	
8	-252	=5/9*(-252-32)	
9	-251	=5/9*(-251-32)	
10	-250	=5/9*(-250-32)	
11			

Figure 5. Middle school students' first attempt at solving the temperature problem.

	A	B
1	Find the temperature that is the same in both Celsius and Farenheit Scales	
2		C=5/9*(F-32)
3	Temperature in degrees Farenheit (Domain)	Temperature in degrees Celsius
4	-256	=5/9*(A4-32)
5	-255	=5/9*(A5-32)
6	-254	=5/9*(A6-32)
7	-253	=5/9*(A7-32)
8	-252	=5/9*(A8-32)
9	-251	=5/9*(A9-32)
10	-250	=5/9*(A10-32)
11		
12		
13		

Figure 6. Temperature spreadsheet formulas after students recognize how to use variables.

	A	B
1	-48	-44.44444444
2	-47	-43.88888889
3	-46	-43.33333333
4	-45	-42.77777778
5	-44	-42.22222222
6	-43	-41.66666667
7	-42	-41.11111111
8	-41	-40.55555556
9	-40	-40
10	-39	-39.44444444
11	-38	-38.88888889
12	-37	-38.33333333
13	-36	-37.77777778
14	-35	-37.22222222
15	-34	-36.66666667
16	-33	-36.11111111
17	-32	-35.55555556
18		

Figure 7. The temperature spreadsheet showing the correct solution.

References

Esty, W. W., & Teppo, A. R. (1996). Algebraic thinking, language and word problems. In P. Elliot (Ed.), *Communication in mathematics, K–12 and beyond* (pp. 45–53). Reston, VA: National Council of Teachers of Mathematics.

Hoyles, C. (1992). Mathematics teaching and mathematics teachers: A meta-case study. *For the Learning of Mathematics, 12*(3), 32–44.

Hoyles, C., & Noss, R. (1992). A pedagogy for mathematical microworlds. *Educational Studies in Mathematics, 23*, 31–57.

Mason, J., & Pimm, D. (1984). Generic examples: Seeing the general in the particular. *Educational Studies in Mathematics, 15*, 277–289.

Sutherland, R., & Rojano, T. (1993). A spreadsheet approach to solving algebra problems. *Journal of Mathematical Behavior, 12*, 353–383.

WORD PROCESSING, DESKTOP PUBLISHING, AND GRAPHICS *in the Mathematics Classroom*

By James H. Wiebe

When teachers consider computer tools and the mathematics classroom, few think of word processors, desktop publishers, or graphics packages. After all, word processors and desktop publishers are designed to manipulate words, and graphics packages are used to create pictures, while in mathematics we manipulate symbols and numbers, right?

Think again! Mathematics involves words, numbers, symbols, and pictures, all of which can readily be produced and manipulated by word processors, desktop publishers, and graphics tools. In students' hands, these products become powerful tools for organizing and representing mathematical information.

Student Uses of Word Processors and Desktop Publishers

These tools are most useful to students in the areas of problem solving and mathematical applications. But they can also be of benefit to students in activities designed to develop understanding of mathematical concepts.

A reasonable way to help students become better problem solvers is to have them learn and use general problem-solving strategies (heuristics) (Polya, 1973). Strategies used may include having students:

- paraphrase a problem situation.
- make lists of information known and/or needed about the problem.
- make tables or charts, brainstorm solution strategies.
- solve a parallel problem.

For many of these steps, a word processor, desktop publisher, and/or graphics program is helpful. You can give problems to students on paper or in a file that contains the necessary problem-solving steps and hints. They then use a word processor, desktop publisher, or graphics package to solve the problem, eventually printing out their work. Students submit their complete printout so that you can evaluate the strategies students use as well as their final answers. Many students will be less threatened by the prospect of entering their ideas and solution attempts into the computer than by writing or drawing them by hand. With a computer graphics package as an aid, students can focus on solving the problem instead of on the by-hand production of elaborate and perfect drawings.

Students can also use the word processor to write their own problems or problems for their peers to solve. To help students, you may want to establish some parameters. For example, you could post a list of rules regarding the problems to be written. (e.g., "The problems must involve addition of more than three whole numbers, and the problem must contain extra information not needed in order to solve it.") Students then work at the computer, perhaps in groups of two, three, or four, to create a set of problems. The teacher or other students then check the problems for appropriateness before giving them back either to those who wrote the problems, or to other students in the class to solve. The students work at the computer and type the strategies they use, their partial results as they attempt to solve the problems, and, finally, their answers.

Another type of problem-solving activity is for you to use the word processor to create a set of problems in which the result or answer is given but the numerical values in the problem or part of the description of the problem are either incorrect or missing (Figure 1). The students enter the missing information or correct the incorrect information using the editing features of the word processor. The teacher can thus create problems that have an interesting range of correct answers.

With certain problems, it might also be helpful to allow students to use a spreadsheet or database manager. For example, students familiar with integrated packages such as *AppleWorks* or *Microsoft Works* could use the spreadsheet to do their computations, then move the data and/or resulting graphs back to the word processor (Figure 2).

You can also create electronic worksheets for practicing mathematical procedures or operations. The worksheets are saved on disk, and students answer the questions on-screen and save or print out the results. Word processors or desktop publishers with good graphics capabilities or stand-alone graphics packages can be used by students or teachers to represent concepts or operations graphically. Figure 3 shows part of a screen-based worksheet designed to help students

Sample Problem

Insert a sentence in the blank spot to complete this problem:

Giant Department Store is having a sale. The regular price of a radio is $32.00. ——————————————.The sale price is $24.00.

Figure 1. Word processor-based math problem.

understand the meaning of addition of fractions before they learn the algorithm.

Graphics Use of Graphics Packages

Graphics packages or the graphics components of word processors or desktop publishers can be sometimes useful in helping students solve problems that involve pictures or models. For example:

> Ashe Street runs East and West. Alice, Beth, Carl, and Doris all live in different houses on the same side of Ashe Street. Their house numbers are 28, 32, 34, and 36. Alice and Carl live to the East of Doris. Beth's house number is higher than Doris'. Carl's house number is the lowest. What are the street numbers of each person's house?

Sample Problem

At the Stamp Collector's Swap Meet, John bought several sets of old stamps from different countries. The sets contained the following stamps: 6 U.S. stamps costing ____ each, 8 Canadian stamps costing ___ each, 21 Mexican stamps at ___ each, 14 French stamps at ___, and 7 Cuban stamps at ___. The total cost of the stamps was exactly $10.00 and single stamps from the different countries cost 17, 10, 21, 12, and 50 cents (the prices are not in order). What was the cost of single stamps from each country? Is there more than one possible set of costs? Type your answers here and print out the results.

1. Restate the problem in your own words.
2. List everything you know about the problem.
3. Set up a spreadsheet like the one below. Experiment with the stamp prices until the prices are correctly matched with the countries (the total should be $10.00—there should be a formula at cell D7 that finds the total of Column D).
4. Copy the completed spreadsheet back into the word processor. Enter the missing values in the problem above, and print out the whole thing, including the problems and the spreadsheet.

	A	B	C	D
1	Country	Stamp Price	No. Stamps	Total/Country
2	U.S.A.		6	
3	Canada		8	
4	Mexico		21	
5	France	$.17	14	$2.38
6	Cuba		7	
7				Total
8				
9			Total Cost	$10.00

Figure 2. Math problem using an integrated software package.

Name: Date:

Directions: Use graphics to find the sums shown below. First, divide and color circles to show each fraction. Next, put the fractions together in a third circle to show their sum. Finally, write a fraction which you think is close to the sum. Print out your work.

Example: 1/2 + 1/4

Estimated Answer: 3/4

1. 1/2 + 1/8

Estimated Answer:

2. 3/4 + 2/3

Estimated Answer:

Figure 3. Solving fractions problems with graphics. (Since the students are not calculating the answer, a range of reasonably close answers is acceptable.)

Students use computer graphics to represent the problem—two lines to represent the street, and numbered boxes to represent the houses. Next, the "houses" are drawn and identified, then moved around on the screen to solve the problem (Figure 4).

Computer tools that allow students to create, manipulate, and print text, symbols, and graphics are useful in representing and solving a variety of mathematical problems and situations. In setting up a laboratory or purchasing software for use in mathematics classrooms, it is worthwhile to consider purchasing an integrated package with word processing/desktop publishing and graphics capabilities, along with spreadsheet and database subprograms. The cost of an integrated package is much less than purchasing each application separately, and provides the convenience of a common set of commands for all the subprograms and of being able to move data easily from one application to another.

Reference

Polya, G. (1973). *How to solve it* (2nd ed.). Garden City, NY: Doubleday.

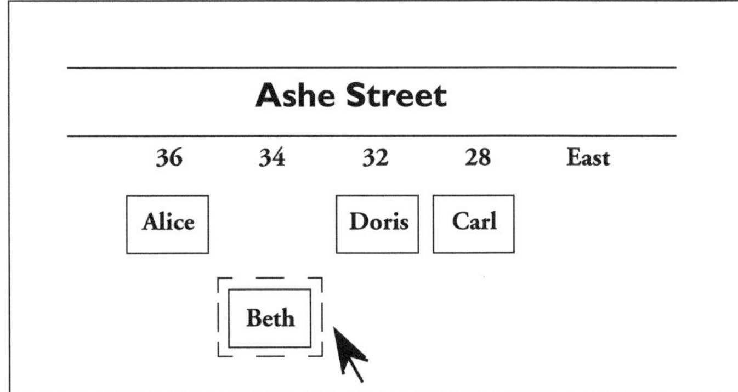

Figure 4. Macintosh screen using SuperPaint.

Do Vampires Exist?

Using Spreadsheets to Investigate a Common Folktale

Using basic mathematical concepts and spreadsheet skills, students investigated the existence of vampires. The spreadsheet provides a computational tool that facilitates the investigative nature of the activity and allows students to solve a mathematical problem in an open-ended, exploratory manner.

Photo: CORBIS/Bettman

By Hollylynne Stohl

Subject: Math, Number Sense, Algebra

Grade Level: 3–9 (Ages 8–14)

Technology: Spreadsheets (e.g., Microsoft Excel or Works, AppleWorks)

Standards: *NETS* 3. (See http://cnets.iste.org for more information on the NETS project.) *NCTM* 2. (See http://standards-e.nctm.org for more information on the updated NCTM standards.)

Do vampires really exist? This question prompted a lively discussion and investigation with a class of third graders. Of course, the students had heard some of the mysterious folktales about vampires and even seen representations of them in cartoons, movies, and television shows. However, in this lesson, we used basic mathematics skills (addition and multiplication by 2) and simple spreadsheet capabilities (formulas, fill down, and graphing) to actually "prove" that vampires could not exist. In this investigation, the students learned simple spreadsheet skills, applied their knowledge of addition and multiplication, and were exposed to some powerful mathematical concepts.

Introducing the Investigation

To begin our investigation, I asked students to discuss what they had heard about vampires. Ideas they shared included: vampires can come out only at night, they sleep in coffins, they can turn into bats, they live forever, they drink other people's blood to survive, you become a vampire if you are bitten by one, and garlic can keep them away. To frame our mathematical investigation, we narrowed the characteristics to two: vampires must drink human blood to survive and a person becomes a vampire when bitten by one. From this, we made three assumptions that would guide our investigation.

1. One vampire currently exists.
2. A vampire must bite someone and drink his or her blood once a week to survive.
3. Once a person is bitten by a vampire, he or she becomes a vampire.

Beginning the Investigation

To prepare for the lesson, I created a worksheet in Microsoft Excel (Figure 1) and displayed it through the television monitor for the entire class to view. Depending on students' ages and experience using spreadsheets, this investigation could be done as a whole class with one computer and a projection system or in a lab where students are working in pairs to create and extend the pattern using formulas.

Using our three assumptions, students recognized that by the end of the first week the one existing vampire would bite someone and that person would become a vampire. Thus, at the end of week one, there would be two vampires in the world. By the end of the second week, both of these vampires had to bite another person, making a total of four vampires. We continued this process up to week eight (two months) and recorded the results in the appropriate cells of the worksheet (Figure 2).

At this point, I asked students to look for any patterns in our table of numbers. They quickly noticed that the total number of vampires doubled each week. To give them a visual picture of the doubling process, I created a line graph of the total number of vampires (Figure 3). From looking at the graph, we connected the doubling process in the table of numbers to the rapidly growing curve. I introduced the term *exponential* and explained that when a sequence of numbers or a graph increases very rapidly, it is often said to be growing *exponentially*. In this case, the total number of vampires in the world is growing exponentially.

	A	B	C
1	Week	Number of People Bitten	Total Number of Vampires
2	0	0	1
3	1		
4	2		
5	3		
6	4		
7	5		
8	6		
9	7		
10	8		

Figure 1. Format of spreadsheet to begin investigation.

	A	B	C
1	Week	Number of People Bitten	Total Number of Vampires
2	0	0	1
3	1	1	2
4	2	2	4
5	3	4	8
6	4	8	16
7	5	16	32
8	6	32	64
9	7	64	128
10	8	128	256

Figure 2. Results after eight weeks.

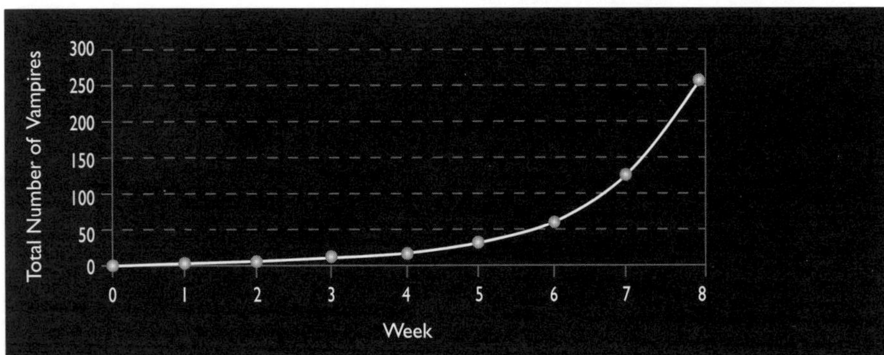

Figure 3. Graph of the total number of vampires through week eight.

Extending the Pattern

The spreadsheet gave students a computational tool they could use to continue modeling the growth of the vampire population. I asked the students to analyze how many vampires were bitten each week as well as the total number of vampires at the end of each week and describe how the vampire population was growing. Two popular observations were:

1. In both the "number of people bitten" and the "total number of vampires" columns, each entry is double that of the previous one.
2. The "number of people bitten" in a week is the same as the "total number of vampires" at the end of the previous week.

To continue the pattern, students used the capabilities of the spreadsheet to describe the process used to calculate the next week's vampire values. To continue the weeks, students added one to the previous week (e.g., = A10 + 1).

Focusing on observation one, a student selected cell B11, typed the "=" symbol, clicked in cell B10 (this action copied the cell reference into the formula) and typed "*2" (the formula =B10*2 now appeared in cell B11). Another student recognized that doubling a number could also be represented by adding the number twice. The student entered the formula =B10+B10 in cell B11. Because both of these formulas resulted in 256 people bitten in week nine, we were able to discuss and reinforce the concept of multiplication as repeated addition. We used similar formulas in cell C11 (=C10*2 or =C10+C10) to model the doubling process in the "total number of vampires" column.

Focusing on observation two, to continue the pattern for week 10, the formula =C11 was used in cell B12. This formula used the value appearing in cell C11 (the total number of vampires from week 9) for cell B12 (the number of people bitten in week 10). Continuing with this observation, the students used the formula =C11+B12 in cell C12 to calculate the total number of vampires for week 10. This formula models the number of vampires from the previous week each biting one person in the current week and calculates the new total number of vampires for the current week.

After we modeled the growth of the vampire population several ways, we decided to use the formula for multiplying the previous entry in a column by 2 to obtain the current entry (e.g., =B10*2). We then used the Fill Down command to quickly calculate the numbers in our table and model the growth of the vampire population through the 35th week. By the 35th week (about 9 months), there were 34,359,738,368 vampires in the world. The students were amazed at the number and we actually discussed the place value of each digit and practiced saying the number (thirty-four billion, three hundred fifty-nine million, seven hundred thirty-eight thousand, three hundred sixty-eight).

Gathering Population Data

To continue our "proof," I asked the students how many people are in the world. The students gave estimates (most were either way too low, "a million," or extremely high, "a gazillion"), and we briefly discussed the effect of births and deaths on the world population. One student reasoned that "there would have to be more births than deaths every minute for the population to grow and if the number of deaths was larger, our numbers would decrease." To get an estimate of the current world population, we visited the world population clock at the Census Bureau's Web site (**www.census.gov/main/www/popclock.html**). This site gives updated estimates every second! For example, on 4/21/99 at 12:23:15 PDT the total population of the world was projected to be 5,981,294,020.

We calculated that there would be 34,359,738,368 vampires in the world after nine months. I pointed out that this is clearly impossible because the world population is a little less than 6 billion people! If our assumptions were true, everyone in the world would quickly become a vampire. I told the students, "I know that I am not a vampire, and I don't think any of you are really vampires." Thus, using mathematics and continuing a pattern, we established a contradiction (something that cannot be true). As this could not be true, then, based on our assumptions, vampires could not exist! Mathematicians often use this method called "proof by contradiction" to prove theories. To do a proof by contradiction, you must start with a set of reasonable assumptions and prove that one or more of them lead to a contradiction.

As an extension to this activity, students could be challenged to find how many vampires would need to exist at the start for everyone in the world (about 6 billion people) to become a vampire in 6 months (about 24 weeks). Students could also use a spreadsheet and formulas to model the growth of the vampire population under different conditions (e.g., starting with three vampires, each vampire biting two people each week to survive).

Summary

This investigation used a well-known folktale to allow students to use their addition and multiplication skills to model the growth of the vampire population; incorporate powerful mathematical ideas such as patterning, variables, formulas, exponential growth, and proof by contradiction; and teach students about basic spreadsheet capabilities. Using the spreadsheet facilitated the process of extending a mathematical pattern and allowed the students to explore the pattern numerically, algebraically (through formulas), and graphically. Thus, instead of just learning *about* a spreadsheet, the students *used* the spreadsheet to do mathematics and learned basic spreadsheet skills while solving an interesting mathematical problem.

My students thoroughly enjoyed this investigation, and many used their new knowledge to persuade their friends and family that vampires don't exist. This lesson won an award in 1998 in a contest held by Microsoft and the Association for Supervision and Curriculum Development. The lesson plan is posted on Microsoft's Web site (**www.microsoft.com/education/k-12**), along with many other lessons using Microsoft products. Furthermore, Microsoft used a version of this lesson in the *Productivity in the Classroom* workbook (Fall/Winter volume 4) that it distributes free to educators.

Exploring Difference Equations With Spreadsheets

When students use a spreadsheet to explore real-world problems involving a periodic change, they can observe what happens at each period, generate a graph, and observe how changing the starting quantity or constants affects the results. Although author Thomas Walsh used Microsoft Excel for these exercises, any spreadsheet will work. These are also good exercises for introducing the power of spreadsheets in a computer applications or business class.

By Thomas P. Walsh

A difference equation calculates many values of a specific quantity that changes over time, such as a population or money in an interest-bearing bank account. When solving first-order difference equations, the present value is usually calculated based upon the previous value, initial conditions, and (perhaps) a small constant.

A computer-based spreadsheet program, such as Excel, makes it easy to calculate many different values for a single equation. Each successive value shows on the computer screen, and by moving to the cell that contains the value, students can view the equation that calculated the value.

My students always enjoy the lessons developed on the following pages. A lot of steps are involved in creating the spreadsheets, but most students like the challenge of carrying out the tasks on the computer. When they have finished entering the spreadsheet data, they enjoy generating different types of graphs, including pie, stacked bar, or 3-D bar, and then moving on to the next lesson so that they can create a new spreadsheet and make more graphs. They also like to experiment by entering different numbers into the difference equations to see how the results change.

The lessons provided here address the National Council of Teachers of Mathematics (NCTM) Standards (1989). Specifically, they address two standards for grades 9–12—Standard 1: Math Problem Solving ("to apply the process of mathematical modeling to real-world problems") and Standard 12: Discrete Mathematics ("represent and solve problems using linear programming and difference equations").

Reference

National Council of Teachers of Mathematics. (1989). Curriculum and evaluation standards for school mathematics. Reston, VA: Author.

Exploring Difference Equations With Spreadsheets

Part 1: Population Growth

Town of 800 People

Simple population growth can be modeled in terms of a difference equation, because a good approximation of the next time period's population is dependent almost entirely upon the previous time period's population. First, let's investigate the population in a town with 800 people. Let's assume that the population growth there is 8% per year. Open the Excel spreadsheet and follow these steps:

1. Type the headings as shown in Figure 1.
2. Enter the numbers 1–10 in cells A7-A16 in the "Year" column.
3. In cell B7 of the "Population" column, enter the number 800.
4. In cell B8 of the "Population" column, enter the formula: = (B7 * .08) + B7, where B7 is the cell containing the initial Population. When you hit the Return key, you should get 864.00 in B8. You may have to adjust the precision to two decimal places.
5. Use the Copy and Paste keys to copy the contents of cell B8 into each of cells B9–B16 adjacent to the numbers 3–10 in the "Year" column.
6. The spreadsheet should now resemble Figure 1.

	A	B	C
1			
2	Population growing at a rate		
3	of 8% per year. Population begins		
4	with 800 people.		
5			
6	Year	Population	
7	1	800	
8	2	864.00	
9	3	933.12	
10	4	1007.77	
11	5	1088.39	
12	6	1175.46	
13	7	1269.50	
14	8	1371.06	
15	9	1480.74	
16	10	1599.20	
17			

Figure 1.

The difference equation for this spreadsheet is: $P_{t+1} = (P_t * .08) + P_t$, where P_{t+1} is the population of the present year and where P_t is the population of the previous year.

Try this: Adjust the "Population" column to allow integers only. How does it affect the final population?

To graph the data from the spreadsheet, follow these steps:

1. Use the pointer to select all the numbers 1–10 in the "Year" column and all the numbers 800–1599.20 in the "Population" column.
2. Choose the Chart Wizard icon in the Window menu.
3. Move the arrow down to the spreadsheet, and notice that it changes to the Chart Wizard icon and a cross. Drag the icon to the area in the spreadsheet where you wish to place the chart, click the mouse button, and the Chart Wizard dialog box appears. At the top, it says "Chart Wizard Step 1 of 5." Follow the steps. The range should read $A$7:$B$16 if you chose just the numbers, and $A$6:$B$16 if you chose the column headings as well. Click the Next button to go to the next step.
4. In the second step of Chart Wizard, you must choose a chart type. Choose Line and click the Next button to go to the next step.
5. In the third step of Chart Wizard, choose a chart format. Choose the first formatting option (upper left corner) and click the Next button. Now the Chart Wizard should show you a picture of your chart (see Figure 2).

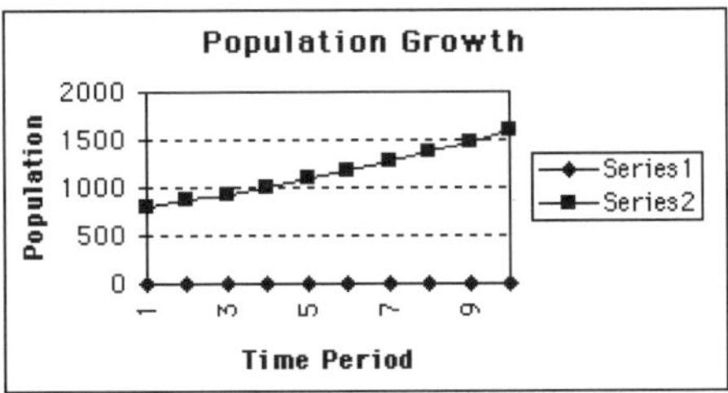

Figure 2.

6. Click Next to add a legend to your chart. Add "Population Growth" as the title of your chart.

Population Growth Considering Death Rate

The previous example showed simple population growth. It did not account for deaths in the population. Next, we'll look at a population that grows but also has a certain death rate associated with it, and see how the population is affected. We'll start with 1,000 people and set a growth rate of 30% and a death rate of 0.02% per year. You'll be surprised at the results!

In Excel, follow these steps:

1. Type the headings as shown in Figure 3.
2. Enter the numbers 1–20 in the "Time period" column.
3. In cell B8 of the "Population" column, enter 1000.
4. In cell B9 of the "Population" column, enter the formula: = (1.3 - (.0002 * B8)) * B8, where B8 is the cell containing the initial Population. When you hit the Return key, you should get 1100 in cell B9.

	A	B	C
1			
2	A population rate that reaches		
3	equilibrium. We start with		
4	1000 people, a growth rate of		
5	30%, and a death rate of .02%		
6			
7	Time period	Population	
8	1	1000.00	
9	2	1100.00	
10	3	1188.00	
11	4	1262.13	
12	5	1322.18	
13	6	1369.20	
14	7	1405.02	
15	8	1431.71	
16	9	1451.26	
17	10	1465.41	
18	11	1475.55	
19	12	1482.76	
20	13	1487.87	
21	14	1491.48	
22	15	1494.02	
23	16	1495.81	
24	17	1497.06	
25	18	1497.94	
26	19	1498.56	
27	20	1498.99	
28			

Figure 3.

5. Use the Copy and Paste keys to copy the contents of B9 into all other cells B10–B27 in the "Population" column.

Use the steps for creating a graph from spreadsheet data in the first example to make a graph for this example (see Figure 4).

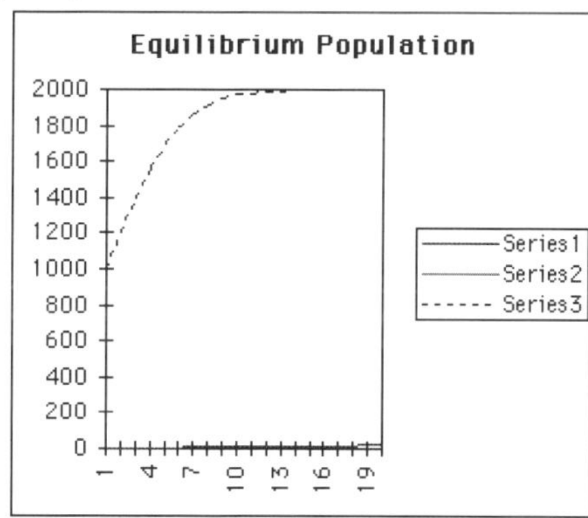

Figure 4.

You wouldn't expect a population that grows at a rate of 30% and that has a mortality rate of .02% to level off at 150% of the original population, but that's exactly what happens. The difference equation for this spreadsheet is $P_{t+1} = (1.3 - (.0002 * P_t)) * P_t$, where P_{t+1} is the population of the present period and P_t is the population of the previous period.

Actually, this is not quite a simple death rate, as you may suspect from the equation. We're multiplying by the square of the previous population, and that ensures that this simulation will level off at 150% of the original population.

Try this: Change the growth rate to 35% and run the simulation. Does anything change? Change the death rate to .04% and run the simulation (with a growth rate at the original 30%). Does anything change?

Part 2: Simple and Compound Interest

Simple Interest

Let's use a difference equation on a spreadsheet to calculate how much the Native Americans, who accepted items worth about $24 from the Dutch settlers in the Hudson Valley for the right to own Manhattan Island, would have made if they had exchanged the items for money and then put that money into an interest-bearing account. We'll calculate simple interest, then compound interest. The first year (1626) they earn no interest, having deposited their money at the end of that year. The interest rate is 5.5%

In Excel, follow these steps:

1. Type headings as shown in Figure 5.

	A	B	C
1			
2	Let's see what would have happened if		
3	the Manhattan Native Americans had		
4	invested the money they received from		
5	the Dutch by selling Manhattan Island.		
6	They received $24.00. The bank		
7	account pays 5.5% simple interest.		
8	We start at the year they sold the		
9	island: 1626.		
10			
11	Year	Interest	Net Worth
12	1626	$ -	$ 24.00
13	1627	$ 1.32	$ 25.32
14	1628	$ 1.32	$ 26.64
15	1629	$ 1.32	$ 27.96
16	1630	$ 1.32	$ 29.28
17	1631	$ 1.32	$ 30.60
18	1632	$ 1.32	$ 31.92
19	1633	$ 1.32	$ 33.24
20	1634	$ 1.32	$ 34.56
21	1635	$ 1.32	$ 35.88
22	1636	$ 1.32	$ 37.20
23	1637	$ 1.32	$ 38.52
24	1638	$ 1.32	$ 39.84
25	1639	$ 1.32	$ 41.16
26	1640	$ 1.32	$ 42.48
27	1641	$ 1.32	$ 43.80
28	1642	$ 1.32	$ 45.12
29	1643	$ 1.32	$ 46.44
30	1644	$ 1.32	$ 47.76
31	1645	$ 1.32	$ 49.08
32	1646	$ 1.32	$ 50.40
33	1647	$ 1.32	$ 51.72
34	1648	$ 1.32	$ 53.04
35	1649	$ 1.32	$ 54.36
36	1650	$ 1.32	$ 55.68
37			

Figure 5.

2. Enter the years 1626 through 1650 in the "Year" column.
3. In cell B12 of the "Interest" column, enter 0.00.
4. In cell C12 of the "Net Worth" column, enter 24.00.
5. In cell B13 of the "Interest" column, enter the formula = $C12$ * .055, where C12 is the cell containing the initial net worth. When you hit the Return key, you should get $1.32 in cell B13.

6. In cell C13 of the "Net Worth" column, enter the formula: = B13 + C12, where B13 is the initial entry in the "Interest" column and C12 is the initial entry in the "Net Worth" column.
7. Use the Copy and Paste keys to copy the same formula in B13 into cells B14–B36 in the "Interest" column. The spreadsheet will want to copy relative numbers, that is, the cell previous to the one being copied. You must ensure that in each cell of the "Interest" column you have copied the formula exactly as it appears in Step 5.
8. Use the Copy and Paste keys to copy the contents of C13 into cells C14–C36 in the "Net Worth" column.

Use the steps for creating a graph from spreadsheet data in the first example to make a graph for this example (see Figure 6).

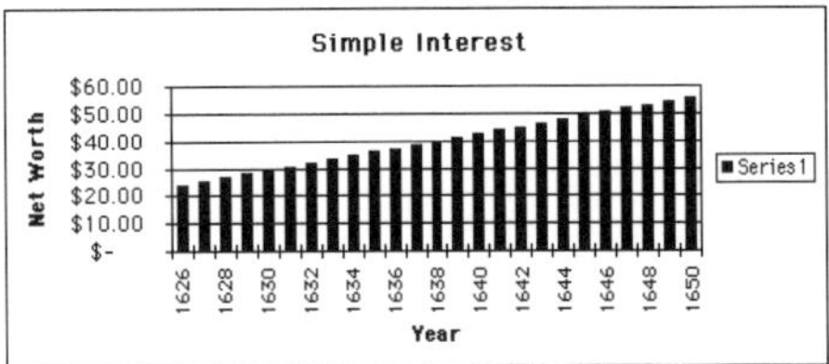

Figure 6.

The difference equation for this spreadsheet is: $NW_{t+1} = (NW_1 * .055) + NW_t$, where NW_{t+1} is the net worth of the present year, NW_t is the net worth of the previous year, and NW_1 is the original principal deposited

Try this: Change the interest rate to 6% and run the simulation.

Compound Interest

Now, let's calculate the interest compounded annually. When interest is compounded annually, the interest from the previous year is added to the present year's principal, and interest is calculated on the sum.

In Excel, follow these steps:

1. Type the headings as shown in Figure 7.
2. Enter the years 1626–1650 in the "Year" column.
3. In cell B12 of the "Interest" column, enter 0.00.
4. In cell C12 of the "Net Worth" column, enter 24.00.
5. In cell B13 of the "Interest" column, enter the formula = C12 * .055, where C12 is the initial entry in the "Net Worth" column. When you hit the Return key, you should get $1.32 in cell B13.
6. In cell C13 of the "Net Worth" column, enter the formula = B13 + C12, where B13 is the initial entry in the "Interest" column and C12 is the initial entry in the "Net Worth" column.

	A	B	C	D
1				
2	Let's see what would have happened if			
3	the Manhattan Native Americans had			
4	invested the money they received from			
5	the Dutch by selling Manhattan Island			
6	They received $24.00. The bank			
7	account pays 5.5% compound interest.			
8	We start at the year they sold the			
9	island: 1626.			
10				
11	Year	Interest	Net Worth	
12	1626	$ -	$ 24.00	
13	1627	$ 1.32	$ 25.32	
14	1628	$ 1.39	$ 26.71	
15	1629	$ 1.47	$ 28.18	
16	1630	$ 1.55	$ 29.73	
17	1631	$ 1.64	$ 31.37	
18	1632	$ 1.73	$ 33.09	
19	1633	$ 1.82	$ 34.91	
20	1634	$ 1.92	$ 36.83	
21	1635	$ 2.03	$ 38.86	
22	1636	$ 2.14	$ 41.00	
23	1637	$ 2.25	$ 43.25	
24	1638	$ 2.38	$ 45.63	
25	1639	$ 2.51	$ 48.14	
26	1640	$ 2.65	$ 50.79	
27	1641	$ 2.79	$ 53.58	
28	1642	$ 2.95	$ 56.53	
29	1643	$ 3.11	$ 59.64	
30	1644	$ 3.28	$ 62.92	
31	1645	$ 3.46	$ 66.38	
32	1646	$ 3.65	$ 70.03	
33	1647	$ 3.85	$ 73.88	
34	1648	$ 4.06	$ 77.94	
35	1649	$ 4.29	$ 82.23	
36	1650	$ 4.52	$ 86.75	
37				

Figure 7.

7. Use the Copy and Paste keys to copy the contents of B13 into cells B14–B36 in the "Interest" column. Now, let the spreadsheet copy relatively and you'll see the interest grow.
8. Use the Copy and Paste keys to copy the contents of C13 into cells C14–C36 in the "Net Worth" column.

Use the steps for creating a graph from spreadsheet data in the first example to make a graph such as the one shown in Figure 8.

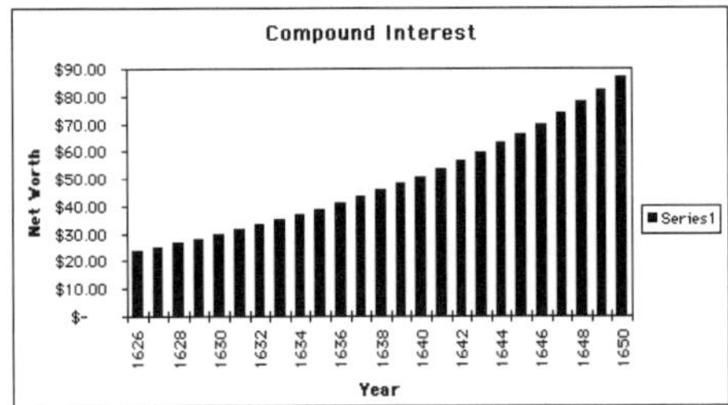

Figure 8.

The difference equation for this spreadsheet is: $NW_{t+1} = (NW_t * .055) + NW_t$, where NW_{t+1} is the net worth of the present year and NW_t is the net worth of the previous year. Note that it takes the Native Americans just 13 years to double their money with compound interest, whereas it takes 19 years for them to double their money when they earn simple interest.

Try this: Change the interest rate to 6% and run the simulation again.

Part 3: An Epidemic Model

In this lesson we will look at a "normal" epidemic. No epidemic is normal, of course, but most epidemics follow a general pattern. The number of people who contract the disease is usually small. For example, the E. bola virus that struck a small town in Zimbabwe was traced to a single man: a charcoal maker in the bush who then infected a dozen or so of his relatives. As these people came in contact with other people, the disease spread quickly through the population. In epidemics, however, as the number of infected people goes up, the number of newly infected people goes down because most of the population is already infected. Finally, the epidemic fades and might even be eradicated. The general pattern, then, is similar to a bell curve, starting out small, getting very big, and tapering off. In this simulation, we'll start with an initial population of 1,000 people. Initially, 10 people will be infected with the disease, and the disease has a contraction rate of .05%.

In Excel, follow these steps:

1. Type the heading for this spreadsheet as shown in Figure 9.
2. Enter the numbers 0–23 in the "Time period" column in cells A8-A31. The first time period is 0 because initially no one is infected.
3. In cell B8 of the "Infectives" column, enter the number 1000 (for reference); and in B9, enter 10. This is the number of people who first contracted the disease.
4. Enter 0 in cell C8 of the "New Infectives" column, and in C9, enter 10 (again, the number of people who first contract the disease).
5. In cell B10 of the "Infectives" column, enter the formula: = (1 + (.0005 * $B$8) - (.0005 * B9)) * B9, where B8 is the initial value in the "Infectives" column and B9 is the second value. When you hit the Return key, you should get 14.95 in cell B10. You may have to adjust the precision to two decimal places.
6. In cell C10 of the "New Infectives" column, enter the formula: = (B10 - C9), where B10 is the third value in the "Infectives" column and C9 is the second value in the same column. When you hit the Return key, you should get 4.95 in cell C10. Again, you may have to adjust the precision to two decimal places.

	A	B	C	D
1				
2	A disease epidemic model. We start			
3	with 1000 people, a contraction rate of			
4	0.05%, and 10 people initially infected			
5	with the disease.			
6				
7	Time period	Infectives	New Infectives	
8	0	1000.00	0.00	
9	1	10.00	10.00	
10	2	14.95	4.95	
11	3	22.31	7.36	
12	4	33.22	10.91	
13	5	49.28	16.06	
14	6	72.71	23.43	
15	7	106.41	33.71	
16	8	153.96	47.55	
17	9	219.09	65.13	
18	10	304.63	85.54	
19	11	410.55	105.92	
20	12	531.55	121.00	
21	13	656.05	124.50	
22	14	768.87	112.82	
23	15	857.73	88.85	
24	16	918.74	61.02	
25	17	956.07	37.33	
26	18	977.07	21.00	
27	19	988.27	11.20	
28	20	994.07	5.80	
29	21	997.02	2.95	
30	22	998.50	1.49	
31	23	999.25	0.75	

Figure 9.

7. In cell C11 in the "New Infectives" column, enter this formula: = B11 - SUM ($C$8 : C10).
8. Use the Copy and Paste keys to copy the contents of B10 and C10 into cells B11-B31 and C11-C31, respectively.

Take the usual steps to create a graph like the one shown in Figure 10.

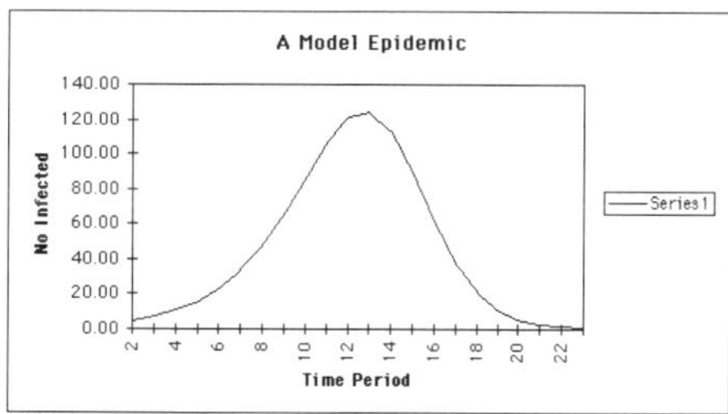

Figure 10.

There are two difference equations for this simulation. For "Infectives" we have: $I_{n+1} = (1 + (.0005 * 1000) - (.0005 * I_n)) * I_n$. The difference equation for the New Infectives is: $NI_{n+1} = (I_{n+1} - NI_n)$, where I_{n+1} is Infectives of the present time period, NI_n is New Infectives of one time period previous to the present one, and NI_{n+1} is New Infectives of the present time period.

Try this: Change the contraction rate to .06%.

CHAPTER 4

Reasoning and Proof

The Conversation Continues

Anne: A lawyer in a court of law seeks to convince the jury of the validity of his points, sequentially presented, to secure an adoption of his position—guilty or innocent. In mathematics, many of us quickly recall the multitude of theorems that were available for proving a point.

Ivan: But my students say, "Big deal! So what! Why do we need to prove those theorems? What difference will it make in my life?"

Anne: Students need to develop the skills to sequentially present evidence to prove or disprove a hypothesis. Individuals who have moved society forward have a well-established background in sequential thinking. Through the development of these skills, the students acquire a background for effectively drawing conclusions in other disciplines. Consider the adage "The proof is in the pudding." When you eat the pudding, you decide, based on taste, whether it is good or not. You also make a decision about whether you want it again or whether you may want to request the recipe.

Ivan: As you read the articles that follow, look for information that will help you find creative ways to help your students build these reasoning skills.

THEORY INTO PRACTICE

Proofs

Examples With Geometry and Technology

By Anne Raymond

The National Council of Teachers of Mathematics identifies reasoning and proof as one of the five process standards that should permeate the mathematics curriculum at all levels. What constitutes a proof varies depending on the audience (and age level) to which you are trying to prove a mathematical idea. The level of rigor of a proof is also dependent on the level of the mathematics concept you are trying to prove.

NCTM Reasoning and Proof Standard

Reasoning and Proof: Instructional programs from prekindergarten through Grade 12 should enable all students to

- Recognize reasoning and proof as fundamental aspects of mathematics
- Make and investigate mathematical conjectures
- Develop and evaluate mathematical arguments and proofs
- Select and use various types of reasoning and methods of proof

Some proofs are elementary in nature and can be developed in a variety of ways—for example, proving that the sum of the interior angles of any triangle is 180 degrees.

For an elementary student, one might have students draw and cut out a wide range of triangles. Then, working from the given assumption that a straight line measures 180 degrees in an arc formed at any point on the line, we have students tear their triangle corners apart and place the three corner angles together (see Figure 4.1). In all cases, the students will see that the three corners, when placed together, form a straight line. Thus, they can conclude that the sum of the interior angles of any triangle is 180 degrees.

Although this is a very basic proof, one can conclude that if the reasoning is clear enough to convince the problem solver, then the proof is successful. Technology can assist in a proof such as this. When students have access to geometric shape builders via computer software, they can build and manipulate many triangles in an effort to develop numerous "cases" from which they can draw their conclusions about the angle measures of triangles.

A much more sophisticated proof, although accessible to young students, involves housing a triangle between parallel lines in a plane and using knowledge of angles formed by transversal lines cutting across parallel lines (Figure 4.2). In this proof, students can show that for triangle ABC, the sum of angles A, B, and C is 180 degrees.

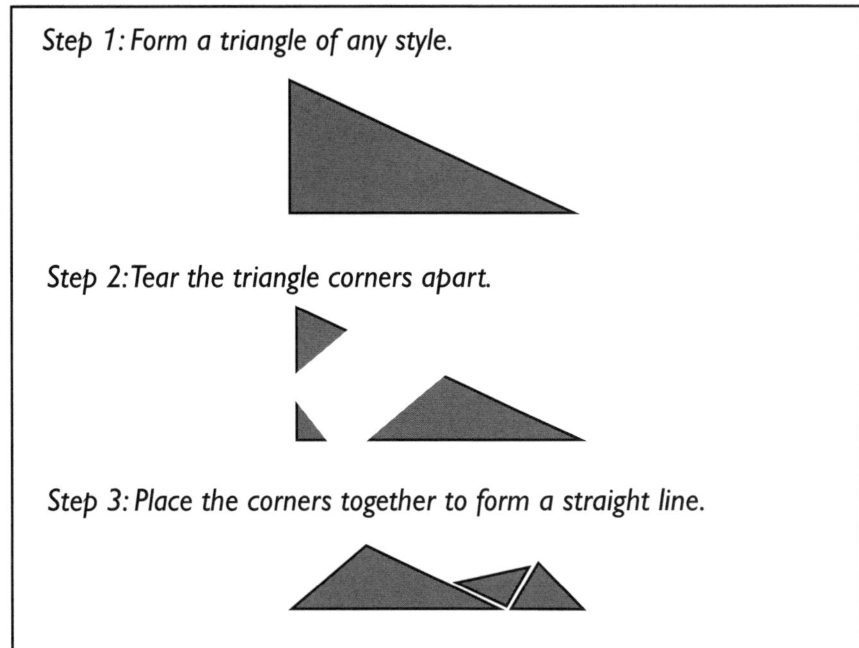

Figure 4.1. Proof of triangle interior angles through paper tearing.

The crux of the proof lies in the known assumption that alternate interior angles, such as angles 1 and B, are equal. Based on this assumption, we know that angles 1 and B are equal and angles 2 and C are equal because they are pairs of alternate interior angles. We also know that angles 1, A, and 2 form a straight line, thus determining that the sum of those three angles is 180 degrees. Consequently, since angle 1 equals angle B and angle 2 equals angle C, we can conclude that the sum of angles A, B, and C equals 180 degrees, and the proof is done.

This more advanced proof illustrated through Figure 4.2 can also be verified more solidly through geometric manipulations using computer software. Using technology, students can verify that alternate interior angles are equal. They can even go further to learn other proofs that stem from the same figure involving alternate exterior angles, corresponding angles, and vertical angles. Thus, through technology, students can experience a range of proofs, solving problems both visually and logically.

A second dimension in the presentation of the proofs occurs when the student prepares a document (word processed or Web site) in which the proof is presented and elucidated.

Two of the National Educational Technology Standards for Students have relevance here.

NETS•S Technology Productivity Standard

3. **Technology productivity tools**
   - Students use technology tools to enhance learning, increase productivity, and promote creativity.
   - Students use productivity tools to collaborate in constructing technology-enhanced models, preparing publications, and producing other creative works.

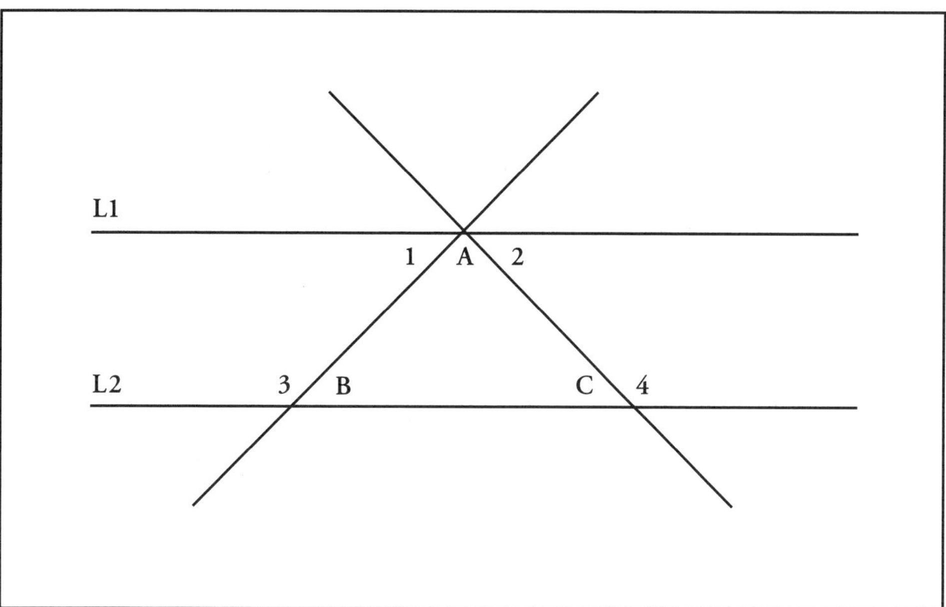

Figure 4.2. Triangle ABC set between parallel lines L1 and L2.

Under the above standard, students will use the computer and drawing or CAD software to construct models that illustrate the concept they seek to prove.

NETS•S Technology Communications Standard

4. **Technology communications tools**
    - Students use telecommunications to collaborate, publish, and interact with peers, experts, and other audiences.
    - Students use a variety of media and formats to communicate information and ideas effectively to multiple audiences.

THEORY INTO PRACTICE

Reasoning, Constructivism, and Skemp's Continuum of Understanding

By Anne Raymond

Reasoning in mathematics depends heavily on how people understand and ultimately come to construct their own knowledge of mathematics. Constructivists support the notion that children must be active participants in the development of their own understanding (Fosnot, 1996; Schifter, 1996). Schifter and Fosnot argue that "no matter how lucidly and patiently teachers explain to their students, they cannot understand for their students."

In light of work by Hiebert and Carpenter (1992), Van de Walle (2001) broadly defines understanding as "a measure of the quality and quantity of connections that an idea has with existing ideas." He further explains that "understanding depends on the existence of appropriate ideas and on the creation of new connections" (p. 28).

In the spirit of trying to understand the notion of constructivism and how it relates to mathematical understanding, we may recall the debate sparked by Skemp (1978) in his landmark discussion of instrumental versus relational understanding. Skemp describes instrumental understanding as an understanding of ideas that are mostly or completely isolated. For example, rote learning or the memorization of a procedure would fall into the category of instrumental learning. On the other hand, relational understanding is the understanding of ideas in connection with other concepts and procedures. Thus, for example, one might understand the additional algorithm relationally if approached through the use of base 10 block manipulation, making visual and kinesthetic connections to the algorithm itself.

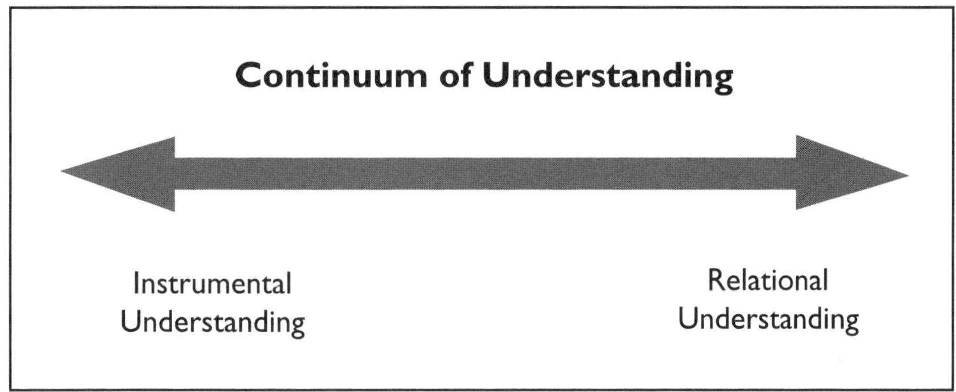

Figure 4.3.

Hiebert and Carpenter (1992) suggest a continuum of understanding where Skemp's relational understanding and instrumental understanding are on opposites ends of the continuum (Figure 4.3). Although constructivists would suggest that understanding closer to the relational end of the spectrum is preferred, it is clear that over the years many students have learned and understood mathematics more in an instrumental way.

When appropriate technology is infused into the mathematics curriculum, teachers have the potential to affect the process of understanding so that it moves the individual away from the instrumental understanding toward a more relational balance. For example, when students have the opportunity to experience mathematics visually through interaction with graphics and spreadsheet manipulations, they come at mathematical concepts from more than one perspective, making vital connections.

References

Fosnot, C. T. (Ed.). (1996). *Constructivism: Theory, perspectives, and practice.* New York: Teachers College Press.

Hiebert, J. C., & Carpenter, T. P. (1992). Learning and teaching with understanding. In D. A. Grouws (Ed.), *Handbook of research on mathematics teaching and learning* (pp. 65–97). Old Tappan, NJ: Macmillan.

Schifter, D. (1996). A constructivist perspective on teaching and learning mathematics. *Phi Delta Kappan, 77,* 492–499.

Skemp, R. (1978). Relational understanding and instrumental understanding. *Arithmetic Teacher, 26*(3), 9–15.

Van de Walle, J. A. (2001). *Teaching mathematics developmentally* (4th ed.). New York: Addison Wesley Longman.

INSIGHT

Inductive and Deductive Reasoning

By Anne Raymond

There are many types of mathematical reasoning students ought to develop. Forms of reasoning include, among others, inductive reasoning, deductive reasoning, algebraic reasoning, proportional reasoning, and spatial reasoning. In problem-solving situations, multiple forms of reasoning may be used in partnership with each other to find an appropriate solution.

Inductive and deductive reasoning are often thought of as key approaches to forming persuasive mathematical proofs. Today's technology, in the form of spreadsheets, charting, and graphic options, supports the development of inductive and deductive reasoning.

Inductive and deductive reasoning are often seen as *opposite* forms of reasoning. Deductive reasoning is reasoning employed when one begins with a set of given general truths or assumptions and, through process of elimination or some sort of comparative analysis, comes to a very specific conclusion. In brief, deductive reasoning involves moving *from general to specific* conclusions.

Conversely, inductive reasoning is thinking that begins with specific examples and, through developing patterns, yields a generalization. Therefore, inductive reasoning involves moving *from specific to general* conclusions. Although viewed as opposite forms of thinking, inductive and deductive reasoning often complement each other and can provide two related means of solving a problem.

More than deductive reasoning, inductive reasoning can be easily played out using programs such as Excel to develop patterns and attempt to generalize the patterns to formulas. Consider the following problem that can be charted and solved using Excel:

> ***The Reunion:*** *At a family reunion, the first family member to show up knocks on the door and enters the party hall alone. After that, every time there is a knock at the door of the party hall, two more family members enter than entered on the knock before. After the door has been knocked on and opened 20 times, all of the family members are there. How many family members showed up at the reunion?*
>
> **Solution**: A solution can be determined by merely drawing a chart and filling in the columns, establishing patterns when possible. However, by using Excel, the problem solver can experiment with possible generalizations for each of the columns (see Figure 4.4, developed using Excel). The "guess and check" strategy of problem solving, combined with the available technology, helps facilitate students' ability to induce generalizations.

THE REUNION PROBLEM		
Door Knock	Number That Enter	Total in the Room
1	1	1
2	3	4
3	5	9
4	7	16
5	9	25
6	11	36
7	13	49
8	15	64
9	17	81
10	19	100
.		
.		
.		
20	39	400
.		
.		
.		
n	2*n-1	n*n

Figure 4.4: An Excel chart solution to the Reunion Problem.

INSIGHT

Hands-On Random Sampling

By Ivan W. Baugh

Estimation is an essential life skill. Whether developing a conclusion or evaluating an answer, the ability to have a sense of the closeness of an answer to the computed value will facilitate an individual's work, often referred to as reasonableness of results. This enables the user to determine if the result produced reasonably approaches the expected answer.

Polls use a statistical form of estimation called random sampling. Many individuals find the concept of random sampling difficult to understand. Because of the abstract nature of the concept, you may want to have your students complete a hands-on activity to help them comprehend what occurs in random sampling.

Inflatable globes provide a tool for experiencing hands-on random sampling. First, have the participants load a spreadsheet. Set up the spreadsheet so that it will look like the one shown in Figure 4.5.

◇	A	B	C	D	E
1		A	B	C	D
2	1		Guess	Random Sample	Actual
3	2	Land			
4	3	Water			
5	4				
6	5				
7					

Sheet1
Ready Sum=0 SCRL CAPS NUM

Figure 4.5

Next, hold the inflated globe in plain view for all to see. I slowly rotate the globe, asking each of them to estimate the percentage of land and the percentage of water. We enter that information in a spreadsheet under the heading Guess.

Then, ask each student to take an erasable pencil or pen and place a dot on the end of the thumbnail of one hand. Have the students arrange themselves in a circle. As they toss the globe, they will look to see if the dot on the thumbnail is over land or water. Should it happen to be on a line, report water if it is mostly over water or land if it is mostly over land. If it happens to be evenly divided, report water. As students announce their results, tally the results. Upon completion of the desired number of globe tosses, have the students enter the results in the spreadsheet under the heading Random Sample.

For the final step, assign selected students the responsibility to research the actual figures (an electronic encyclopedia works well). Enter the information in the spreadsheet under the heading Actual.

Have the students create pie charts to enable them to compare the results. How they do this will depend on the software they are using. In AppleWorks, select cells A1 through D3 and it will create three separate pie charts. In Excel, select each data set separately and make the chart. To select nonadjacent data, press the CTRL key while you select the cells you want to chart. For example, select the cells containing data in column A, then press the CTRL key as you select the cells containing data in column C. Then use the Chart Wizard to complete one chart. Repeat the process for each data set.

After completing the charts, discuss the findings. Then pose the question, "What difference would it make if we used a larger globe or a smaller globe?" After discussing students' points of view, complete a second globe toss using either a larger or a smaller globe. Compare the results.

The first time I did this activity, we used 24-inch globes. The workshop had more than 60 participants. We made two groups and used two globes. Each group tossed a globe 100 times. One group had the results 65 water, 35 land. The other group posted 67 water and 33 land. I have done this with a 12-inch globe with similar results. When I used a globe the size of a tennis ball, the results frequently moved toward 60 land, 40 water. One time, using the smallest globe, we got 50 water, 50 land. Upon repetition, we got different results, with more water than land.

I found my students developed a better understanding of random sampling through completing this hands-on activity. One student raised the question, "Could we not control the results by tossing the globe without allowing it to spin?" We discussed the potential impact this could have. I then challenged them to toss it without it spinning. They had very limited success. You will want to make sure each student catches the globe at least one time.

One Worthy Concept,

Many Worthwhile Lessons

By Clint Mason Luscombe

I recently fell in love with both the power and potential of spreadsheet software as a major tool to help me teach high school math students. My students enjoy both re-creating tables using spreadsheets and converting those spreadsheets into graphical form. In the process, they understand how math tables are created and why a table for the same function should be the same in all cases. Turning math tables into graphs adds tremendous meaning to each table, because it becomes more than a "list" of numbers when brought to life in graph form. The newly created graph often takes on a noticeable and predictable form that allows students better to appreciate and understand each table they chart as they experience the software's powerful feedback.

One set of tables that we re-created and graphed was the six trigonometric functions. Students were asked to create a spreadsheet table of the six trigonometric functions using decimal degrees. "Lesson for Trigonometric Functions" shows the formulas I used to make the spreadsheet. My students, however, were not given such details, because I wanted them to work their way through the process as I had.

Using Spreadsheets to Re-Create the Tables

With minimal instruction, students can be shown how to use formulas to re-create the selected table easily and powerfully. Figure 1 shows a formulas table and its corresponding values table. Notice that once the initial degree of "1" was selected, the formulas refer back to "A2," its row-and-column position.

Lesson for Trigonometric Functions

1. Create a spreadsheet table of the six trigonometric functions using decimal degrees. Here are the formulas you will need:

Degrees	Sine	Cosine	Tangent
A2 + 1	= SIN(A2*PI()/180)	= COS(A2*PI()/180)	= TAN(A2*PI()/180)

Cosecant	Secant	Cotangent
= 1/B2	= 1/C2	= 1/D2

2. Analyze your trigonometric table by using your program's chart-making capabilities.
3. Predict what the graphs would be like if the table were extended to higher values.
4. Be prepared to demonstrate any time-saving "maneuvers" you use in setting up the tables and graphs.
5. Print the charts, the values table, and the formulas table.

Initially, students were challenged with converting decimal degrees to radian measure because Excel required such units. They were encouraged to recreate the degree table rather than the radian table because they might have better success making this table given that they were a little more accustomed to degree measure. The students recalled how familiar the software notation "PI()/180" was to their own notation. Because degrees and radians are related by the formula 180 degrees = p radians, the students noticed that Excel simply multiplied the degree measure by p/180. Also, students learned they could choose the number of decimal places displayed in the table by going to "Format," choosing "Category," and then selecting the number of decimal places they desired.

Using a Spreadsheet's Graph-Making Capabilities to Analyze the Table

I asked students to use the spreadsheet to create only a significant part of the table and then use the graph-making capabilities of their software to represent their tables. A "significant part" is one that yields enough information to predict the rest of the curve safely, so each pair of students had different significant parts that they felt comfortable using as predictors. Some students simply duplicated a large enough portion of the table to make predictions of its extension easy. They found that each table has its unique significant region that needs to be compiled before its entire graph can be understood. For example, each trigonometric function can be better pictured if a 360-degree region is converted into a graph. Looking at the sine and cosine curves as well as their reciprocals, cosecant and secant, the students saw that every "width" of 360 degrees was repeated. When prompted, they realized that any 360-degree width was wide enough to contain the whole "repeated pattern." The tangent and cotangent curves have a pattern that is repeated every 180 degrees, but students felt more comfortable verifying that fact with at least 360 degrees' worth of data.

Most curves, however, were not initially of proper lengths, so the tables had to be enlarged to guarantee that a significant region was being graphed. Because making a table is easy, most students will be able to enlarge their tables if they have to find their tables' significant regions.

Figure 1.

DEGREES	COSINE	SINE
1	= COS(A2*PI()/180)	= SIN(A2*PI()/180)
= A2+1	= COS(A3*PI()/180)	= SIN(A3*PI()/180)
= A3+1	= COS(A4*PI()/180)	= SIN(A4*PI()/180)
= A4+1	= COS(A5*PI()/180)	=SIN(A5*PI()/180)
= A5+1	= COS(A6*PI()/180)	= SIN(A6*PI()/180)
= A6+1	= COS(A7*PI()/180)	= SIN(A7*PI()/180)
= A7+1	= COS(A8*PI()/180)	= SIN(A8*PI()/180)
= A8+1	= COS(A9*PI()/180)	= SIN(A9*PI()/180)
= A9+1	= COS(A10*PI()/180)	= SIN(A10*PI()/180)

DEGREES	COSINE	SINE
1.0000	0.9998	0.0175
2.0000	0.9994	0.0349
3.0000	0.9986	0.0523
4.0000	0.9976	0.0698
5.0000	0.9962	0.0872
6.0000	0.9945	0.1045
7.0000	0.9925	0.1219
8.0000	0.9903	0.1392
9.0000	0.9877	0.1564

Predicting the Graphs from the Tables

From that point, I asked students to predict what the rest of the graph would look like if the table were extended. Once a sufficient region was graphed, the horizontal axis had to be adjusted to improve the graph's readability (i.e., they either had to reformat their table or put a legend on their chart).

When the sine and cosine curves were superimposed on one chart, the students could readily see their wavelike motion and, when prompted, explain their 360-degree periods. Students were prompted to use their fingers to trace the curve on the computer screen; once the movement of their fingers reached the point that it moved in the same direction as it did from the starting point, they had found the period (see Figure 2).

While analyzing the four other trigonometric curves—tangent, cosecant, secant, and cotangent—the students found vertical asymptotes and were encouraged to explain why they appeared; this led to a discussion of limits and infinity. The students noticed, for instance, that when the cosecant curve was graphed with the sine curve, its asymptotes were at the *x*-intercepts. They were then prompted to compare the definition of sine with cosecant. I

Figure 2.
Sample Sine and Cosine Chart

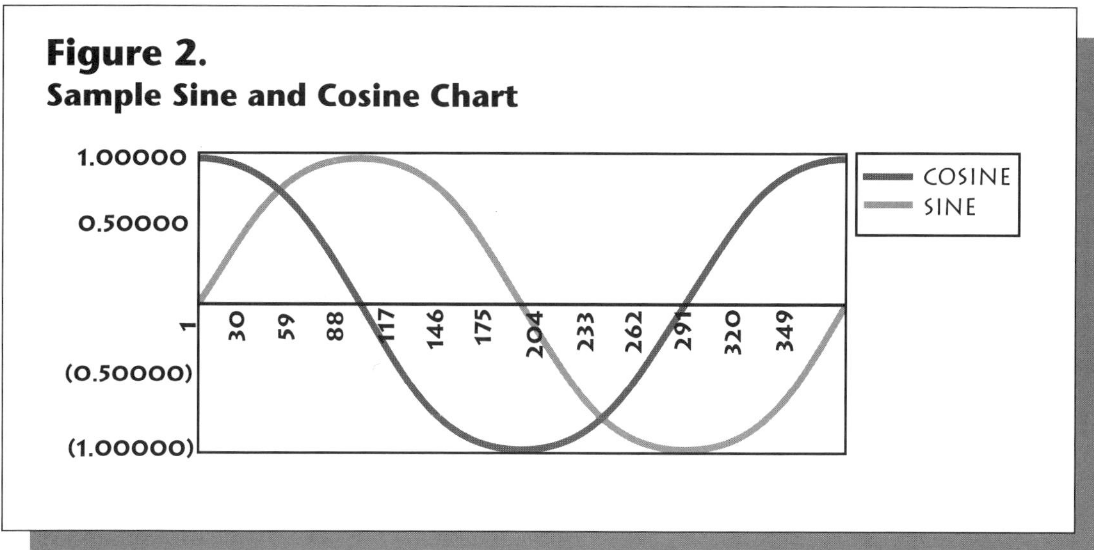

asked them, "What is the value of the sine curve at its x-intercepts?" Because they were reciprocals, their values at each angle should also be reciprocals. The students declared, "The reciprocal of zero is undefined, so there is no graph at that location!" To see how limits pertained to the asymptotes, the students were asked to study the values of the cosecant curve at degree measures close to each asymptote. They observed the cosecant approaching either positive or negative infinity as it approached each asymptote.

These tables are filled with patterns, so developing the student's ability to recognize them is critically important in improving his or her understanding of math. When asked, the students explored these patterns: At which points do the curves intersect? At which points are they the maximum distance apart? Is there a definite starting point for the curves' repeating pattern? How are they similar and dissimilar? Does the curve have a maximum or minimum? Does the curve have a positive or negative slope or both?

As more and more tables were analyzed, students learned to appreciate such concepts as periodic curves, limits, infinity, critical points, and much more. Each lesson was filled with opportunities for the students to make and check hypotheses.

Sharing the Shortcuts

Finally, students shared some neat shortcuts they discovered that added to the power of the software. While pressing "Control" and the left single quotation mark, Excel changes the values version of the worksheet display to the corresponding formulas display. The students also used the "fill handle" in the lower right corner of the active cell by dragging it over the desired cells to complete their charts. To return to the top left corner of the work-sheet, they used "Control" plus "Home," and to go to the lower right corner of the worksheet, they used "Control" plus "End."

Figure 3.
Sample Values Table

Degrees	Sine	Cosine	Tangent	Cosecant	Secant	Cotangent
5	.0872	.9962	.0875	11.47	1.004	11.43
10	.1736	.9848	.1763	5.759	1.015	5.671
15	.2588	.9659	.2679	3.864	1.035	3.732
20	.3420	.9397	.3640	2.924	1.064	2.747
25	.4226	.9063	.4663	2.366	1.103	2.145
30	.5000	.8660	.5774	2.000	1.155	1.732

Conclusion

After students gain experience and confidence re-creating tables, they can create one of their own—for example, a greatest integer table. Designing a table is a terrific problem-solving activity.

Most math books have tables in their appendixes that can be re-created using spreadsheet software. The following tables, for example, were found in the appendixes of my own school math books:

- Table of Squares of Integers from 1 to 100
- Table of Square Roots
- Cube and Cube Roots
- Table of Trigonometric Ratios in Decimal Degrees
- Table of Trigonometric Ratios in Radian Measure
- Common Logarithms of Numbers
- Logarithm of Trigonometric Functions
- Natural Logarithms
- e^x
- $e^{(-x)}$

Is There a Pattern?

By Chris Wallace

Editor's Note: *"Students should learn to use the computer as a tool for processing information and performing calculations to investigate and solve problems" (NCTM Curriculum and Evaluation Standards, 1989). This activity is particularly nice because the search for patterns is clearly aided by looking at the data in different ways provided in integrated software. The database provides a mechanism to quickly sort the data to look at subsets. The spreadsheet graphing capabilities beautifully describe the existence of a pattern. Even if you only have one computer, the search for patterns can be done by having groups identify investigations and showing the whole class the results. As the teacher, be sure to facilitate the student discussion with questions, not answers.*

While browsing through old issues of *The Mathematics Teacher*, I read a letter written by Henry Lulli in the January 1982 issue. He described the following exercise and stated an interesting fact that aroused my curiosity.

Think of a nonpalindromic two-digit number (i.e., the two digits are not the same). Reverse the digits to get a second nonpalindromic two-digit number. Subtract the smaller number from the larger. If the result is not 9, repeat the process. Keep reversing digits and subtracting until the difference is 9. Count the number of subtractions. For example, suppose my nonpalindromic two-digit number is 63.

Directions

1. Reverse the digits of your two-digit number.
2. Subtract the smaller of the two-digit numbers from the larger.
3. Repeat steps 1 and 2 on the result until the result is equal to 9.
4. Write down the number of subtractions it took to get a result of 9.
5. Repeat for all 3 two-digit numbers on this card.

	My Numbers		
	10	12	13
Number of subtractions	___	___	___

Name ______________________________

Figure 1. Use the multiple label format of Works to print large-sized labels with all the information except the numbers. Attach labels to the cards and fill in the numbers.

Reverse the digits:	36
Subtract the smaller number from the larger:	63 - 36 = 27
Is the result 9?	No
Reverse the digits:	72
Subtract the smaller number from the larger:	72 - 27 = 45
Is the result 9?	No
Reverse the digits:	54
Subtract the smaller number from the larger:	54 - 45 = 9
Is the result 9?	Yes

The number 63 requires three subtractions to get a difference of 9. It turns out, Mr. Lulli stated, that, at most, five subtractions are required to reduce any two-digit nonpalindromic number to 9. I set out to prove his statement and arrived at an activity ideally suited to an algebra class with access to computers and an integrated software package such as Microsoft Works on the Macintosh.

Day 1

Before class, write 3 nonpalindromic two-digit numbers on a 3 x 5 card for each student (see Figure 1.) There are 81 such numbers. If there are more than 27 students, assign 2 rather than 3 numbers per student. You may also want to write the directions on the card.

In class, describe the subtraction procedure and do at least one example on the board. Pass out the 3 x 5 cards and have students find the number of subtractions required to reduce their original numbers to nine. Do not tell them that the maximum number will be five, let them discover that on their own later. Collect the cards in preparation for Day 2.

Day 2

Before class, record all the data in a database. Figure 2 is a picture of the screen displaying a single record of the database using Works on a Macintosh. Note there are only two fields in the record, "Nonpalindromic two-digit number" and "Number of subtractions."

Copy the database onto a disk for each computer station. Ideally there will be two students per computer. By compiling the data beforehand, the students spend their class time working with the data rather than spending half the class time entering the data.

Display an overhead of Table 1 and ask students what they notice about the data. They probably will not notice the pattern when the data is in table form, but they probably will notice that all the nonpalindromic two-digit numbers require from one to

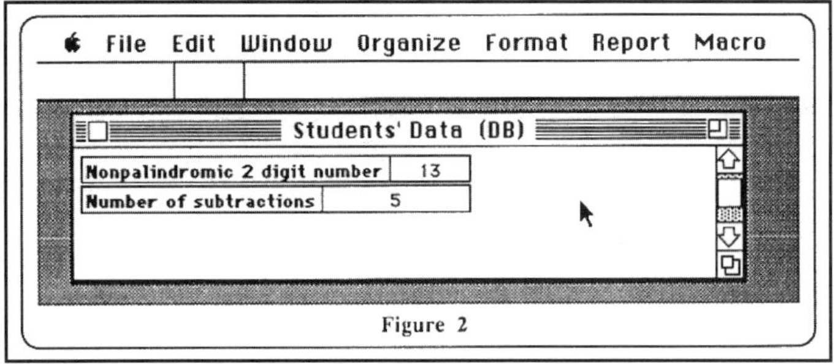

Figure 2.

five subtractions. Ask the class for possible explanations. Ask if anyone sees a pattern. Suggest that a graph might be more enlightening than a table of numbers. To graph the data using Works, each group of students must copy the data from their database to a spreadsheet and use the graphing utility under the Chart menu. Go through the graphing procedure step by step so that everyone will have the same line graph with the same scale.

The students generate a graph similar to that shown in Figure 3. (Mine was cleaned up with a paint program.) Suddenly, everyone will see that there is a pattern. Discuss the pattern. Ask the students how many points are in each cycle (10). Is the cycle from 12 to 21 symmetric? (Yes, the left side is the mirror image of the right side.) Are there any other symmetries in other cycles? (Yes, for example, the cycle from 17 to 27 is also symmetric.)

Pose the problem of finding out the reason for the pattern. One way to investigate this problem is to separate the two-digit nonpalindromic numbers into five groups according to the number of subtractions. For instance, consider only those numbers that require just one subtraction. What do they all have in common? Is there a pattern to the sequence? To make this easier, have the students go back into the database and do a search

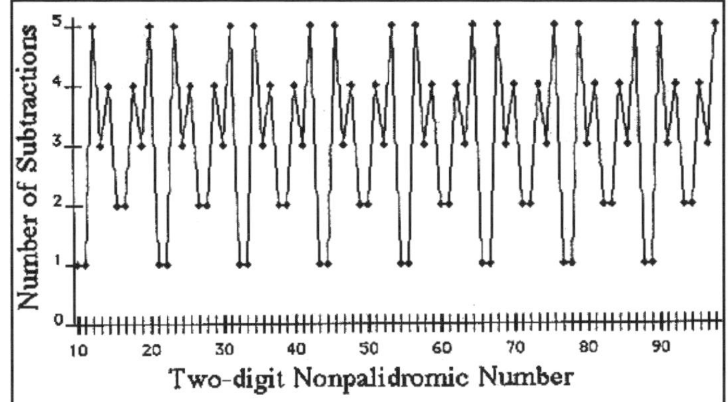

Figure 3.

for all records with 1 in the "Number of subtractions" field. In Works, you select Record Selection under the Organize menu. Then you highlight "Number of subtractions" and "Equals" and enter 1. Table 2 was created by doing five separate searches in the "Number of subtractions" field in the database and copying the data into columns in a spreadsheet. Afterward, the whole table was copied to the word processor.

Each group's goal is to discover and explain as many patterns as it can. They should write a report containing graphs and tables of their findings. Give them at least two class periods to complete the report.

Some of the observations that students might make include the following:

1) Subtract the 1st term from the 2nd, the 2nd from the 3rd, the 3rd from the 4th, ... For the sequence of numbers in the one subtraction group this becomes:

12 - 10 = 2
21 - 12 = 9

Number of subtractions for two-digit nonpalindromic numbers

#	Number of Subtractions	#	Number of Subtractions	#	Number of Subtractions	#	Number of Subtractions	#	Number of Subtractions	#	Number of Subtractions
10	1	26	4	41	3	57	5	72	2	87	1
12	1	27	2	42	5	58	3	73	4	89	1
13	5	28	2	43	1	59	4	74	3	90	5
14	3	29	4	45	1	60	2	75	5	91	3
15	4	30	3	46	5	61	2	76	1	92	4
16	2	31	5	47	3	62	4	78	1	93	2
17	2	32	1	48	4	63	3	79	5	94	2
18	4	34	1	49	2	64	5	80	3	95	4
19	3	35	5	50	2	65	1	81	4	96	3
20	5	36	3	51	4	67	1	82	2	97	5
21	1	37	4	52	3	68	5	83	2	98	1
23	1	38	2	53	5	69	3	84	4		
24	5	39	2	54	1	70	4	85	3		
25	3	40	4	56	1	71	2	86	5		

Table 1.

1 sub	2 sub	3 sub	4 sub	5 sub
10	16	14	15	13
12	17	19	18	20
21	27	25	26	24
23	28	30	29	31
32	38	36	37	35
34	39	41	40	42
43	49	47	48	46
45	50	52	51	53
54	60	58	59	57
56	61	63	62	64
65	71	69	70	68
67	72	74	73	75
76	82	80	81	79
78	83	85	84	86
87	93	91	92	90
89	94	96	95	97
98				

Table 2.

$$23 - 21 = 2$$
$$32 - 23 = 9, \text{ etc.}$$

Each group produces it own repeating sequence when this algorithm is followed. These sequences are shown below.

1 subtraction group 2, 9, 2, 9, 2, 9, 2, 9, 2, 9, 2, 9, 2, 9, 2, 9.

2 subtractions group 1, 10, 1, 10, 1, 10, 1, 10, 1, 10, 1, 10, 1, 10, 1.

3 subtractions group 5, 6, 5, 6, 5, 6, 5, 6, 5, 6, 5, 6, 5, 6, 5.

4 subtractions group 3, 8, 3, 8, 3, 8, 3, 8, 3, 8, 3, 8, 3, 8, 3.

5 subtractions group 7, 4, 7, 4, 7, 4, 7, 4, 7, 4, 7, 4, 7, 4, 7.

2) In each group, if XY is an element of the group (where X is the tens digit and Y is the ones digit), then YX is also in the group. This result must be the case by the nature of the subtraction algorithm. For example, 32 is in the group requiring just one subtraction. The number 23 is also in that group. In the one subtraction group these numbers are next to each other in the sequence.

10, (12, 21) (23, 32) (34, 43) (45, 54) (56, 65) (67, 76) (78, 87) (89, 98)

3) Another more difficult pattern to see is that all the numbers in the first column of Table 2 have a difference of 1 between their two digits. Likewise the numbers in the second column have a difference between digits of either 5 or 6. Table 3 summarizes this new pattern.

	Digit difference
1 subtraction group	1
2 subtractions group	5 or 6
3 subtractions group	3 or 8
4 subtractions group	4 or 7
5 subtractions group	2 or 9

Table 3.

The results in Table 3 can be shown algebraically by the following:

$$\begin{array}{r} x\,(x-1) \\ \underline{-(x-1)\,x} \\ 9 \end{array} \qquad \begin{array}{r} x\,(x-2) \\ \underline{-(x-2)\,x} \\ 18 \end{array} \qquad \begin{array}{r} (x-3) \\ \underline{-(x-3)\,x} \\ 27 \end{array}$$

$$\begin{array}{r} x\,(x-4) \\ \underline{-(x-4)\,x} \\ 36 \end{array} \qquad \begin{array}{r} x\,(x-5) \\ \underline{-(x-5)\,x} \\ 45 \end{array} \qquad \begin{array}{r} x\,(x-6) \\ \underline{-(x-6)\,x} \\ 54 \end{array}$$

$$\begin{array}{r} x\,(x-7) \\ \underline{-(x-7)\,x} \\ 63 \end{array} \qquad \begin{array}{r} x\,(x-8) \\ \underline{-(x-8)\,x} \\ 72 \end{array} \qquad \begin{array}{r} x\,(x-9) \\ \underline{-(x-9)\,x} \\ 81 \end{array}$$

All two-digit numbers with a difference of 1 between digits require just 1 subtraction to get to 9.

All two-digit numbers with a difference of 5 or 6 between digits result in 45 or 54 after the first subtraction. Since 45 and 54 have a difference of 1 between digits, it will take just one more subtraction to get 9. So these numbers require a total of two subtractions.

All two-digit numbers with a difference of 3 or 8 between digits result in 27 or 72 after the first subtraction. Since 27 and 72 have a difference of 5 between digits, it will take just two more subtractions to get 9. So these numbers require a total of three subtractions.

All two-digit numbers with a difference of 4 or 7 between digits result in 36 or 63 after the first subtraction. Since 36 and 63 have a difference of 3 between digits, it will take just three more subtractions to get 9. So these numbers require a total of four subtractions.

All two-digit numbers with a difference of 2 or 9 between digits result in 18 or 81 after the first subtraction. Since 18 and 81 have a difference of 7 between digits, it will take just four more subtractions to get 9. So these numbers require a total of five subtractions.

Summary

It is important that the students take part in the data collection step. Presenting Table 1 would not be as meaningful as the students generating the data themselves. This activity is rich with patterns and reasoning skills. The students draw graphs in the spreadsheet program, execute searches in the database program, and assemble everything in the word processor to complete the assignment. Remind them how stumped they were when all they had was a table of all the data. Discuss how tedious it would have been to graph the data and do all the searches by hand. Finally, discuss the patterns and their explanations that everyone sees.

CHAPTER 5

Representation

The Conversation Continues

Ivan: Graphical representations can contain very convincing information when you're trying to influence someone's decision or opinion on a topic.

Anne: Graphs can also be very misleading. By simply adjusting the vertical scale on a line graph or bar graph, you can distort the image of the information being represented by the graph. Depending on how you depict the information, you can persuade people to think differently about an issue.

Ivan: Here's an example of what you're talking about. Consider the graph in Figure 5.1.

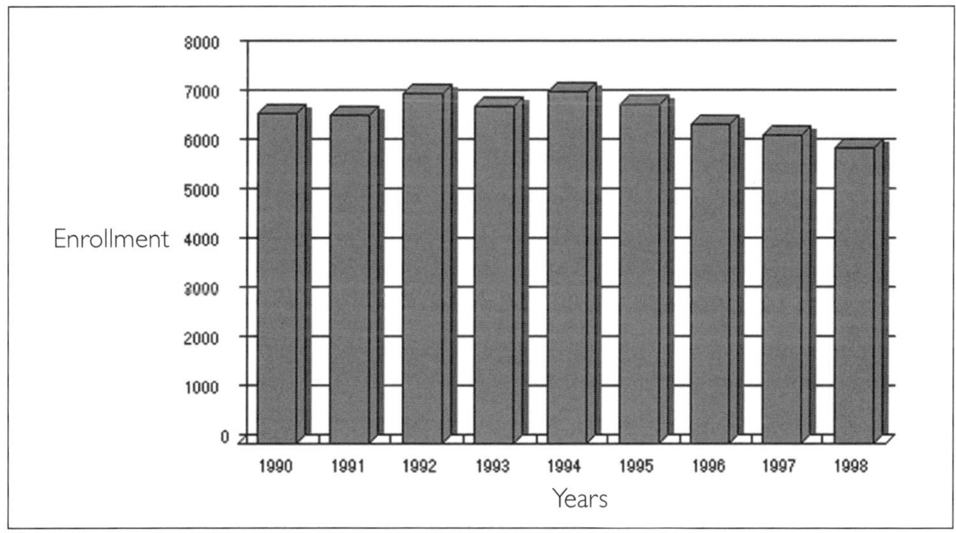

Figure 5.1.

The enrollment management team at Learn-A-Lot College may use the graph to illustrate that enrollment was fairly steady throughout the 1990s. On the other hand, the vice president of financial management of the same college may use the graph in Figure 5.2 to illustrate how enrollment has dropped considerably toward the end of the 1990s, thus explaining the need for financial cutbacks.

Both graphs illustrate enrollment between the years of 1990 to 1998 as being 6,700, 6,680, 7,100, 6,850, 7,150, 6,900, 6,500, 6,250, 6,000. However,

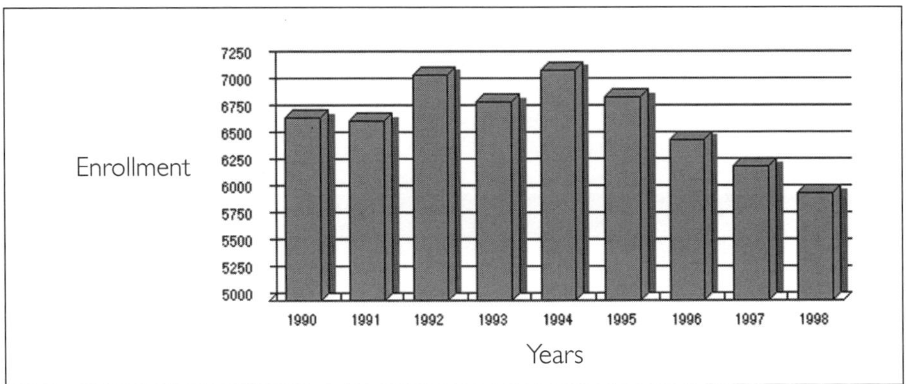

Figure 5.2.

each representation paints a different picture of how enrollments have or have not fluctuated over the decade. The only difference between the graphs is the choice of scale along the vertical axis. This single difference can really change the view of the situation at hand.

Anne: The articles in this chapter on the representation standard focus on different ways to represent mathematical data. For example, in the article "Choose the Right Graph," by Christopher Moersch, students are challenged to consider which types of graphical representations are appropriate for the given data.

THEORY INTO PRACTICE

Representation and Technology

By Ivan W. Baugh

How we represent numbers makes a difference in the results one obtains. You have likely heard of sleight of hand. Let's do some sleight of mind. Did you know that half of 12 is 7? Now you immediately think the author has gone off the deep end. But write the number 12 in Roman numerals:

XII

Draw a horizontal line dividing the number in half:

~~XII~~

Now erase the part below the horizontal line and tell me the number you see:

VII

Here's another one for you. Did you know that 2 + 1 = 4? Write the number 2 with an elongated line at the bottom:

2__

Now draw a vertical line across the extended line:

2+

The result is the number 4. Neither of these representations agrees with the mathematical principals underlying the original question, but they do accurately represent a result.

NCTM Representation Standard

Representation: Instructional programs from prekindergarten through Grade 12 should enable all students to

- Create and use representations to organize, record, and communicate mathematical ideas
- Select, apply, and translate among mathematical representations to solve problems
- Use representations to model and interpret physical, social, and mathematical phenomena

One early use of representations involves developing basic mathematical skills. I observed a primary teacher working with her students to develop a working knowledge of basic mathematics facts. She drew a 2 x 3 matrix on the board.

She guided her students as they learned the many different ways they could represent the information they saw in the matrix: 1 + 1 + 1 + 1 + 1 + 1; three groups of two horizontally, leading them to 2 + 2 + 2, which led them to 3 x 2; groups of three vertically, teaching them 3 x 2; and then moving to the converse of multiplication and showing them division, 6 ÷ 2 = 3 and 6 ÷ 3 = 2. Each presentation shows various ways of representing the information given in the matrix.

With the above analysis available for student use, give the students work sheets on which they will apply the basic mathematical facts studied above. Also provide them with manipulatives they can use as they solve the problems. As they use the information on the board and the manipulatives in solving the problems on the work sheet, they begin internalizing the processed information.

For homework, assign the students the task of explaining the completed work sheet to their parents. Encourage them to use the matrix or manipulatives as they explain what they have learned.

NETS•S Productivity Standard

3. Technology productivity tools

- Students use technology tools to enhance learning, increase productivity, and promote creativity.
- Students use productivity tools to collaborate in constructing technology-enhanced models, preparing publications, and producing other creative works.

The next day in class, continue with the matrix and examples available for student use. Give students another work sheet, where they will apply the studied principles. Rather than grading their papers, make use of a spreadsheet on a computer, such as the one shown in Figure 5.3, which the students will use to evaluate their work.

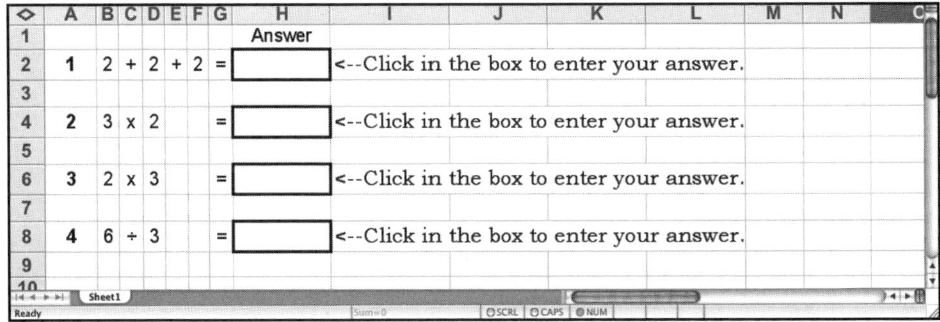

Figure 5.3.

The formula in I2 reads:

```
=IF(ISBLANK(H2),"<--Click in the box to enter your answer.",IF(H2=B2+D2+
F2,VLOOKUP(RAND()*5+1,$N$1:$O$5,2),"Use the manipulatives to check
your work."))
```

In another cell on the row, you could put a formula that would tally the number of correct answers on the first try. In column M2 off the screen to the right, you could have the spreadsheet tally the number of times it took the student to arrive at the correct answer: =if(isblank(G2),1,0). This checks to see if an answer has been entered in G2.

In columns V and W, the spreadsheet keeps track of how successfully the student solved the problems. When no answer has been entered, column V contains a zero. Column W adds 1 to column V when the entered answer is incorrect. This helps the teacher identify problems the student had difficulty solving correctly. The teacher can use this information to plan the next instructional activity. This could include grouping those who need additional instruction on a skill and moving those who mastered the skill ahead to a new skill.

For assignments you will record in your grade book, look at the formula in column Z that checks for the correct answer. If the student enters the correct answer, a 1 shows in the cell on the row with the problem. Cell Z1 uses a formula to total the correct answers and divide by the number of problems to determine the grade on the assignment or test. Note that I put this formula in a column far to the right of the area where the students will work. They will not see it. Each student will save the file with the student's name and date. The teacher will open each file; use CTRL + G to access the Go To window; and, to see the grade, enter the destination cell (Z1) where the grade is calculated. By using the spreadsheet in this manner, the teacher has more time for planning and for helping students who have difficulty with a skill.

Note: *On the companion Web site for this book (http://education.bellarmine.edu/baugh/), you will find set up for your use a spreadsheet named "Math Facts Worksheet."*

INSIGHT

Multiple Representations—A Big Idea

By Anne Raymond

It is important for students to obtain a sense of the "big ideas" in mathematics. Big ideas are themes that permeate multiple areas of mathematics and help provide a big picture of the underlying themes. According to Masingila, Lester, and Raymond (2002), big ideas are unifying notions that

- Help students make connections among what may for them seem like unrelated mathematical topics
- Help students make connections between the world of mathematics and the "real world"
- Provide ample opportunities for students to acquire a greater appreciation for, and understanding of, the beauty and subtlety of mathematics and mathematical activity

Big ideas can fall into several categories. Some big ideas may be more content-oriented in nature, thus suggesting certain structures that exist in mathematics, such as the ideas of equivalence, function and relation, and space and shape. Other big ideas can be classified more as actions that occur while exploring mathematics. Examples in this category include generalizing, decomposing/composing, verifying, and representing (Masingila et al.).

When teaching students to engage in the process of representing mathematics, it is vital that students learn to recognize multiple representations. Multiple representations can be as simple as illustrating the many ways to represent the numeric value 568, such as in Example 1.

EXAMPLE 1

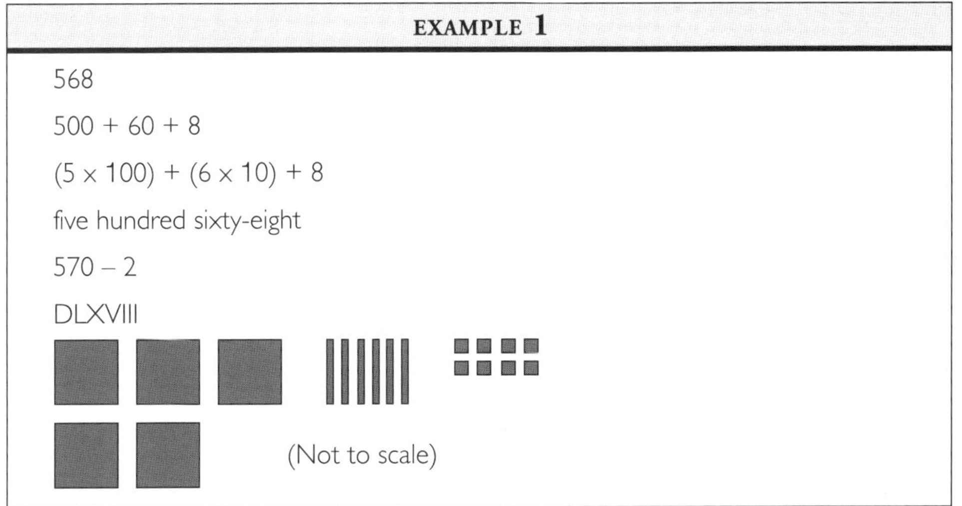

Example 1 shows many ways to represent the amount 568. It is written in conventional Hindu-Arabic form "568," but also in several expanded forms, in written words, as a difference in a subtraction problem, in Roman numerals, and pictorially using base 10 blocks. Students should develop their sense of "number" by being able to represent them in multiple forms. Many problem-solving activities with calculators afford students the opportunity to represent numbers in a variety of ways.

Other examples of multiple representations in mathematics are the following:

EXAMPLE 2

"Light margarine has 1/3 less fat than regular margarine."

"Light margarine has 2/3 the fat of regular margarine."

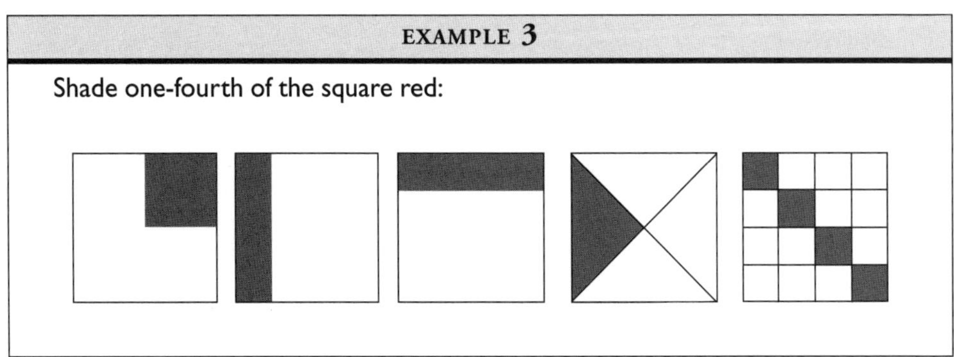

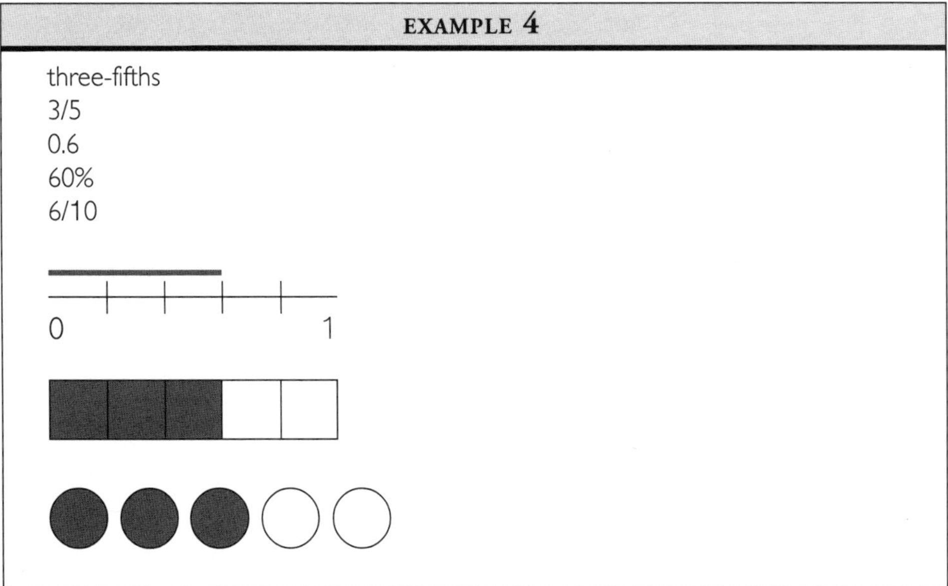

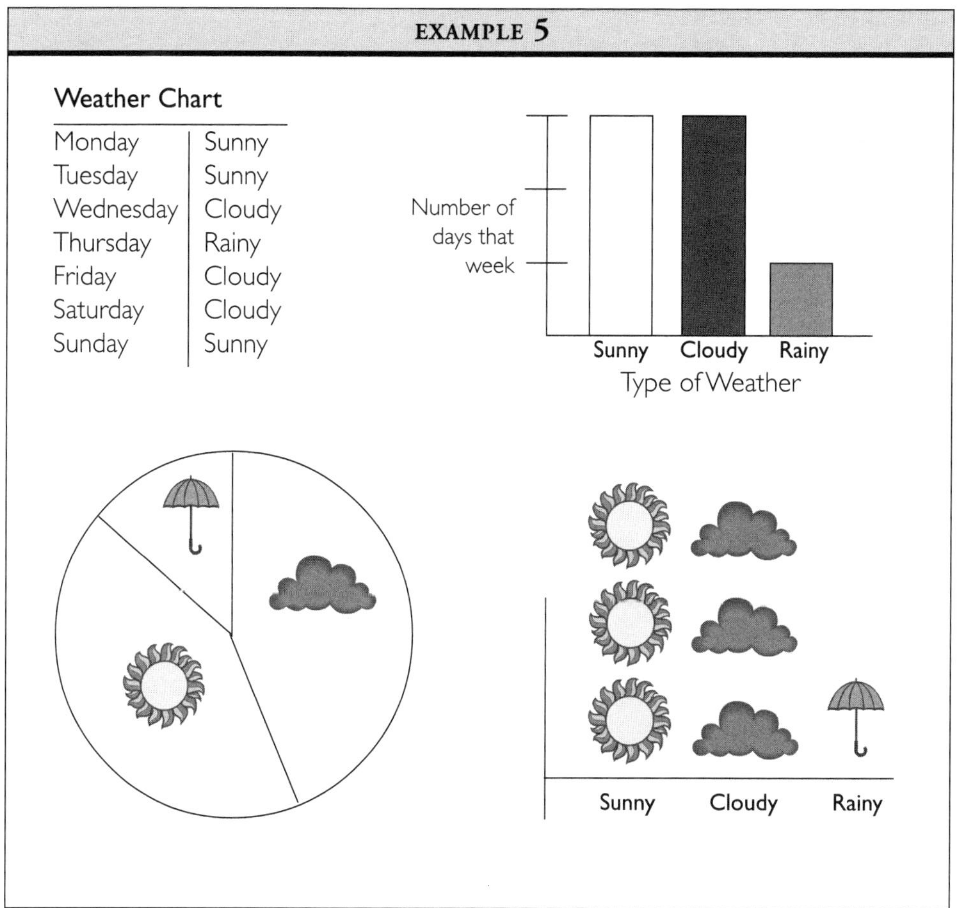

Reference

Masingila, J.O., Lester, F. K., & Raymond, A. M. (2002). *Mathematics for elementary teachers via problem solving: Instructor manual.* Upper Saddle River, NJ: Prentice-Hall.

INSIGHT

A Function Is a Function Is a Function ...

By Anne Raymond

Too often, students of mathematics have a single view of a function. They generally view functions solely in a formulaic way. That is, their image of a function is something of the form: $y = 2x + 8$, or perhaps even $f(x) = 2x + 8$. Nevertheless, there is the sense that a function is an equation that relates an independent variable to a dependent variable.

Students need to have a more diverse understanding of a function. Today's technologies, in the form of graphing calculators and computer algebra systems, provide students many opportunities to experiment with and examine functions in many forms.

Six key ways of representing functions include:

- an equation, such as $y = 7x - 1$
- a graph, such as:

- a table or chart of related values, such as:

x	y
-3	-6
-2	-4
-1	-2
0	0
1	2
2	4

- a "mystery box" or function machine, such as:

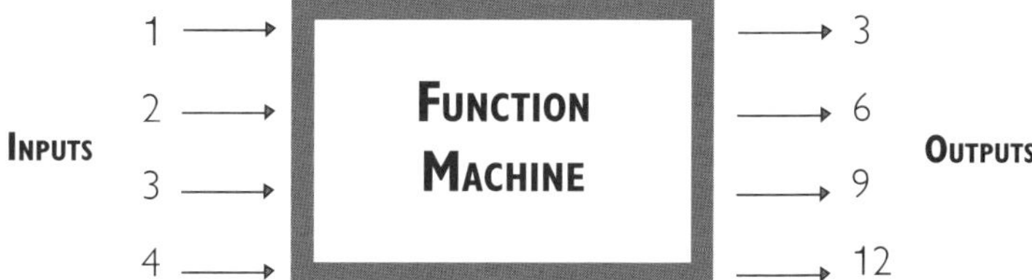

- a mapping, such as:

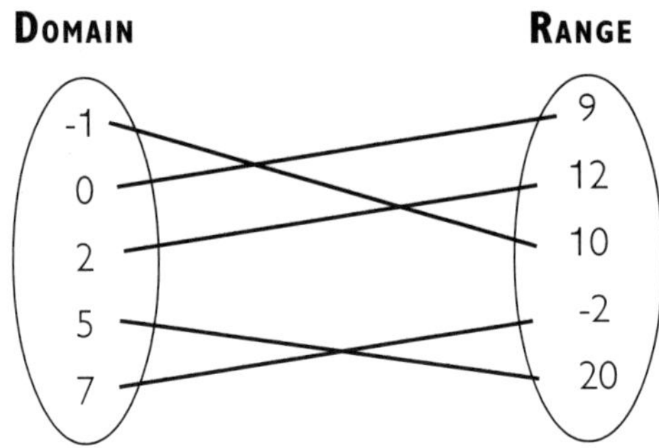

- ordered pairs, such as:

(5, 10), (10, 20), (2, 4), (7, 14), (8, 16)

In all the cases above, one can determine relationships between corresponding values. Students should be taught to think of functions in all these ways and to be able to determine scenarios that may be described by the different types of functions.

Beyond Just Doing It

Making Discerning Decisions about Using Electronic Graphing Tools

Do your students understand the data they are creating just because they can hand in a computer-generated graph of it? Maybe, or maybe not. Catherine Miles Grant describes a project that incorporates both hand-drawn and computer-generated graphs to enrich students' understanding of data and what they represent.

Figure 1. These students chose to represent their entire group's data on a single outfit.

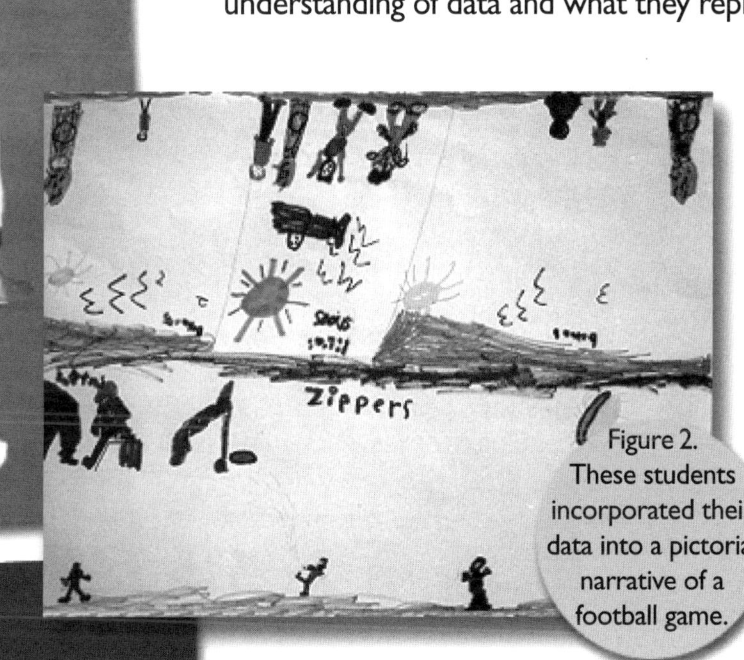

Figure 2. These students incorporated their data into a pictorial narrative of a football game.

By Catherine Miles Grant

Subject: Mathematics, graphing

Grade Level: 2–4 (Ages 7–9)

Technology: The Graph Club (Tom Snyder Productions) or other software with graphing capabilities

Standards: *NETS•S* 3 & 5. (To find out more about the NETS project, go to www.iste.org and select Standards Projects.) *NCTM* 11. (Visit www.nctm.org to find out more about its mathematics standards.)

Online Supplement: www.iste.org/L&L

The work reported in this article was supported by National Science Foundation grant REC-9612905. The opinions, findings, conclusions, and recommendations expressed here are those of the author and do not necessarily reflect the views of the funder.

Schools are flocking to software publishers to equip their newly acquired computers with the latest programs. Electronic graphing tools are an important component of any computer tool kit. It is not unusual to see student projects with computer-generated graphs lining school hallways. Parents are delighted that their children are working with data and using computers to produce neat, professional-looking products that incorporate graphs along with the text, graphics, and tables they have come to look for.

Yet the fact that students are graphing with computers doesn't, in and of itself, mean that they are developing rich understandings of data and of the subtleties of data representation. The ease with which computers can generate graphs and other mathematical representations must not allow teachers to keep young students from developing their own representations. Children's own often-innovative representations can connect them with the nuances of the data with which they are working, providing opportunities to understand the ways in which choices of representational schemes can highlight different features of data.

I co-taught with classroom teachers in Department of Defense schools in Hanau, Germany. We worked together to facilitate mathematical investigations. (*Editor's Note:* Read more about Cathy's work in Hanau in "Beyond Tool Training" in *L&L* vol. 27 no. 3. And read "The Tool Kit" in *L&L* vol. 26 no. 5 to find out more about technology use in Hanau schools.) The second-grade activity I describe here illustrates how children's thinking can be highlighted in an extended investigation that uses a range of tools.

Clothes Fasteners

It was springtime in Hanau, and temperatures fluctuated from day to day, from moment to moment, and from recess (outside) to the classroom (inside). The clothes children wore used a number of fasteners, displaying ingenious ways for adding and removing layers of clothing easily. Students collected data about the different types of clothes fasteners and analyzed them using different tools.

First, the class brain-stormed types of clothing fasteners they knew of (zippers, snaps, buttons, hook and loop fasteners, laces, belts, etc.), and then decided to focus on zippers, snaps, and buttons. They considered which of the three types of clothes fasteners were most common in their classrooms.

Students asked questions that helped define the question and clarify the terms: "Is this about counting how many zippers, snaps, and buttons there are or how many people there are with zippers, snaps, and buttons?" "Is this about what we're wearing now or what's in our cubbies, too?" "What's the difference between a button and a snap?" Coming to a common understanding of a question is enormously important in reading data from a graph, yet students are rarely given the opportunity to clarify their understandings of just what is represented in any given graph.

Students jotted down their individual predictions and then shared their reasoning with each other.

Generating Representations by Hand

Students gathered in groups of four to count the number of zippers, snaps, and buttons and to make representations of the data for their group. Each group worked with poster board, construction paper, markers, scissors, and glue.

The variety of representations was striking. Many were picture graphs, in which items were drawn and clustered in categories. Some involved variations on standard graphical representations (bar and circle graphs). One group showed a prototype "child" wearing the requisite total number of zippers, snaps, and buttons from the group of four on a single outfit (Figure 1). And, one showed a pictorial narrative of a football game in which each player represented a zipper, snap, or button (Figure 2).

From this work we get a sense of the importance, for many children of this age, of incorporating detail and of setting their representations in a meaningful context. Most of the representations communicated the functions of the items—although there were almost no labels, looking at the drawings would allow the viewer to know what items were being counted and often their functions as well (i.e., their association with clothing). Several groups chose to go with what "looked good," rather than with what would best represent the actual data. For example, the creators of Figure 1 ingeniously resolved the dilemma of how to include the right number of fasteners on a figure while still making it "look right." They decided that each symbol would represent two items. The two "snaps" on the ears represent four snaps, and the five buttons represent 10 buttons. Thus, they had invented for themselves a way to use scales representing other units. Another group made a color-coded key to accompany their drawing of an outfit on a hanger. There had been no instructions to label the drawings, yet in seeing a need for what they called "instructions" this group displayed an awareness of audience and an understanding of the importance of clarity for that audience.

The representations that children developed by hand did not necessarily follow the conventions of graphing that adults would use. Yet they powerfully communicated the meaning of the data they collected in their small groups. Textured discussions about the data and where the next steps in the investigation might be grew from these experiences.

Making Graphs with the Computer

When their hand-drawn representations were well under way, students used The Graph Club to create further representations of their data. They had explored the tool previously, getting to know its features and familiarizing themselves with the interface and its options.

Predictably, the tool provided plenty of "flash appeal" to these second graders. They were drawn to the professional look of their graphs and proudly made copies for each team member. As an adult viewer interested in children's mathematical thinking, I was struck by how much more sterile these graphs were than the rich representations they had generated by hand. Had they not had prior experience with the data, these second graders might have been pulled further from the original question and their findings by the use of electronically generated graphs, rather than closer to it. Yet introducing students to standard representations of data with which they had already worked and become familiar made it possible for them to make some fresh connections.

For example, the creators of the representation in Figure 2 had become so carried away with their football narrative that they had lost track of the items they were representing. Entering the data in the table on the computer brought them back to the original question. They were astonished to see their data represented first as a picture graph and then as a bar graph (the two were displayed side by side). They moved back and forth between the two graphs, first counting the symbols in the picture graph, then noting that the top of the stack of symbols was the same height as the bars on the bar graph. There followed a moment of "Aha!" when a student noticed that the bar graph had numbers on the vertical (Y) axis, from which one could tell how many items there were in each category without referring to the picture graph: "The numbers on the side show how many we counted!" This was an exciting connection with a standard feature of graphical representations. Nevertheless, their attachment to their original football narrative was strong: This group still labeled their computer graph "Football," even though the categories were Zippers, Snaps, and Buttons.

Comparing Groups' Understanding

Two groups' graphs show similarities in end products but differences in how they arrived there. Generally, circle graphs (or pie charts) aren't introduced until the upper-elementary grades when students have had experience with proportions, fractions, and percentages. Yet many children see data represented in this way in other places, such as television news, newspapers, and books. It is to be expected that, when invited to make their own representation of data, some students will choose to work with a representation they have seen previously.

Two of the groups represented their data with circle graphs. One group grappled at length with the issue of proportionality as they tried to accurately show eight buttons, two snaps, and two zippers on their hand-drawn circle graph. There were heated exchanges about whether the sections were too large or too small. Their argument was mathematical: Were they showing the proportion of buttons, snaps, and zippers accurately? There was a good deal of discussion, revision, and re-coloring before they eventually produced a graph they could all live with. Then, as though recognizing how lost the actual numbers had become in their graph, they created a picture graph to accompany it (Figure 3).

For this group, using The Graph Club to generate a graph of this same data was a thrilling confirmation of what they had settled on. They were amazed to see how closely the proportions on the computer-generated graph and their hand-drawn graph matched (Figure 4). They then became intrigued with the different options for labeling the categories (by number, name, fraction, or percent).

A second group also decided to represent their group's data (three zippers, three snaps, and three buttons) with a hand-drawn circle graph. But their representation suggests they didn't consider the relative size of the different sections, which, in this case, should be equal (Figure 5). For them, entering the same data in an electronic program and seeing it graphed electronically was eye-opening: the three sections were all equal in size! It provided an invitation to discuss the concept of equal portions in a circle graph of this data set.

Had the students gone straight to the computer to graph their findings, the teachers might not have seen the difference in the mathematical thinking of the students in the two groups. Each would have come up with a crisp circle graph to display the data. Yet the combination of student and computer-generated representations provided a fruitful context in which to further both students' understanding and the teacher's perspective on that understanding.

Supporting Children's Thinking through Mathematical Talk

The sometimes-heated discussions about how to represent group data constituted genuine mathematical discourse. As children talked their way through the challenge they had set out for themselves, they articulated their own thinking, considered their teammates' ideas, and found a way to come to agreement about their final representation. In the process, they clarified their own thinking about the importance of relative proportions in this problem and furthered their burgeoning understanding of key components of graphical representations. Children need real problems in which an answer is not immediately apparent to engage them in thoughtful discussion about mathematical ideas.

Once again, students made a prediction about data for the whole class: All together, are there more zippers, snaps, or buttons in this class? Their predictions were informed by results shared by each group. At that point the computer made it possible to quickly graph the cumulative findings of all the groups put together. (Buttons were the winners.)

One class then wondered whether the results would be the same if they asked the same question of family members at home. Students surveyed family members at home, and then used The Graph Club to graph their findings.

Here the neatness of electronically generated graphs helped students draw comparisons between home and school data. Though buttons were the "winners" among both second graders and family members, the relative sizes of the categories in the bar graphs demonstrated that there was a higher proportion of buttons to other kinds of fasteners among family members than among second graders. A class discussion centered around speculation that many family members were wearing military uniforms, which have a lot of buttons.

This class had moved beyond a view of graphs as an end product, some-

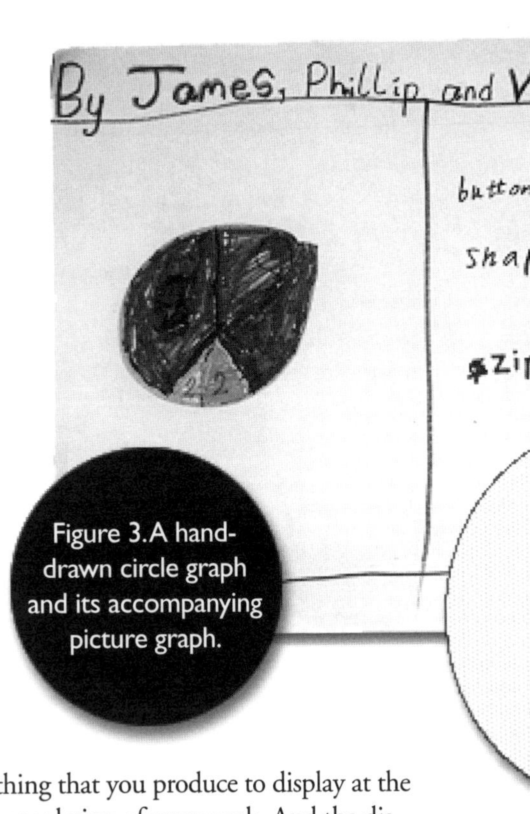

Figure 3. A hand-drawn circle graph and its accompanying picture graph.

Figure 4. The computer-generated graph of the data in Figure 3.

Figure 5. This graph shows equal data in unequal proportions.

thing that you produce to display at the completion of your work. And the discussion went beyond the mere labeling of the "highest" and the "lowest." Rather, both hand-drawn and computer-generated graphs were used as part of a continuous and ongoing process, where data displayed in a variety of ways highlight key elements, point to hypotheses, and suggest further avenues and questions to explore. With guidance, children referred back to their original question, to what their findings showed, and to further investigations.

Supporting the Development of Children's Understandings

With the proliferation of data in our society, it is no wonder that educators are drawn to the tools that make the incorporation of data representations easy. The key is to be discerning about the roles we choose for hand- and computer-generated graphs to support the progressive development of children's understanding of data.

One of the strongest arguments for having children create representations of data by hand is that they provide teachers with clues about these understandings. Rich examples of student work (generated both by hand and electronically) provide a powerful means of communicating to teachers the subtleties of children's work with data. Student work informs teachers about their engagement with the question itself, their conclusions about the data they collect, and their understanding of key features of data representation as a means of communicating information.

Effective work with children on data collection, representation, and analysis requires that we: work from questions that are of genuine interest and will yield meaningful data; give students the opportunity to develop their own representations of data; consider when and how to introduce standard representations and their uses; develop in students a grounded sense of when they need to clarify their questions and collect further data; help students make sense of their findings and describe them in rich ways; and encourage students to consider the data in light of their original questions and pose follow-up ones that will uncover more of the story.

From their earliest years in school, children can engage in rich explorations of their world through the analysis of data about authentic questions. In these investigations, computers can serve as powerful tools for supporting young people's mathematical thinking. The value of their presence as a mathematical tool will depend, however, on teachers' recognition of the strengths and limitations of their use in the classroom.

Acknowledgments

With grateful acknowledgment to Nancy Bogensberger, Nettie Eddie, Anna Heuer, and their students, who made this study possible.

Thanks to Elizabeth McNamara, Judith Davidson Wasser, and Andee Rubin for their review and suggestions.

Resource

The Graph Club is available from Tom Snyder Productions. Visit www.tomsnyder.com, call 800.342.0236 or 617.926.6000, or e-mail ask@tomsnyder.com for information or a free demo version.

Choose the Right Graph

"Getting information from a table is like extracting sunlight from a cucumber."

—Farquhar & Farquhar in *Economic and Industrial Delusions: A Discourse for the Case of Protection,* New York: Putnam, 1891.

By Christopher Moersch

The Farquhars' insightful commentary on the importance of graphic presentation in the 19th century has equal significance today. Contrary to popular belief, interpreting and generating graphs related to real-world phenomena is not limited to some distant research laboratory or federally funded think tank. It is very much a lifelong skill. Though the science of constructing bar charts, pie charts, and box plots has been "taught" for years, the practical application or art of graphing has been relatively neglected in classrooms nationwide. Why else would television commercials and newspaper ads continually market their products and services using graphs that often lack a clearly-defined scale or accurate labeling of the x- and y-axes?

Both the *NCTM Standards* and *Benchmarks for Science Literacy* reference the importance of collecting, tabulating, and graphing data as outcomes for a literate citizen. According to *Benchmarks*, "Students should practice using tabular, graphical, and symbolic representations of functions and translating among them — and they should be called upon to describe their tables, graphs, and equations in clear English." The art and science of graphing provides a succinct, visual representation for any data set, whether it originates from a classroom survey, experiment, outdoor simulation, database, or external probe.

The plethora of available spreadsheet and graphing software packages, including *LabQuest*, *Excel*, *ClarisWorks*, and *Cruncher* for the Apple, MS-DOS, and Macintosh environments have made graphing a relatively simple and painless process. Yet knowing what to graph and which graph to use (called an individual's graphicacy) requires some fundamental understanding of a few basic, yet essential, graphing conventions, idiosyncrasies, and prerequisites.

Conventions

Graphing conventions include proper labeling of the x/y axes, scaling of the graph to promote the maximum level of clarity and understanding, and most importantly, selecting the right graph to display tabular data. Whether to plot data as a bar chart versus a box and whisker graph, or a line graph versus an x/y graph are types of decisions that need to be conceptualized by students early in their formal education and applied liberally throughout their educational experience. In short, the conventions used in graphing should empower the reader to infer patterns, irregularities, and so forth from the data that were not apparent from its tabulated counterpart.

Idiosyncrasies

Most graphs can be partitioned into several dichotomous categories such as x-axis dependent (e.g., normal curve graph, line graph) versus non-x-axis dependent (e.g., pie chart, stem and leaf graph), or few data points (e.g., pie chart, bar chart) versus many data points (e.g., box plot, stem and leaf graph, normal curve). However, several of these graphs contain idiosyncrasies that make them truly unique. A pie chart, for example, is the only graph that provides a percentage breakdown of data points when compared to the total. The box plot is similarly unique because it is able to separate data points into quartiles. Lastly, the bar graph is the only chart with a dual identity—a histogram (if you are tracking frequencies) or a bar chart (if you are making comparisons).

Prerequisites

Parker and Widmer (1992) and Eng (1988) describe some practical prerequisites that should precede the actual graphing experience. These prerequisites include framing a question and organizing data in a worksheet or spreadsheet. All graphs should be in response to a significant, relevant question posed by the student. In this manner, the graphing experience mir-

SUGAR.WKS

◇	A	B	C
1	Food	Size	Sugar (tsp)
2	Gum	1 stick	0.5
3	Fudge	1 oz square	4.5
4	Candy	4 oz	20
5	Jello	1/2 cup	4.5
6	Jelly	1 Tbsp	6
7	Shake	10 oz glass	5
8	Cake	4 oz piece	10
9	Soda	8 oz bottle	5
10			

Figure 1. Spreadsheet

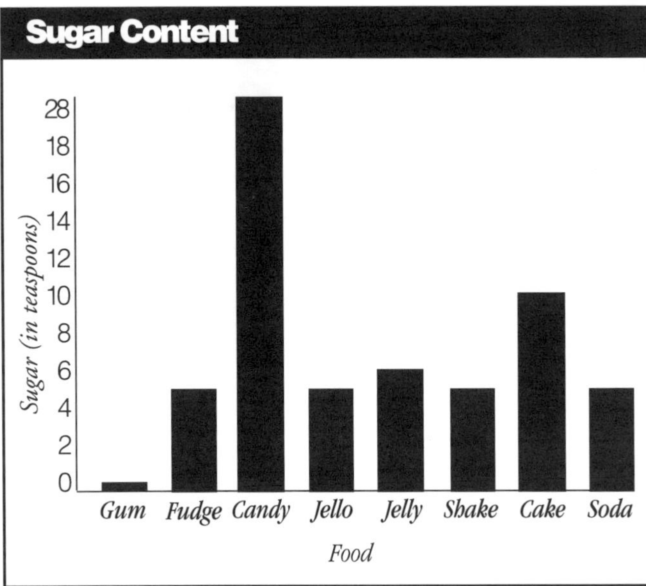

Figure 2. Graph

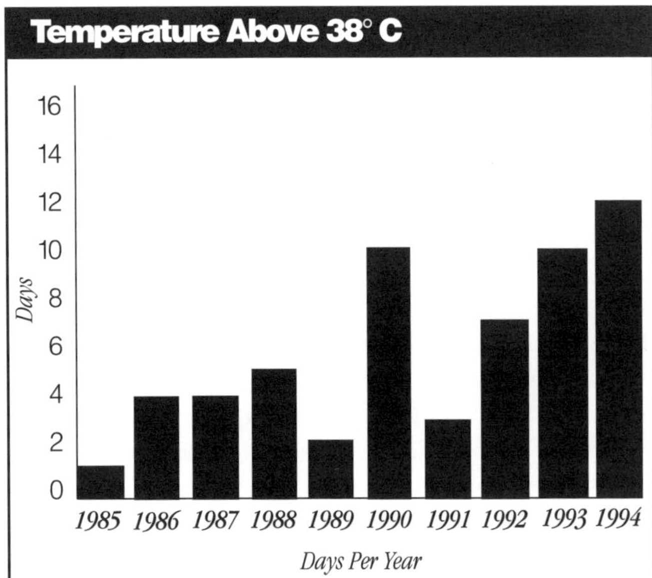

Figure 3. Bar Graph

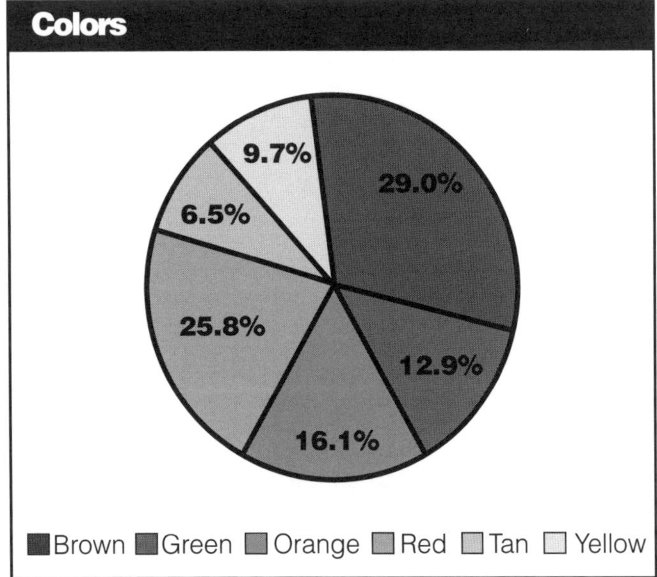

Figure 4. Pie Chart

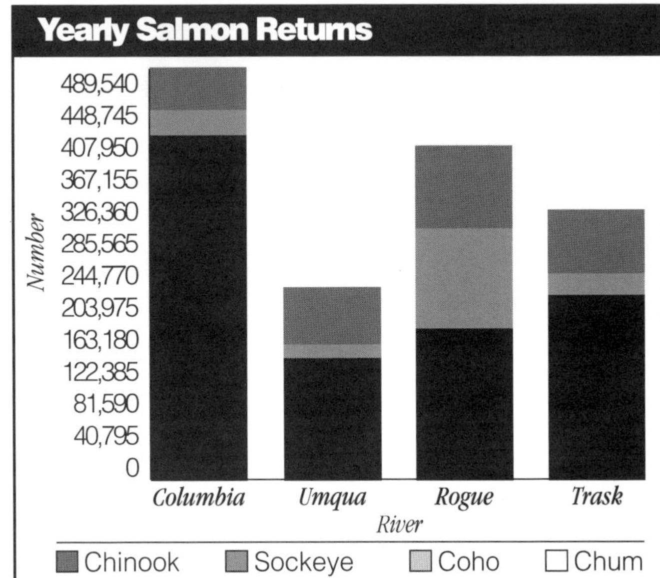

Figure 5. Stacked Bar Graph

rors the real world process of acquiring and analyzing data and finding solutions to authentic problems. Without such context, the graph itself becomes an esoteric expression void of any lasting meaning and relevancy to students.

Regardless if the data originates from a survey, experiment, census report, almanac, or newspaper, the strategy used to organize it in a spreadsheet can greatly enhance the ease to which it can be displayed in graphic format. The spreadsheet and graph in Figures 1 and 2 using *LabQuest®* contain data comparing the sugar content for different foods. In most spreadsheets, the data (including labels and values) are organized by columns. Column A is typically reserved for labels or values appearing along the x-axis while Columns B, C, D, and so forth contain values that are represented on the y-axis. Notice that Cell C1 contains the sugar label (which appears as the legend in the bar chart) while Column A contains the actual food types (which appear along the x-axis in the bar chart).

A thorough discussion of graphing prerequisites would be incomplete without also emphasizing the necessity for students to translate tabulated data into a graph (e.g., bar, line, stem and leaf) on grid or line paper prior to using a computer. Demonstrative success with graphing by hand will ensure that students have first conceptualized the numerous mathematical concepts and processes (e.g., ordered pairs, x/y axes, frequency distributions) embedded in each graph type. A description of 10 different graph types found in *LabQuest* and some practical tips regarding their use follow.

Bar Graph

A bar graph or histogram is one of the most useful, simple, and popular techniques in graphic presentations. Bar graphs use vertical or horizontal bars to make comparisons between one or more ranges of data or to report frequencies (i.e., histogram). The length of each bar or of its components is proportional to

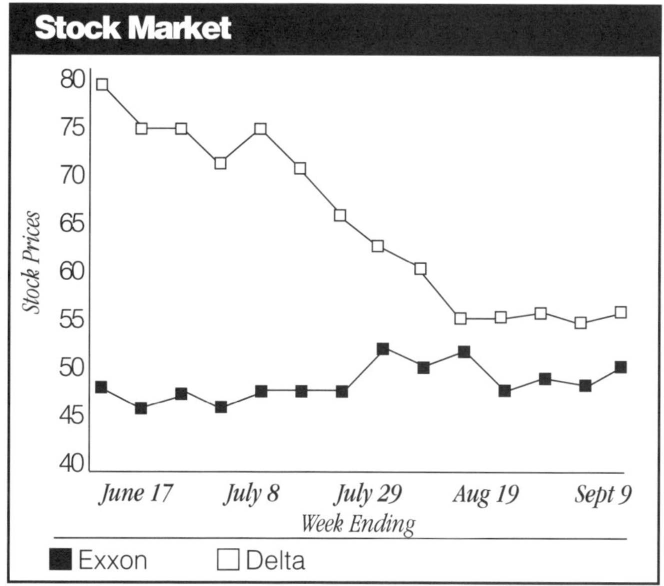

Figure 6. Line Graph

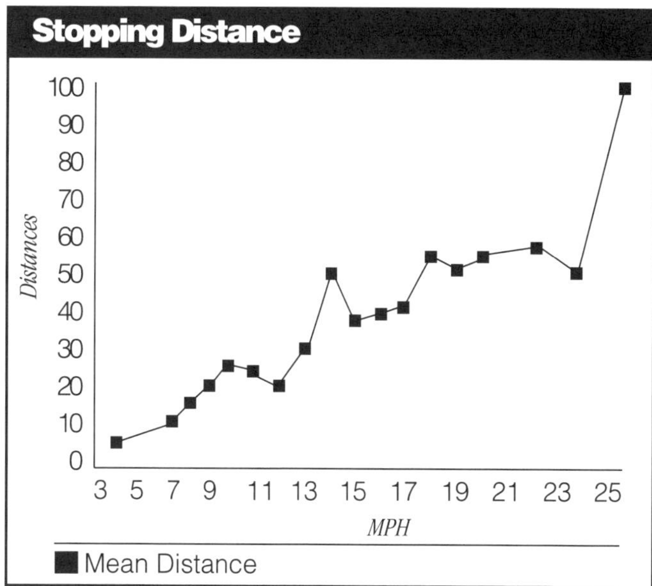

Figure 7. X/Y Graph

the quantity or amount of each category represented. Bar graphs are most useful for comparing the magnitude or size of coordinate items or of parts of a total.

The simple *LabQuest* bar graph in Figure 3 compares the days per year with a temperature above 38 degrees Celsius for a selected city in the United States. Keep in mind that the main emphasis and purpose of simple bar graphs is to focus on comparative magnitudes. As with all graphs, the scale, scale lines, and ordering of categories should be considered based on the intent of the graph to communicate a particular message (e.g., persuade an audience, support a position, report a finding).

Pie Graph

The most recognizable graph type is the pie graph. A pie graph is a graphic form used to impart information about percentages. The various "parts" of the pie chart portray component parts of an aggregate or total. The pie graph in *LabQuest* will display the percentage breakdown of each component as well as a legend that defines each component.

The *LabQuest* pie graph in Figure 4 displays (compares) the color distribution found in a bag of M&M's. When using a pie graph, it is advisable to limit the number of categories (e.g., colors found in a bag of M&M's) to perhaps five or six; otherwise, it may be confusing to differentiate the relative values portrayed on the graph. When capturing data from the spreadsheet, *LabQuest* will automatically display the first column or row as the legend and the next column or row as the individual components of the pie chart.

Stacked Bar Graph

A stacked bar graph uses multiple bars (colors and/or patterns) to make comparisons within and among data sets. The total length of each bar (regardless of the number of individual bars) allows you to make comparisons between different ranges or categories of data. The individual bars allow you to make comparisons within each range or data set.

The *LabQuest* stacked bar graph in Figure 5 makes two comparisons: (1) it *compares* the total number of salmon found in each river and (2) it *compares* the variety of salmon found in each river. Don't be misled by the scaling on the stacked bar graph. Keep in mind that the y-axis is adding all of the individual bars for each data set.

Line Graph

A line graph is used to show the amount of change. Lines plotted on line graphs are typically related to real or observable events, but do not contain or imply a cause and effect relationship between the x- and y-axes, such as daily weather patterns or stock market prices over a period of time.

The *LabQuest* line graph in Figure 6 displays the stock prices for Exxon and Delta Airlines over several weeks. In *LabQuest*, the first column captured will always appear on the x-axis as a label regardless if it is a label (e.g., Monday) or a value (e.g., 25). The remaining captured columns will appear along the y-axis.

X/Y Graph

An x/y graph shows the degree of relationship (i.e., cause and effect relationship) between two or more variables such as changes in pH when incremental amounts of vinegar are added to a solution or the changes in deer population from a Project Wild simulation. Points on an x/y graph are connected by lines. If you select an x/y graph in *LabQuest*, the computer will automatically display your first column of values along the x-axis as values. If your first column of captured data are labels, *LabQuest* will skip your first column and plot the second column (assuming that they are values) along the x-axis as values (Macintosh version only).

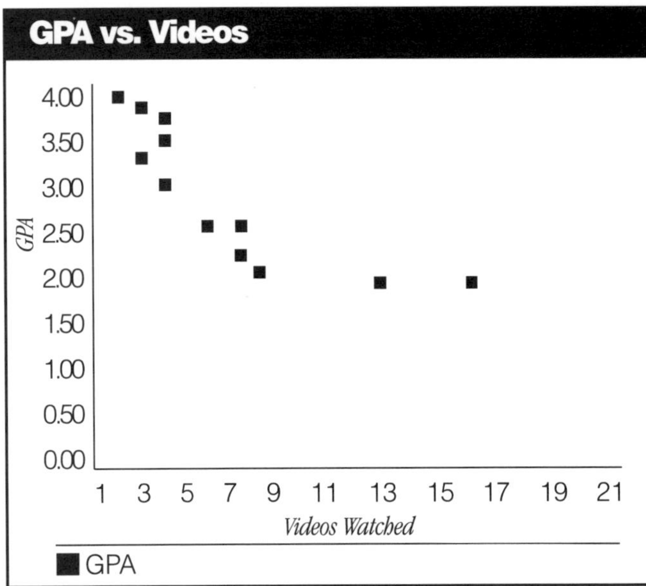

Figure 8. Scatter Graph

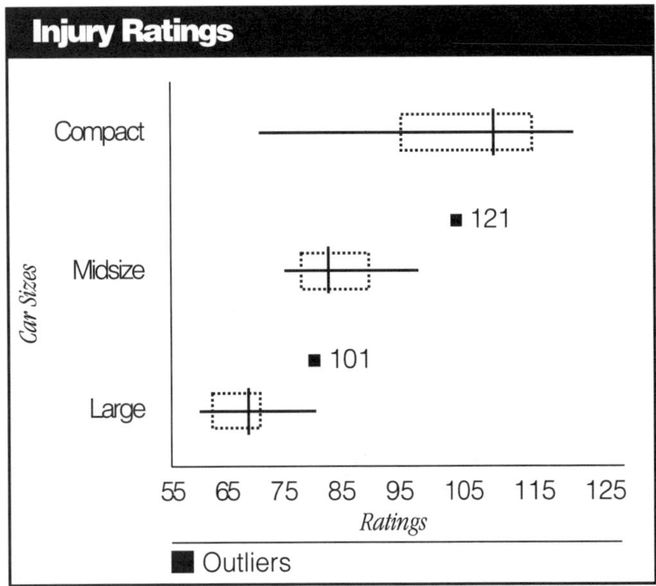

Figure 9. Box Plot

The x/y graph in Figure 7 displays the relation of speed of automobiles to stopping distance. The x/y graph is also very useful for graphing linear equations using the standard equation, $y = mx + b$.

Scatter Graph

A scatter graph is a two-dimensional plot of points that shows the degree of association or correlation between two variables. Each point represents the paired measurements of the two variables. The degree of correlation referred to as the correlation coefficient can range from -1.0 to +1.0 inclusive. The sign of the coefficient indicates the direction of the relationship; the absolute value of the coefficient indicates the magnitude of the relationship.

The *LabQuest* scatter graph in Figure 8 displays the degree of association or correlation between grade point average and the number of videos watched by students each week. Don't panic—a negative correlation exists in this graph!

You need not get into a discussion of correlation coefficient or the Pearson Product-Moment Correlation to use the scatter graph. It is a very practical, easy-to-understand, and entertaining graph if it is presented as a way to discover patterns or associations between different real world phenomena.

Building Heights

LA	STEM	SF
	00	
	01	
	02	
	03	
55531	04	
71	05	222355667
922	06	00
53	07	7
5	08	5
0	09	

90/0 = 900

Figure 10. Stem and Leaf Graph

Box Plot

A box plot or a box and whisker plot is a graphic form used to plot the median, minimum, maximum, quartiles, and outliers for one or more data sets. In *LabQuest*, the program will automatically generate the quartiles, the median, and the presence of any outliers. An outlier is any number more than 1.5 interquartile ranges above the upper quartile, or more than 1.5 interquartile ranges below the lower quartile. The interquartile range represents the difference between the upper quartile and the lower quartile. A box plot allows you to focus on the relative position of different sets of data and thereby compare them more easily. Box plots are best used when the set of data contains hundreds of numbers. The *LabQuest* box plot graph in Figure 9 displays the injury ratings for compact, midsize, and large cars. Which car size would you purchase?

Note: The cars are rated in relative terms; 100 represents the average for all cars. Lower numbers mean a better safety record. A rating of 121, for example, means 21% worse than the average.

Unlike the stem and leaf graph, the box plot does not display individual points, but rather, concentrates on specific points including the median, quartiles, and extremes. This can be very useful especially if you use large data sets.

Stem and Leaf Graph

A stem and leaf graph is used to plot the frequency distribution

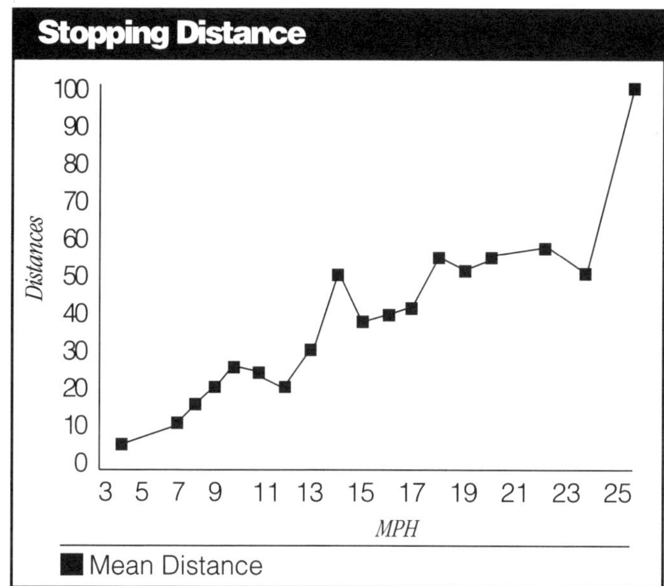

Figure 11. Best Fit Graph

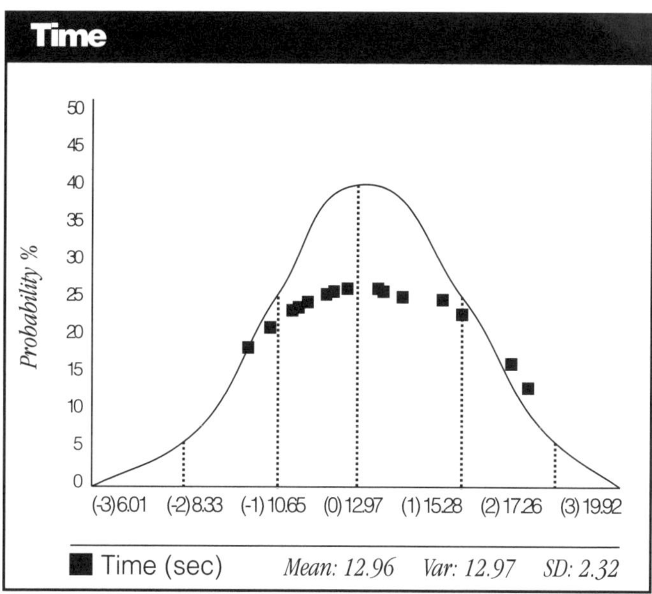

Figure 12. Normal Curve Graph

of a data set. By rotating a stem and leaf graph 90 degrees counterclockwise, you can get a plot that resembles a bar graph or histogram. Stem and leaf graphs are often better than the bar graph or histogram because all the original data values are displayed. The stem and leaf graph enables you to locate clusters, gaps, largest and smallest values, outliers, the relative position of any item important to you, and the general shape of the distribution. In addition, a back-to-back stem and leaf graph lets you compare two sets of data.

The *LabQuest* back-to-back stem and leaf graph in Figure 10 displays the number of buildings above 400 feet in Los Angeles and San Francisco. Before making comparisons with a back-to-back stem and leaf graph, first check to ensure that both sets have about the same total number of values.

Keep in mind that stem and leaf graphs are best used when there are more than 25 pieces of data. The stem and leaf graph is often used as a substitute for the less informative histogram and bar graph.

Best Fit Graph

A best fit graph, commonly referred to as a linear regression line, is actually a scatter graph with a linear regression line imposed on it. The best fit graph is used to predict the mathematical relationship between two variables. To use a linear regression for prediction, *LabQuest* employs the standard equation, $y = mx + b$ where y = the predicted score, m = the slope, and b = the y-intercept. In this setting, the equation of the straight line becomes the regression equation used for prediction.

The *LabQuest* best fit graph in Figure 11 attempts to predict the relationship between the speed of automobiles to stopping distance. In *LabQuest*, it is recommended that students first use the spreadsheet and graph to estimate the best fit line between two sets of data on an x/y graph prior to selecting the computer-generated best fit line. This will encourage estimation of the equation that best represents the mathematical relationship between the two variables.

Normal Curve Graph

A normal curve graph represents the mathematical model of a distribution of normally distributed standard scores with a mean equal to zero (0) and a standard deviation equal to one (1.0). In *LabQuest*, data displayed as a normal curve will display the mean, standard deviation, and variance for the data set along with a graphical display of the data points along the x and y- axes.

The *LabQuest* normal curve graph in Figure 12 displays the 100 meter times for boys and girls at a rural high school in Texas. Obviously, the school must have some excellent sprinters based on the times. The value of a normal curve is that it allows you to compare the position of an individual value relative to the distribution as a whole and against a "normal" set of data.

Students' ability to discern an abundance of statistical data, make inferences, and draw conclusions, as well as construct meaningful graphs, are essential elements for proactive participation in a technology-based democracy. As mentioned earlier, a paramount need exists for all students to acquire these skills early in their educational experiences. Until this approach becomes a reality, the level of subtle distortion using graphical data in newspapers, reports, and advertisements throughout society will continue to exist unchecked.

References

Eng, K. L. (1988, October). Real graphs, real fun, real learning. *Learning.*

Hartley, J. A (1992, June/July). [Postscript to Wainer's "Understanding graphs and tables.] *Educational Researcher.*

Head, J. T. and Moore, D. M. (1988-89). The effect of graphic format on the interpretation of quantitative data. *Journal of Educational Technology Systems, 17*(4).
Hinkle, D. E., Wiersma, W., and Jurs, S. G. (1979). *Applied statistics for the behavioral sciences.* Rand McNally College Publishing Company: Chicago.
Landwehr, J. M., and Watkins, A. E. (1986). *Exploring data.* Palo Alto: Dale Seymore Publications.
Mosenthal, P. B. (1990, March). Understanding documents. *Journal of Reading.*
Parker, J., and Widmer, C. C. (1992, April). Statistics and graphing. *Arithmetic Teacher.*
Schmid, C. F., and Schmid, S. E. (1979). *Handbook of graphic presentation.* New York: John Wiley & Sons.
Wainer, H. (1992, January/February). Understanding graphs and tables. *Educational Researcher.*

Software

LabQuest; Learning Quest, Inc., P.O. Box 61, Corvallis, OR 97339; 800/742-6232.

MODELING MATHEMATICS CONCEPTS

By Connie Widmer and Linda Sheffield

Middle school students enjoy creating things and using technology. You can harness their interest with physical models and technology for mathematics activities that encourage students' in-depth exploration by using physical, calculator, and computer models to teach area and perimeter.

Most students like to make things. And most students like to use calculators and computers. These statements could support crafts and games in the classroom, but they might better support lessons that teach students to investigate concepts on their own. You can, in fact, put students' learning preferences to good use. Introduce concepts by having students build physical models and then ask them to explore the concepts more deeply by using technology. This sequence of experiences promotes mathematical explorations that go far beyond those possible through paper-and-pencil exercises.

U.S. mathematics education has been criticized for being too broad, and the results of the Third International Mathematics and Science Study (TIMSS) support more in-depth mathematics teaching. Eighth-grade TIMSS achievement data, in fact, revealed that U.S. students scored below average in mathematics when compared with students in other participating countries. One analysis of the report observed that the mathematics curriculum in the United States seems to be "a mile wide and an inch deep" (Schmidt, McKnight, & Raizen, 1996, p. 1). "Less is more" is becoming a familiar cry as teachers recognize the need to help students truly understand concepts rather than just acquire rote skills.

Recently, Northern Kentucky University, in conjunction with the Eisenhower Mathematics and Science Education Program, sponsored a mathematics institute for middle school teachers. During the institute, we modeled ways of using simple problems and situations to probe deeply into the underlying concepts. We also encouraged the teachers to construct responses to problems by using physical models, calculators, and computers. In other words, they were to solve problems the same way students would. They were also learning a method of teaching mathematics that they could take back to their classrooms. The following problem, which involves area and perimeter, is one such example.

The Problem

Latisha is designing quilt sections made of square pieces. When the quilt section is finished, Latisha will finish the edges by sewing a wide ribbon around the outside. For example, if Latisha uses six squares, she might decide to put them together like this:

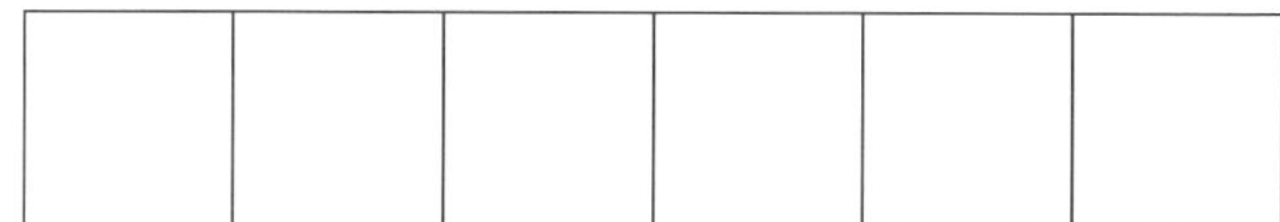

or like this:

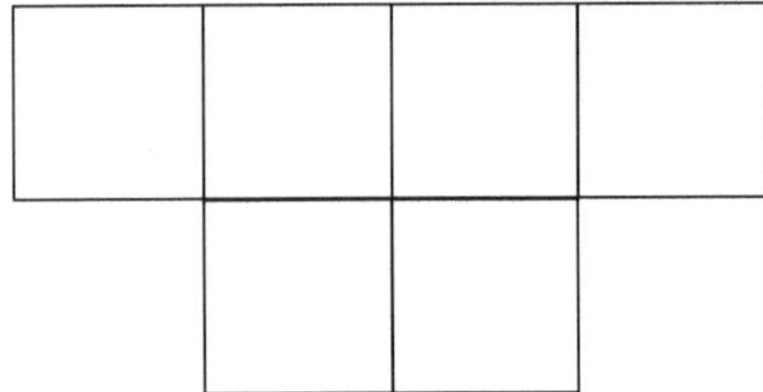

Would the same amount of ribbon be needed to go around each quilt? If not, which quilt would take more?

If each quilt square is one square foot, then how much ribbon would be needed to go around each quilt (ignore the overlap of ribbon needed on the corners)? What other perimeters are possible using the same six squares? Each square must be attached to at least one other square on a full side.

Extending the Problem

After reading and discussing this problem, the teachers were assigned to create various arrangements using one-square-inch plastic tiles. They were then encouraged to ask questions that extended the original problem for deeper exploration. They were given the following list of question starters to help them with this task.

1. Why?
2. Why not?
3. How is this like ... ?
4. How is this different than ... ?
5. What if?
6. What if not?
7. Is that always true?
8. Will that ever work?
9. Can you do that another way? How many ways might you ... ?
10. What is the largest? The smallest? How many different answers can you find?
11. What patterns do you notice?
12. What predictions or generalizations can you make?
13. Convince me. (Prove it. Show me.)

Although the teachers generated many follow-up questions, they decided to explore the following: What is the maximum perimeter? the minimum perimeter? What would happen if the number of squares changed? Can a pattern be detected?

The teachers then used a chart to organize their data so that they could search for patterns and generalize their results (see Figure 1).

They began by exploring rectangular shapes for the quilts. They soon realized that a rectangular quilt would give the maximum perimeter for each area, but not always the minimum. The number of squares used to create the shapes determines which shape will give the minimum perimeter. If you use an even number of quilt squares, then a rectangle will give the minimum perimter. But if you use an odd number of squares (greater than three), then a nonrectangular figure such as the following would give the least perimeter:

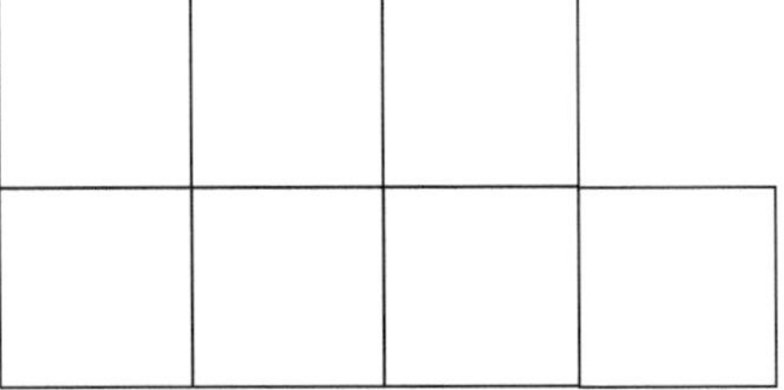

Looking for Patterns

After they completed their charts, the teachers noted that the maximum perimeter increased by two each time another square was added to the quilt. Putting all the tiles in a straight line always gave a rectangle with the largest perimeter, but several other shapes also gave the same perimeter. Teachers expressed this pattern in several ways, including:

- The maximum perimeter is two more each time you add another square to the quilt.
- The maximum perimeter is the number of squares times two plus two.
- If you add the length and the width and double the total, you get the maximum perimeter if the width is 1.

The pattern for the minimum perimeter was not as obvious. The teachers noted that whenever the number of tiles was a square number, a square (and only a square) gave the minimum perimeter. For example, given 9 tiles, a 3 × 3 square gives the minimum perimeter of 12 units. However, the pattern for other numbers of tiles was not always evident, especially for prime numbers. You might want to stop at this point and attempt to discover some of the patterns for minimum perimeter.

Using Spreadsheet Tables and Graphs

The teachers decided that a graphic representation of both the maximum and minimum perimeters might help them generalize their results to any number of squares. The participants discovered that the largest perimeter occurs when the tiles are simply put in a straight line to form the longest possible rectangle (see the first quilt diagram). The participants chose to enter their data into a spreadsheet. The values for area and perimeter could either be entered into their respective columns or generated by the spreadsheet by using formulas. In this problem, the area always equals the numbers of tiles, so the teachers did not create a separate column for area (see Figure 2).

Computing maximum perimeter in terms of the number of tiles takes a bit more reasoning. With the tiles arranged in a single line to form the longest rectangle, the teachers observed that the height of the rectangle formed is one unit. Hence, the formula for perimeter becomes [2 × (number of tiles)] + 2]. If the number of tiles is represented in Column A, then the formula becomes [2 × A4] + 2, [2 × A5] + 2, and so on (see Figure 3). Formulas allow anyone to explore beyond the given data; large numbers can be inserted and the resulting values easily calculated.

The graph of maximum perimeter for a given area reveals that the largest possible perimeter increases linearly as

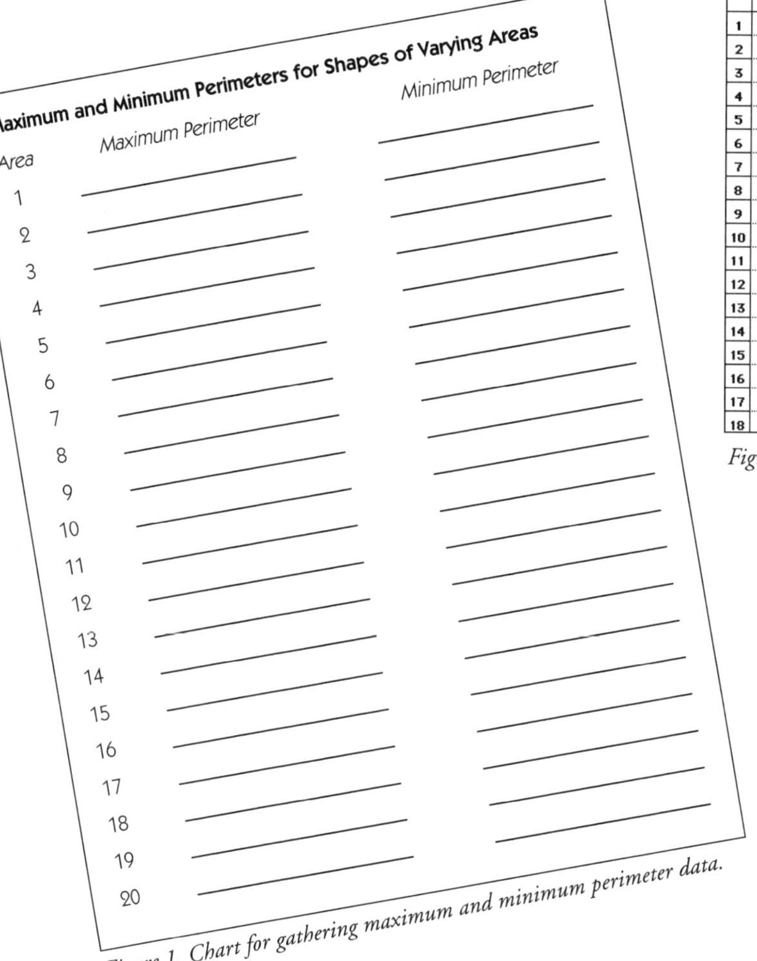

Figure 1. Chart for gathering maximum and minimum perimeter data.

	A	B
1	Number of Tiles	Maximum Perimeter
2	(= Area)	
3		
4	1	4
5	2	6
6	3	8
7	4	10
8	5	12
9	6	14
10	7	16
11	8	18
12	9	20
13	10	22
14	11	24
15	12	26
16	13	28
17	14	30
18	15	32

Figure 2. Spreadsheet showing maximum perimeter.

	A	B
1	Number of Tiles	Maximum Perimeter
2	(= Area)	
3		
4	1	=2*A4+2
5	=A4+1	=2*A5+2
6	=A5+1	=2*A6+2
7	=A6+1	=2*A7+2
8	=A7+1	=2*A8+2
9	=A8+1	=2*A9+2
10	=A9+1	=2*A10+2
11	=A10+1	=2*A11+2
12	=A11+1	=2*A12+2
13	=A12+1	=2*A13+2
14	=A13+1	=2*A14+2
15	=A14+1	=2*A15+2
16	=A15+1	=2*A16+2
17	=A16+1	=2*A17+2
18	=A17+1	=2*A18+2

Figure 3. Formulas for calculating maximum perimeter.

the area increases (see Figure 4). Some participants expected the minimum perimeter data to behave the same way. They entered data to find the least possible perimeter of each fixed area from 1 to 16 square units. Then they selected Line Graph to view the data (see Figure 5).

Many teachers were surprised by the results: The graph did not form a straight line! This discovery led to interesting investigations: Why do maximum and minimum perimeters behave so differently? Is there a pattern to this graph? Does the minimum perimeter always increase by 2? Why do some areas yield the same minimum perimeter? What is happening to the figure at the values for which the minimum perimeter changes? Several teachers noted that the perimeter always increased immediately after the number of tiles was a square number. For example, the mimimum perimeter for 13, 14, 15, and 16 tiles is 16, but for 17 tiles the minimum perimeter increases to 18. Upon further investigation, teachers realized that the perimeter also gets larger following a number of tiles that can be arranged into a rectangle with a length one unit longer than the width. For example, 12 tiles can be arranged into a 3 × 4 rectangle with a perimeter of 14. For 13 tiles, the minimum perimeter increases to 16. Several other patterns begin to emerge as the problem is explored more fully.

Using Graphing Calculators

After working with spreadsheets on the computer, the teachers knew the formula for the maximum perimeter for any number of tiles. Because they had noted that the maximum perimeter was always double the number of tiles plus 2, it was fairly easy for them to translate this formula to a graphing calculator; they used the formula $y = 2x + 2$. By using the trace function, they found that tracing the graph gave them several points that were not on the chart, such as the point with coordinates (2.3, 6.6). Discussion followed about the meaning of such points and whether they would occur with tiles. The teachers then investigated what would happen if other fractions and decimals were used and if the shapes were not restricted to squares or rectangles. They quickly noted that an area of 2.3 square units was not possible when using the 1" square tiles. However, if they did allow an area of 2.3, a rectangle of dimensions 2.3 units × 1 unit would have a perimeter of 6.6 units. One teacher noted that if dimensions were not constrained to whole number units, then a rectangle with dimensions of 6 units by 1/2 unit would have an area of 3 square units and a

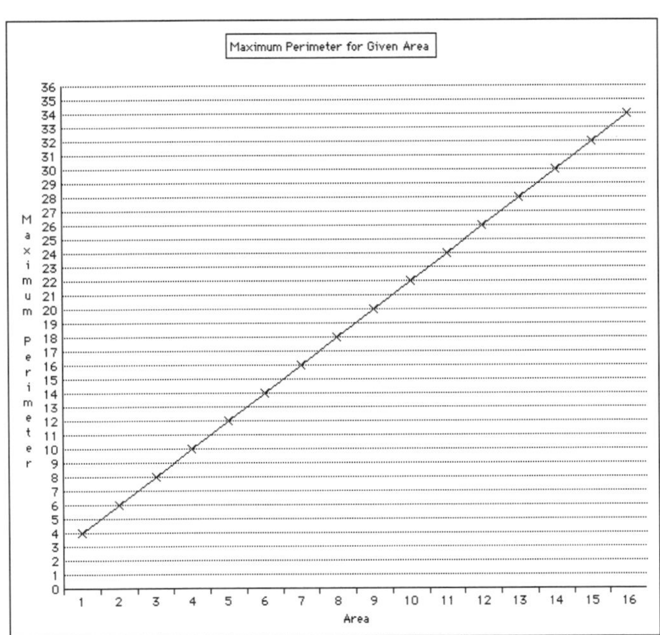

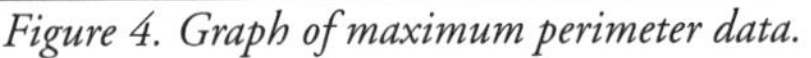

Figure 4. Graph of maximum perimeter data.

	A	B
1	**Number of Tiles**	**Minimum Perimeter**
2	**(= Area)**	
3		
4	1	4
5	2	6
6	3	8
7	4	8
8	5	10
9	6	10
10	7	12
11	8	12
12	9	12
13	10	14
14	11	14
15	12	14
16	13	16
17	14	16
18	15	16
19	16	16
20		
21		
22		
23		
24		
25		
26		
27		
28		

Figure 5. Spreadsheet and graph of minimum perimeter.

perimeter of 13 units, a much larger perimeter than that of a 3 × 1 rectangle, which has a perimeter of only 8 units. Several teachers also noted that each answer to one question brought several other questoins to mind; the problem solving seemed to have no end.

Final Remarks

These teachers learned an important lesson: Problem solving often leads to more problem solving. Problems involving area and perimeter, for example, lead naturally to problems that involve maximum and minimum volume and surface area. These types of investigations, facilitated by physical models and technology, encourage in-depth explorations of concepts. Having engaged in such investigations, these teachers are ready to model and lead similar experiences in their own classrooms. In this way, their students will be encouraged to view learning as a process of investigation rather than as memorization.

Reference

Schmidt, W. H., McKnight, C. C., & Raizen, S.A. (1996). *Splintered vision: An investigation of U.S. mathematics and science education: Executive summary.* Washington, DC: U.S. National Research Center.

INTERLUDE

Developmental Learning and Mathematics

By Anne Raymond

Developmental learning theory has had a strong impact on mathematics teaching for years. The emphasis of mathematics teaching has gone through many stages, from rote teaching and memorization inspired by behaviorist learning theories to a problem-solving approach related to Piaget's stages of learning. As we learn more about the developmental processes of learning, while simultaneously proposing national standards for the learning and teaching of mathematics, we need to be cognizant of meeting all students' needs, from the special needs learner, to the "at grade level" learner, to the gifted learner.

It is an impossible struggle to attempt to meet the needs of all learners with a single approach in the classroom. Consequently, much has been studied regarding teaching according to different "learning styles" (Kolb, 1976, 1984) and by tapping into students' natural intelligences (Gardner, 1993).

Kolb's work on learning styles defines two separate learning activities: perception and processing. These activities can be divided into opposites. For example, some people

Kolb's Learning Styles

- The **Type I Learner** is primarily a "hands-on" learner. He primarily relies on intuition rather than logic. He likes to rely on other people's analysis rather than his own. He enjoys applying learning in real life situations.
- The **Type II Learner** likes to look at things from many points of view. She would rather watch than take action. She likes to gather information and create many categories for things. She likes using her imagination in problem solving and is sensitive to feelings when learning.
- The **Type III Learner** likes solving problems and finding practical solutions and uses for learning. He shies away from social and interpersonal issues and prefers technical tasks.
- The **Type IV Learner** is concise and logical. Abstract ideas and concepts are more important to her than people issues. Practicality is less important to her than a good logical explanation.

Gardner's Multiple Intelligences

- **Linguistic Intelligence:** Students think in words and love to read, write, and tell stories.
- **Logical-Mathematical Intelligence:** Students think by reasoning and love to question, experiment, and calculate.
- **Musical Intelligence:** Students think through rhythms and melodies and love to hum, whistle, sing, listen, and tap to the beat.
- **Spatial Intelligence:** Students think in images and pictures and love to draw, design, doodle, and visualize.
- **Kinesthetic Intelligence:** Students think through physical sensations and love to move, dance, run, touch, gesture, and build.
- **Naturalistic Intelligence:** Students think about structure and nature and love to sort, classify, examine, and collect.
- **Interpersonal Intelligence:** Students think by bouncing ideas off others and love to lead, relate, mediate, and organize.
- **Intrapersonal Intelligence:** Students think inside themselves and love to meditate, plan, dream, be quiet, and set goals.

perceive information using concrete experiences, while others best perceive information abstractly. Similarly, some people process information by active experimentation, while others perceive best by reflective observation. Thus, the two sets of "opposites" of his model: Concrete Experience versus Reflective Observation, and Abstract Conceptualization versus Active Experimentation. These opposites create four quadrants of learning behavior, and the quadrants define four types of learners.

Just as Kolb's model suggests that people learn best according to their own particular learning styles, Gardner's theory of multiple intelligences implies that people have natural strengths and preferences in terms of their approach to learning. He refers to these as "intelligences" and has determined eight distinct intelligences that can be developed and tapped into during the learning process to meet individual students' needs.

In addition to work in the areas of learning styles and intelligences, studies in the areas of equity in education (Secada, Fennema, & Adajian, 1995), gifted and talented students (Sheffield, 1999), inclusion (Bley & Thorton, 1995; Thorton & Bley, 1994), tracking, and heterogeneous versus homogeneous grouping (Johnson & Johnson, 1989) have all informed both the necessity to meet and the ways to meet all students' needs. In mathematics education, research in these areas has resulted in the call for mathematics classrooms that allow for concrete experiences that lead to abstract understanding through the use of mathematics manipulatives and other models. Results also include increased emphasis on cooperative learning in the mathematics classroom, increased contextualization of mathematics problems that are culturally diverse and related to real-life issues, and increased engagement of students in higher-level problem-solving experiences.

The introduction of technology into the curriculum provides another tool for helping students at each phase of the mathematics learning process. The calculator is a tool that supports the development of skills and the problem-solving activities. The computer provides opportunities for a wide range of visualization and for organiza-

tion of mathematical ideas. Technology, therefore, offers new avenues for remediation, practice, and challenges in mathematics.

References

Bley, N. S., & Thorton, C. A. (1995). *Teaching mathematics to students with learning disabilities* (3rd ed.). Austin, TX: Pro-Ed.

Gardner, H. (1993). *Multiple intelligences: The theory in practice.* New York: BasicBooks.

Johnson, D. W., & Johnson, R. T. (1989). Cooperative learning in mathematics education. In P. R. Trafton (Ed.), *New directions for elementary school mathematics* (pp. 234–245). Reston, VA: NCTM.

Kolb, D. A. (1976). *The learning style inventory: Technical manual.* Boston: McBer.

Kolb, D. A. (1984). *Experiential learning.* Englewood Cliffs, NJ: Prentice Hall.

Secada, W. G., Fennema, E., & Adajian, L. B. (Eds.). (1995). *New directions for equity in mathematics education.* New York: Cambridge University Press.

Sheffield, L. J. (Ed.). (1999). *Developing mathematically promising students.* Reston, VA: NCTM.

Thorton, C. A., & Bley, N. S. (Eds.). (1994). *Windows of opportunity: Mathematics for students with special needs.* Reston, VA: NCTM.

CHAPTER 6

Geometry

The Conversation Continues

Ivan: Did you ever think about how geometry, and shapes in particular, impact our world?

Anne: Well, the definition of geometry is "earth measure," so it makes sense that geometry is connected to describing the world around us. In fact, a favorite activity of many teachers at the beginning of a geometry unit is to take the students on a "geometry walk" outside so that students can identify geometric shapes in their environment.

Ivan: It's fascinating to think about how geometry influences not only our day-to-day life, but also such arenas as space exploration. I mean, when space shuttles connect with space stations or when satellites have to follow an exact trajectory, geometric ideals are the means through which aerospace experts plan for these events.

Anne: Yes, even the simplest geometric ideas have had an impact on the space program. Recall the moment during the Apollo 13 dilemma when the aeronautical engineers had to figure out how to fix the air filtering system on board the spacecraft by fitting a round tube into a square opening with only a limited set of materials available to construct the connection. This is an example of how simple geometric ideas can play a role in solving very complicated and important problems.

Ivan: Considering the range of interesting possibilities of study that the area of geometry poses, I have been frustrated to learn from surveys with my preservice teachers that many of their least favorite mathematics school memories relate to their experiences in high school geometry classes. Students expressed that they felt disconnected from the geometry because they were merely trying to recreate abstract proofs that had very little meaning to them.

Anne: I think messages stated by the National Council of Teachers of Mathematics in the geometry standard imply that the study of geometry needs to be more hands-on and relevant to students' lives. Fortunately, the examples of the articles in this chapter provide interesting opportunities for students to see the value of geometry in their world. For example, in the article "Does Cincinnati Need a New Bridge?" students meld social studies and mathematics to answer a real-life question.

THEORY INTO PRACTICE

Geometry and Multiple Intelligences

By Anne Raymond

One of the content areas emphasized in the National Council of Teachers of Mathematics' *Principles and Standards for School Mathematics* is geometry. The descriptors of the geometry standard imply that students must have an active role in developing geometric reasoning. The standard also suggests that students must approach geometry from a variety of learning perspectives.

NCTM Geometry Standard

Geometry: Instructional programs from prekindergarten through Grade 12 should enable all students to

- Analyze characteristics and properties of two- and three-dimensional geometric shapes and develop mathematical arguments about geometric relationships
- Specify locations and describe spatial relationships using coordinate geometry and other representational systems
- Apply transformations and use symmetry to analyze mathematical situations
- Use visualization, spatial reasoning, and geometric modeling to solve problems

Many theorists have expressed that individual students learn in different ways and that students must construct their own understanding. Gardner (1993) suggests that students can learn from many perspectives. He is noted for developing what he calls "multiple intelligences." In his original work *Frames of Mind,* Gardner (1983) suggests "the existence of a number of different intellectual strengths, or competencies, each of which may have its own developmental history."

Originally, Gardner proposed that there were seven distinct intelligences:

- Linguistic
- Logical-Mathematical
- Spatial
- Bodily-Kinesthetic
- Musical
- Interpersonal
- Intrapersonal

Later work has determined that we should also consider an eighth intelligence:

- Naturalistic

Educators purport that we should consider these intelligences when trying to meet all students' learning styles through our teaching. Further, we should attempt to develop areas of intelligences that students currently do not exhibit as strengths.

In the study of geometry, we can work toward developing students' spatial intelligence in conjunction with their logical mathematical intelligence. Armstrong (1994) has determined, for each intelligence, a way to classify how students who have that intelligence think, what they love to do, and what types of materials and experiences they need. He claims that children who are strongly *spatial*

- think in images and pictures,
- love designing, drawing, visualizing, and doodling,
- need art, LEGO toys, videos, slides, imagination games, mazes, puzzles, illustrated books, and trips to art museums.

On the other hand, students who are strongly *logical-mathematical*

- think by reasoning,
- love experimenting, questioning, figuring out logical puzzles, and calculating,
- need things to explore and think about, such as science materials, manipulatives, or trips to science museums.

In teaching geometry, we must tap into spatial sense and logical-mathematical reasoning. Software programs such as Geometer's Sketchpad, Geometric Supposer, and Tessel Mania offer activities for students to engage in manipulation, exploration, reasoning, and designing. These types of activities are tied to the goals of the NETS•S on Technology Productivity Tools.

NETS•S Productivity Standard

3. **Technology productivity tools**
   - Students use technology tools to enhance learning, increase productivity, and promote creativity.
   - Students use productivity tools to collaborate in constructing technology-enhanced models, preparing publications, and producing other creative works.

Drawing or painting software offer opportunities for students to investigate and communicate what they have learned more easily than relying solely on previously available tools. More than ever before, through technology students have many ways to connect the learning of geometry to the multiple intelligences.

Readers interested in learning more about multiple intelligences should refer to the Interlude of this book, "Developmental Learning and Mathematics," for details on all eight of Gardner's intelligences. Readers can examine suggestions for connecting multiple intelligences to teaching in a variety of content areas via technology in McKenzie (2002), *Multiple Intelligences and Instructional Technology: A Manual for Every Mind*.

References

Armstrong, T. (1994). *Multiple intelligences in the classroom.* Alexandria, VA: Association for Supervision and Curriculum Development.

Gardner, H. (1983). *Frames of mind: The theory of multiple intelligences.* New York: Basic Books.

Gardner, H. (1993). *Multiple intelligences: The theory in practice.* New York: Basic Books.

McKenzie, W. (2002). *Multiple intelligences and instructional technology: A manual for every mind.* Eugene, OR: ISTE.

INSIGHT

Geometry and Tessellations

By Anne Raymond

Tessellations are special geometric patterns. The word tessellation comes from the Latin word *tessella,* which was the small, square tile used in ancient Roman mosaics. Usually, a tessellation is a pattern in a plane that is made up of one or more geometric shapes completely covering a surface without any gaps or overlaps (Seymour & Britton, 1989).

There are categories of tessellations. For example, regular tessellations are patterns made using a single regular polygon. Semi-regular tessellations are tessellations made up of two or more regular polygons. Students can learn a lot about properties of geometric shapes as they explore tessellations.

Examples of regular tessellations include those in Figures 6.1 and 6.2.

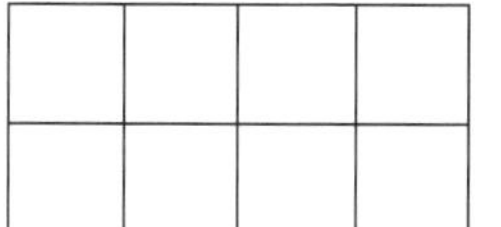

Figure 6.1. Tessellation of a square.

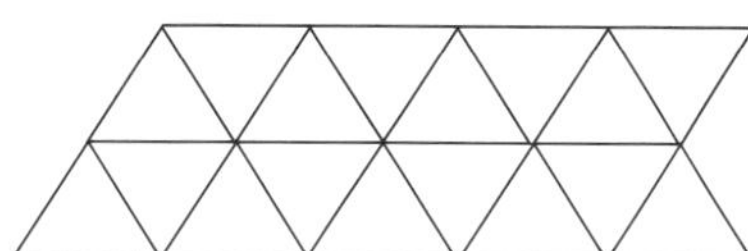

Figure 6.2. Tessellation of an equilateral triangle.

Examples of semi-regular tessellations include those in Figures 6.3 and 6.4.

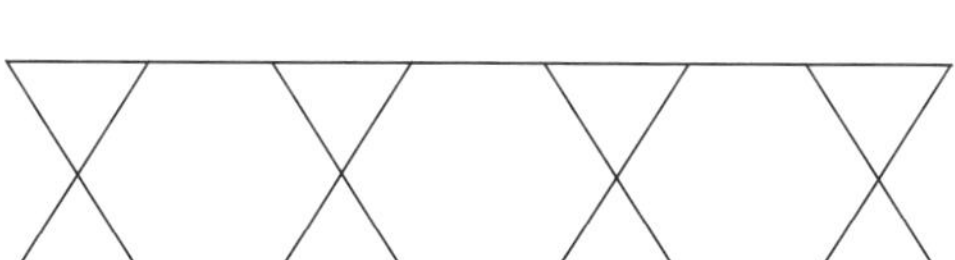

Figure 6.3. Tessellation of an equilateral triangle and a hexagon.

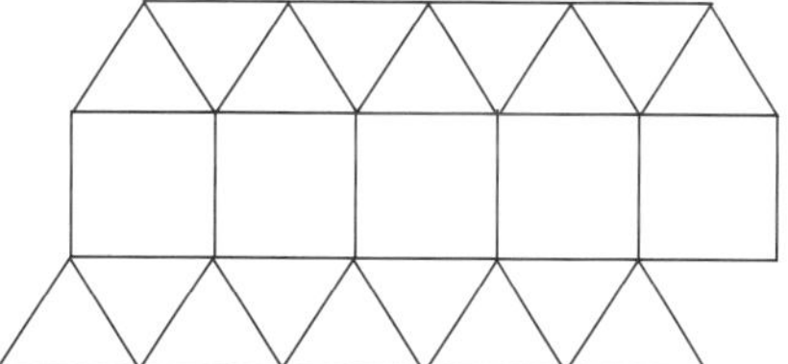

Figure 6.4. Tessellation of an equilateral triangle and a square.

Figures 6.5 and 6.6 show other tessellations.

Tessellations can be seen all around us in such items as wallpaper patterns, floor and ceiling tiles, and clothing materials. Tessellations are also connected to artistic expressions. In particular, many of the works by M. C. Escher can be formed by tessellating shapes in the plane. Read, for example, *The World of M.C. Escher* (Escher, 1971) to learn about Escher art, or go to www.Escher.com to see examples of Escher art.

A number of software programs are available for students to use to explore the geometry of tessellations. One popular program, TesselMania! (MECC, 1995), allows students to explore given tessellations as well as create their own designs. I created

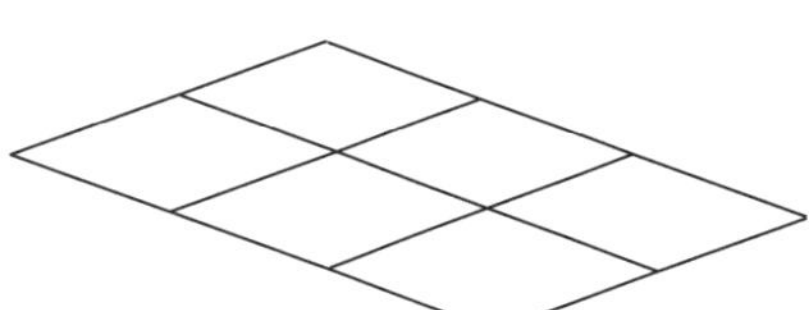

Figure 6.5. Tessellations with a parallelogram.

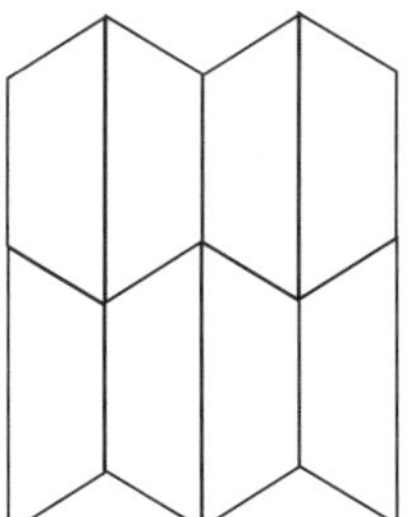

Figure 6.6. Tessellations with a trapezoid.

the following designs using TesselMania! The first tesselation (Figure 6.7) started with a hexagon that I indented and then rotated throughout the plane. The second tesselation (Figure 6.8) started as a parallelogram that was reshaped and "glide-reflected" across the plane.

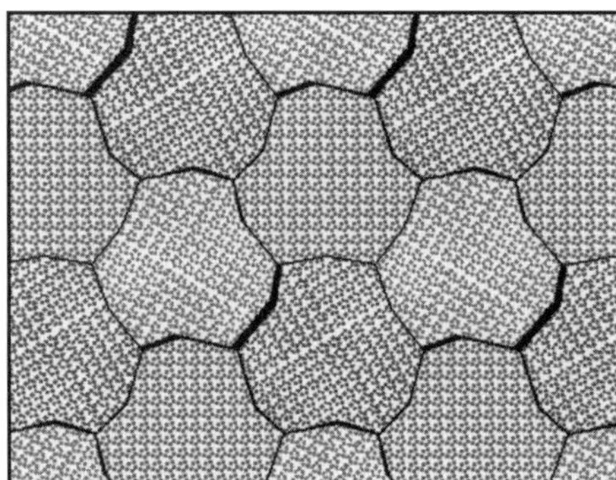

Figure 6.7. Tesselation 1.

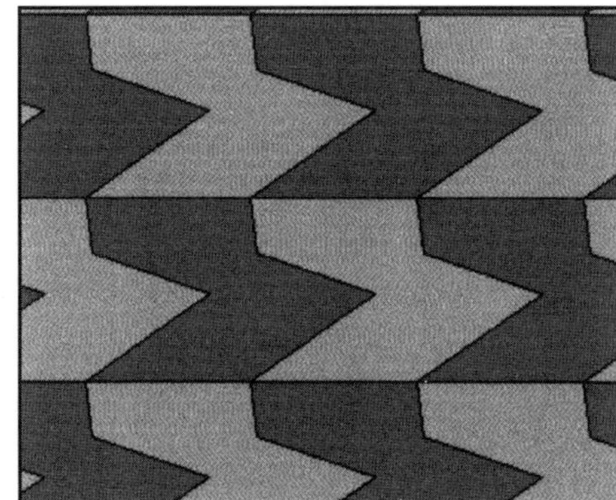

Figure 6.8. Tesselation 2.

On many educational Web sites, teachers and students can find opportunities to build and manipulate tesselations. For example, one of NCTM's Web sites (www.illuminations.com), a virtual manipulative site (matti.usu.edu/nlvm/nav/index.html), and an offshoot of the regular mathforum site (mathforum.org/sum95/suzanne/tess.intro.html) offer students the chance to do such tesselation activities.

Working with tesselations is a key way to build geometric understanding. While creating tesselations, students have to visualize and determine shapes that will fit together without leaving gaps. Therein, students learn about shapes, their angle measures, and their side lengths. Further, when extending patterns formed by tesselations, students learn about symmetries such as reflections, rotations, and translations. Finally, by manipulating tesselations to form new designs, students encounter Big Ideas in mathematics such as decomposition and composition. This is particularly true when students create Escher-type designs. Overall, tesselations offer a wealth of opportunities to see and manipulate geometric ideas in a fun, creative way.

References

Books

Escher, M. C. (1971). *The world of M.C. Escher.* New York: Harry N. Abrams.

Seymour, D., & Britton, J. (1989). *Introduction to tessellations.* Palo Alto, CA: Dale Seymour.

Software

MECC (1995). *TesselMania! The art and design tool.* Cambridge, MA: SoftKey International.

INSIGHT

Geometry and Humor

By Anne Raymond

Question: What did the acorn say when it grew up?
Answer: Gee, I'm a tree (geometry)!

This is one of many jokes related to mathematics that you can find in books and on Web sites. Although jokes are available for almost any category of mathematics, I am particularly fond of the wealth of geometry jokes. When I tell mathematics jokes in class, the students usually find them corny, but they love them. Here are some of my favorite geometry jokes:

Question: What did the woman say when her parrot flew away?
Answer: Polly gone (polygon).

Question: What do you call a man who spent all summer at the beach?
Answer: A tan gent (tangent).

Question: What do you call a crushed angle?
Answer: A wrecked angle (rectangle).

Question: What do you call an angle that is adorable?
Answer: A cute angle (acute angle).

Question: What do you call more than one L?
Answer: A pair of "l" (a parallel).
Comment: Without geometry, life is pointless.

Question: What do you call people who are in favor of tractors?
Answer: Pro-tractors (protractors).

Question: What do you get when you divide the circumference of a pumpkin by its diameter?
Answer: Pumpkin pie (pi).

Question: What's the best way to pass the geometry test?
Answer: Know all the angles.

Resources

Book

Azzoline, A., Silvey, L., & Barnabas, H. (Eds.). (1978). *Mathematics and humor.* Reston, VA: National Council of Teachers of Mathematics.

Web Sites

Fathom Resource Center, Jokes: www.keypress.com/fathom/jokes.html

Geometry Jokes: http://ctap295.ctaponline.org/~prandolp/student/geojokes.html

INSIGHT

A Picture Is Worth a Thousand Words

By Ivan W. Baugh

Visual examples help learners grasp concepts not easily presented textually. Drawing and painting software enables technology-using students to create illustrations that have a professional appearance. Fortunately, one does not need to be an artist to make effective use of these tools.

Why would you want to use drawing or painting software in the study of geometry? Illustrating geometric principles using drawing and painting software promotes learner understanding of the illustrated principles. For example, you want your students to demonstrate their knowledge of a tessellation, and you do not have a software package such as those listed in the tessellations article. Have them draw their shapes using a drawing program, and then copy and paste them to produce the desired number of shapes (this enables having multiple shapes the same size). They can then rotate them, change the perspective of the shapes, fill them with a pattern or color, and arrange them to create a tessellation.

Drawing and painting programs often have a default grid eight pixels apart. Whatever you draw will snap to the grid. When you need to move an object a smaller amount than the default grid, turn off the "snap to grid" feature so that you can move your drawing one pixel at a time instead of jumping to the next unit of the grid.

One advantage of creating computer-generated geometric patterns derives from the fact that you can save your creation as you work. Then, if you add figures that you decide you don't want, you can always close the file without saving and revert to the previously saved version to continue working rather than having to start over. This becomes particularly advantageous when working with painting programs, as they usually allow you to undo only one time.

If you draw a shape with multiple parts, you can select all the parts and group them to combine them into one piece. Shapes drawn in this manner you cannot fill with a color or pattern. If you want to fill shapes with a color or pattern, do not draw a shape, such as a square, with the line tool. You can wrap text around the item, thereby avoiding the text's disappearing behind the illustration.

After students have worked with tangrams, have them design activities using shapes to fill a created shape, such as a rectangle, triangle, or trapezoid. Just as they can rotate and flip the tangrams as they seek to create a specified shape, they can rotate and flip shapes created in a drawing or painting program. Your students can create a puzzle that they will challenge other classmates to solve. Provide opportunities for your students to complete hands-on activities with tangrams before they attempt them with the drawing or painting tools.

Your students have taken a geometry walk. You want them to create a Web page demonstrating the applications of geometry they saw on the walk. Drawing or painting software provides tools that will enable them to prepare illustrations with a professional appearance. It will promote pride in their work. They will strive for a higher level of achievement knowing their work will be available on the Web, not just turned in for the teacher to grade.

If you want your students to draw examples to scale, and your software includes rulers, show them on the screen. If it doesn't offer rulers, use the grid described above as a tool for drawing to scale. This will necessitate showing the grid. Students can use this grid when they complete an activity such as drawing their room at home to scale.

Whatever software package you use, consider how this will enable your students to create illustrations with a professional appearance. If the students will publish the document on the Web, you will want to save the illustrations in one of these formats: GIF, JPG, or PNG. Any of these formats works across platforms (Macintosh and Windows) and in various browsers (Internet Explorer or Netscape Navigator).

Databasing Geometry in the Elementary Classroom

By Marilyn Sue Ford

Teachers have been quick to think of uses of databases to teach social studies and science. Within these contexts databases have also been used to explore mathematical situations. Are there areas in mathematics that can benefit from the organization that a database brings to a set of data?

The attributes of geometric figures might benefit from such organization if the students themselves develop the database and then use it. According to Burger (1988), elementary geometry instruction should reflect the research of van Hiele. Elementary students must be provided with opportunities to identify, analyze, and classify shapes through concrete to abstract reasoning experiences and eventually apply geometric properties to problem situations. An information processing computer environment facilitates and supports Burger's recommendations for a progressive, sequential, and comprehensive geometry curriculum in the elementary grades.

In this teaching unit for intermediate grades, students determine the geometric attributes of a variety of triangles and quadrilaterals and set up a database for these figures and their attributes. Then they use the database to answer geometric questions about the figures. Pictures of the shapes used on display in the room help students connect the shape with its abstract attributes listed in the database.

Preparation

This geometry database unit incorporates both pre-computer (offline) and computer (online) activities that alternate between whole group and small group instruction. The computer activities in the teaching unit are developed with the Apple ProDOS version of Scholastic PFS: File software, but can be adapted for use with any database management software.

Activities are also designed for a self-contained classroom with one computer and approximately 30 students.

> Geometric Characteristics of the Rectangle
>
> 4 angles
> all angles = 90 degrees
> all angles equal
> 4 sides
> 2 pairs of equal sides
> 4-sided polygon
> all right angles
> 4 vertices
> sum of angles = 360 degrees
> opposite sides parallel quadrilateral
> etc.

Figure 1.

For whole group activities, a large demonstration monitor (or computer interfaced with a projection system) and printer are preferred, but optional.

Small group computer activities necessitate use of a computer station, learning center approach. You will need to create a computer station schedule whereby each small group has approximately 20–30 minutes of online time for computer activities.

Materials needed: Ample supply of butcher paper, thick felt tip marking pens, and a package of 8" x 10" lined index cards.

Preparation: For each student, make several paper copies of a rectangle. For each group of three students, make several paper copies of one plane geometry shape. The following geometric shapes will adequately meet the needs for a class of 30 students: isosceles triangle, equilateral triangle, scalene triangle, acute triangle, right triangle, obtuse triangle, square, parallelogram, rhombus, and trapezoid.

Organizing Geometric Observations

The first objective is to determine the characteristics of the plane geometry shapes and then organize these observations. To begin, provide each student with a copy of the "class" shape (i.e., the rectangle). If students are familiar with the basic geometric components of a rectangle, they should readily recognize

Geometry Word Bank Chart

Categories	Descriptors
Types of sides	equal, nonequal, parallel, opposite, intersecting, perpendicular
Number of sides	# (numeric value)
Types of angles	obtuse, acute, right, equal
Number of angles	#
Sum of interior angles	#
Polygon name	triangle, quadrilateral
Shape name	isosceles triangle, equilateral triangle, scalene triangle, right triangle, acute triangle, obtuse triangle, rectangle, square, parallelogram, rhombus, trapezoid

Figure 2.

such characteristics as perpendicular lines, parallel lines, right angles, and opposite equal sides. If necessary, protractors and rulers can be used for verification of side and angle measurements. Encourage students to determine the geometric terms used to describe a rectangle and list them on a sheet of paper. Then make a class list of their observations on the chalkboard (Figure 1).

Provide each small group with several copies of one plane geometry shape that will become their group shape for the duration of the teaching unit. Have each small group select a recorder and repeat the above process with their group shape. Each recorder should label a sheet of butcher paper with the name of their group shape and list the geometric terms the group decides describe the shape. Post the butcher paper lists for easy viewing. Have students organize their observations by looking for similarities among the lists. For example, ask which terms describe line segments. On a large sheet of butcher paper labeled Geometry Word Bank Chart, sort these terms together as DESCRIPTORS and give them a CATEGORY label such as "sides." Continue this process, asking which terms describe angles, general names, specific names, and so forth.

If necessary, further refine categories. Either break them down into two or more categories or relabel them with more precise geometric terminology. The category "sides" could be broken down into two categories called "types of sides" and "number of sides." A category called "general names" could be relabeled as "polygon name." This process will produce a butcher paper Geometry Word Bank Chart similar to the one in Figure 2.

Extending Students' Geometric Knowledge Base

The descriptive, organizational process provides the classroom teacher with an informal assessment of students' geometric knowledge. Any areas not addressed on the Geometry Word Bank Chart become possible topics for instruction. For example, given that students can readily generate a list of geometric terms such as those listed in Figures 1 and 2, then the next objective is to extend their geometric knowledge by exploring the geometric qualities of line and rotational symmetry (Renshaw, 1986) and tessellation. The ultimate outcome is to apply this newly acquired knowledge to each group shape and organize the information on the Geometry Word Bank Chart.

Provide each student with a copy of the "class" shape. Have each student cut out the rectangle. Define "line of symmetry" as an imaginary line that produces mirror images. Using a pencil, have students draw a possible line of symmetry on the rectangle. Fold the rectangle on the marked line. If the halves match, then the shape has line symmetry. Continue this procedure until all possible lines of symmetry have been found. Count each line/fold that produces matching halves as a line of symmetry.

Have the small groups refer to their plane geometry shape and apply the line of symmetry procedures to their shape. Groups should determine if their shape has none, one, or more lines of symmetry. Then have the whole group discuss how line of symmetry information could be added to the Geometry Word Bank Chart. Student discussion should be guided toward a category label such as "lines of symmetry" and descriptor such as "#" which indicates any numeric value equal to or greater than zero.

The preceding sequence of whole group activity with the rectangle, small group activity with group shapes, and whole group discussion related to the Geometry Word Bank Chart should be repeated for exploring, applying, and organizing information on the rotational symmetry and tessellation qualities of geometric shapes. For rotational symmetry, cut out the shape, mark an "X" at the top, and a dot at the center of rotation. Place the cut out shape on top of a tracing of the shape. Using a pencil point, hold the cut out figure on the center of rotation. Rotate the cut out shape about the center until it matches (fits) the shape on the bottom exactly. Keep rotating until the cut out shape returns to the starting "X" position. Count the number of rotational turns that create an exact match. The last turn, which places the cut out shape in the starting position, completes the cycle and does not count.

Cut out several copies of a geometric shape in order to determine the quality of tessellation. Arrange the shapes on a piece of colored construction paper according to the tessellation rule of touching edges. If the edges touch and completely cover the construction paper surface, then the shape tessellates a plane. If the edges touch, but leave spaces so that parts of the construction paper surface are still visible, then the shape does not tessellate a plane.

If further whole group exploration of line and rotational symmetry and tessellation is warranted before small group activities, use a pentagon shape. The pentagon has five lines of symmetry, four turns of rotational symmetry, and does not tessellate a plane surface. These results will provide data that students can compare and discuss in relation to the rectangular outcomes. Once the exploration of symmetry and tessellation is completed for both the "class" shape and the group shapes, the resulting information must be added to the Geometry Word Bank Chart. Students' experiences and group discussions should produce an amended Word Bank Chart with the additions listed in Figure 3.

Creating a Database Record Format

The next step is to design a database record format using the Geometry Word Bank Chart as a guide. Post a large sheet of blank butcher paper next to the Geometry Word Bank Chart. Label the butcher paper "Plane Geometry Shapes File" with a subheading of "Record Format." Substituting database terminology, rename categories as fields and descriptors as data. As a whole group activity, begin to construct a record format. Have the students discuss and list the fields in a logical sequence. For example, begin with name fields first, then side and angle, and end with fields related to other geometric qualities.

The exact phrasing of data is important, as all records must utilize the same data phrases and spellings for efficient and effective use of the database. Due to the nature of the topic, the two fields labeled "types of sides" and "types of angles" will probably be the most cumbersome to deal with. The key is to use the least number of words when selecting descriptive data. Reference to the "types of sides" category on the Geometry Word Bank Chart reveals the words equal, nonequal, parallel, opposite, intersecting, and perpendicular. Student discussion could lead to the following conclusions:

- describing some sides as equal means all other sides must be nonequal; therefore, use only equal,
- since the definition of perpendicular includes the word intersecting, use only perpendicular, and
- because the word opposite is used to define parallel sides, use only parallel.

Using only the data words equal, parallel, and perpendicular, students must determine how to phrase data entry so all shapes can be accounted for. Using each group's shape, students should come to realize their shapes have either all, none, or pairs of equal sides; none or opposite pairs of parallel sides; and/or none or intersecting pairs of perpendicular sides. Thus the data section for "types of sides" will need to incorporate these ideas. The same process will need to be followed for the field "types of angles." Refer to Figure 4 for an example of how this process could result in a standard data format that all groups apply to their shape.

The whole group should now carefully review the format of all the fields and data. Some fields may have too many different kinds of data descriptors. If so, then these data descriptors could become fields of their own. Pursue further refinement into more fields with less data descriptors. This refining process will produce a butcher paper record format such as the one shown in Figure 5.

Using the Record Format chart, have the whole group develop a record for the "class" shape on another large piece of butcher paper. Stress preciseness, because the data descriptors must match those agreed upon for the Record Format chart. This large group activity should result in a record for the rectangle that looks like Figure 6.

Provide 8" x l0" index cards for each small group and have them copy the fields from the Record Format chart onto the card. Have each group refer to its plane geometry shape and complete

Amended Geometry Word Bank Chart

Categories	Descriptors
(Data previously entered)	
Lines of symmetry	#
Turns of rotational symmetry	#
Tessellates	yes, no

Figure 3.

Record Format for Types of Sides and Types of Angles

Field	Data
Types of sides	all or # pairs of equal sides # pairs of opposite parallel sides # pairs of intersecting perpendicular sides
Types of angles	all or # pairs of equal angles # right angles # acute angles # obtuse angles

Figure 4.

Plane Geometry Shapes File Record Format

Fields	Data
Polygon name:	triangle, quadrilateral
Shape name:	isosceles triangle, equilateral triangle, scalene triangle, right triangle, acute triangle, obtuse triangle, rectangle, square, parrallelogram, rhombus, trapezoid
Number of sides:	#
Equal sides:	all, # pair(s)
Pairs of opposite parallel sides:	#
Pairs of intersecting perpendicular sides:	#
Number of angles:	#
Equal angles:	all, # pair(s)
Right angles:	#
Acute angles:	#
Obtuse angles:	#
Total interior degrees:	#
Lines of symmetry:	#
Turns of rotational symmetry:	#
Tessellates:	yes, no

Figure 5. Finalized record format.

Sample Record for the Rectangle

Fields	Data
Polygon name:	quadrilateral
Shape name:	rectangle
Number of sides:	4
Equal sides:	2 pair(s)
Pairs of opposite parallel sides:	2
Pairs of intersecting perpendicular sides:	4
Number of angles:	4
Equal angles:	all
Right angles:	4
Acute angles:	0
Obtuse angles:	0
Total interior degrees:	360
Lines of symmetry:	2
Turns of rotational symmetry:	1
Tessellates:	yes

Figure 6. Sample record.

an index card record for its group shape. Remember to stress that data descriptors must match those designed by the class. It is helpful to post both the Record Format chart and the Sample Record for the rectangle within easy viewing for all groups to use as a reference.

Creating the Computer Database File
The teacher or a knowledgeable student must now use the database software to create a file that incorporates the record format designed by the class. The file could be named "Plane Geometry Shapes." For ease of use, the blank computer record may be laid out differently from the butcher paper Record Format in order to use only one computer screen.

Adding Data to the Geometry File
This data entry activity will require one computer booted with the "Plane Geometry File" data disk. With the whole group demonstrate the data entry process using the "class" shape. In order to standardize the data format, refer to the prominently posted Record Format chart and the Sample Record chart for the rectangle as data is entered into each field. Once the record for the "class" shape is completed, carefully review each field for standard data format and potential spelling errors. If any data is incorrectly entered, retype the correct data before finalizing the adding process.

Post a computer station schedule that provides each group approximately 20–30 minutes of online computer time for data entry. Each small group will follow the same procedures outlined with the "class" shape to add its shape's data to the file. Groups should refer to their index card records while entering data into the computer database. Post the Record Format chart and the Sample Record chart for the rectangle near the computer station while students are entering their data to the file. Keeping the data format the same for all shapes will make it easier to compare and contrast records later on.

Simulating Database Searches
In order to explore a computer database students must understand two concepts.

First, what is a computer database search? Second, how does a particular database accept search criteria? Understanding both ideas can be accomplished by having students simulate how a computer searches a database. Provide each small group with a hard copy printout of their computerized record from the "Plane Geometry Shapes" file. If a printer is not available, have groups use their index card record. Ask the whole group a geometric shapes question. For example, "Which geometric shapes have two equal angles?" Have each small group look at the hard copy or index card record that correlates with its group shape. If the shape answers the question, then someone in the group raises a hand. Have students tell which field they were looking at and what data they were looking for. (The field is "equal angles," and data is "2.")

Demonstrate how the computer database answers the same question. Computerized search criteria must match the data format used to create the file. Refer to the Record Format chart where data is specified as numeric by the number sign, or otherwise as alphanumeric. As the computer indicates a shape's record, have someone from that group stand up. Compare the raise of hands performed earlier to the students standing. The two groups should match. If any discrepancies occur, use the computer records and the hard copy printouts to carefully review the field(s) searched for any data format or spelling errors. Update (change) incorrect records immediately and repeat the exercise. See Figure 7 for sample geometric questions and search criteria strategies using the PFS: File database.

Search Criteria Strategies

Once students understand how to explore a database file, they are ready to apply search criteria strategies to solve other geometric shapes questions. As a whole group activity, formally introduce a step-by-step process for applying search criteria strategies to answer questions about the geometric shapes. First, determine which fields will help answer the question. These fields should be written down. Then analyze which search criteria will be needed to search the field, (i.e., equals, does not equal, is less than, etc.). Remember, search criteria must use the same data format as the Record Format chart. This search criteria should be written beside the appropriate field. The need to write down fields and search criteria becomes evident when a search goes astray. Being able to see what the computer was asked to do helps both you and your students analyze any searches that do not provide a reasonable answer. Finally, students should write down the information that appears on the screen in answer to the question. Other questions that require students to think about their answer and/or carefully scrutinize individual records can also be incorporated (see Figure 8).

Follow-up small group, online computer activities should stress use of this process for applying search criteria strategies. A worksheet like the one on the Copy Me! page will help guide students through this geometric database search activity. The computer station schedule established for data entry purposes can be followed for small group exploration of the database search questions. After groups have answered the teacher-developed questions, encourage them to write and solve their own geometry question.

As a culminating activity, provide students with an opportunity to share their database shape questions that were written during the database search activity. Place questions on index cards at the computer station for groups to answer during online computer time.

Conclusion

According to Olds and Dickenson (1985) there are three stages for learning to use database management software in educational settings. The easiest is to use a database created by someone else while

Geometric Questions and Search Criteria

Exact Match (Alphanumeric data)

Which geometric shapes have only one pair of equal sides?
TAB to the EQUAL SIDES field.
Type: **1 pair(s)**

Partial Match (Alphanumeric data)

Which geometric shapes belong to the quadrilateral family of polygons?
TAB to the POLYGON NAME field.
Type: **quad..**

Numeric Matches (Numeric data)

Which geometric shapes have no turns of rotational symmetry?
TAB to the TURNS OF ROTATIONAL SYMMETRY field.
Type: **0**

Which geometric shapes have two or more lines of symmetry?
TAB to the LINES OF SYMMETRY field.
Type: **> 1**

Which geometric shapes have fewer than four right angles?
TAB to the RIGHT ANGLES field.
Type: **< 4**

Numeric Range (Numeric data)

Which geometric shapes have one to three acute angles?
TAB to the ACUTE ANGLES field.
Type: **= 1..3**

Not Matches (Alphanumeric or numeric data)

Which geometric shapes do not have "angle" anywhere in their names?
TAB to SHAPE NAME field. Type:
1..angle..

Which geometric shapes do not have interior angles which sum to 180 degrees?
TAB to TOTAL INTERIOR DEGREES field. Type: **1 /180**

Figure 7.

The Search Process

Shapes Question
Which shapes have pairs of parallel and perpendicular sides? (Why are none of the shapes triangles?)

Fields and Search Criteria
PAIRS OF OPPOSITE PARALLEL SIDES: >0
PAIRS OF INTERSECTING PERPENDICULAR SIDES: >0

Answer
square, rectangle
(Although the right triangle has one pair of perpendicular sides, triangles do not have parallel sides.)

Figure 8.

learning to search and print data. The next stage is to create and use a database given the record format with fields, but no data. As students collect and enter data they learn the importance of standardizing vocabulary. The most challenging stage demands that students create and use an original database from scratch. This geometry unit outlines a guided discovery process for developing the third and most difficult stage of databasing.

Moreover, this process incorporates a standard curriculum mathematical topic of geometry. While students identify, analyze and classify geometric shapes, they also learn about database management as a powerful computer tool for exploring and managing information. Computer literacy knowledge for today's students must include skills with information processing software. Coupling geometry instruction with a computer environment optimizes instructional time on task through the integration of mathematics and computer education.

Editorial Notes

This month's column provides an interesting use of databases to explore geometrical concepts. Though directed at the intermediate level, mathematics students at most grade levels could benefit from such an exploration. Students in a formal geometry course in high school could be asked to build and expand this database as the year progresses. Another option is to use a product such as HyperCard or LinkWay to provide more powerful options for the activity.

The author briefly mentions the research of the van Hieles, two Dutch mathematicians. In 1959, they suggested that students studying geometry progress through five "thought levels." Each level is identified by its respective geometrical reasoning skills, appropriate vocabulary, and form of symbolism. The levels and a brief overview are:

Level 0: Visualization—The student visually studies geometric shapes, recognizing them as a whole without reference to properties of their components.

Level 1: Description—The student analyzes the shapes on the basis of their components or attributes, then establishes related properties.

Level 2: Informal Deduction/Ordering—The student logically orders the shapes, using the established properties. Definitions are formulated.

Level 3: Formal Deduction—The student develops a collection of theorems within a geometrical system, supporting each theorem with an acceptable proof.

Level 4: Rigor—The student studies the foundations of geometry by altering postulates to create different systems.

The database activity described in this article would fit under both Levels 1 and 2. If students should later prove some of their observed relationships (which really are conjectures), then they would be at Level 3.

If you would like to read further regarding the ideas of the van Hieles (and I strongly suggest that you do), an excellent resource is the September 1985 issue of The Mathematics Teacher. For the real die-hards, Pierre van Hiele has written a text, Structure and Insight: A Theory of Mathematics Education (Academic Press, 1986), that expands their ideas considerably.

This article is perhaps the first to illustrate a non-trivial use of a database to explore mathematics, other than for statistical compilations. Thus, I applaud the author on her efforts.—*Jerry Johnson, Mathematics Editor.*

References

Burger, W. F. (1988, November). One point of view: An active approach to geometry. *Arithmetic Teacher, 36,* 2.

Ford, M. S., & Gregg, L. (1987, March). The rockbound files. *Teaching and Computers, 4,* 10-15.

Olds, H. F., Jr., & Dickenson, A. (1985, October). Move over, word processors—here come the databases. *Classroom Computer Learning, 6,* 46-49.

Renshaw, B. S. (1986). Symmetry the trademark way. *Arithmetic Teacher, 34,* 6-12.

Woldman, E., & Kalowski, P. (1985, February). Make your own dinosaur data base. *Teaching and Computers, 2,* 14-19.

Creating a Mathematical Laboratory

Have you limited your teaching of matrices to advanced high school math classes? Now, technology can help you teach them to middle school students. In this article, the author shows how careful planning and a computer spreadsheet program can foster a highly constructive learning environment.

By Louis Feicht

Subject: Math, Technology

Grade Level: 7–12 (Ages 12–18)

Technology: spreadsheet software (e.g., Microsoft Excel or Works, AppleWorks)

Standards: NETS 3 and 6. (See http://cnets.iste.org for more information on the NETS project.)

Supplement: www.iste.org/L&L

Matrices and their applications have typically been topics that were reserved for advanced algebra courses in high school. But no longer. Now these concepts are accessible to middle school students when teachers use technology effectively in their classrooms. Using carefully planned activities, strategies, assessments, and communication, a teacher can effectively guide students in their understanding of meaningful mathematical concepts. Furthermore, using a computer spreadsheet, a teacher can create an environment for students to transform specific concepts into generalizations and construct formal mathematical meaning. The more opportunities students have to coordinate actions with mental operations, the greater the potential for learning.

In this lesson, an investigative teaching strategy allows students from middle school through college to discover and construct concepts on their own. Students investigate connections between matrix multiplication and addition, and with geometric transformations, creating bonds between algebra and geometry. They develop their understanding of the applications of matrices along with the geometric transformations of reflections, translations, dilations, and rotations; the concepts of function composition and inverse functions; and applications of math and its use in real life.

The concepts, connections, strategies, and communication used in this lesson are consistent with the goals of the National Council of Teachers of Mathematics' *Curriculum and Evaluation Standards for School Mathematics* (1989), *Professional Standards for Teaching Mathematics* (1991), and *Assessment Standards for School Mathematics* (1995). The lesson is the result of several years of refinement in my classrooms of an idea from the Montana Council of Teachers of Mathematics (1996) Systemic Initiative for Montana Mathematics and Science (SIMMS) Project. An online version of this lesson is available at the Microsoft Encarta Lesson Collection at **http://encarta.msn.com/schoolhouse/default.asp** and on my school Web site at **www.silverfalls.k12.or.us/~lou/**.

Setting Up the Spreadsheet

This activity can be used in any classroom that has a computer that can run a modern spreadsheet program (we used Microsoft's Excel). First, students are introduced to matrices and the appropriate terminology. Then, using blackboard and chalk, I show them how matrices can be used to represent points in a coordinate system. I set up the spreadsheet program with a matrix of ordered pairs and graph them on the same worksheet by using a scatterplot with connected points (Figure 1).

I have named our preimage matrix "Flag," because that's what the graph looks like. You might choose to use a different set of ordered pairs to create a different preimage; the concepts of the lesson remain the same. Several more examples like the Flag matrix are used to reinforce the students' understanding of how ordered pairs are represented graphically.

Students are next introduced to matrix multiplication. You could do this either before the spreadsheet activities or after exploration and investigation using the spreadsheet. Another matrix, "Transform," is entered into the spreadsheet to the right of the preimage Flag matrix (Figure 2). In the spreadsheet seen in this example, the gridlines and row and column headings are turned off. These two matrices are then multiplied using Excel's built-in multiplication function (Figures 2 and 3). If your spreadsheet does not have this function, then you can use the algorithm for matrix multiplication and the spreadsheet's normal addition and multiplication functions.

Figure 4 shows how to set up the spreadsheet in ClarisWorks (now AppleWorks), a program that does not have a built-in matrix multiplication function. The resulting product matrix (the image) is graphed on the same set of axes in Excel using the spreadsheet's capabilities (see Figure 5 for directions). Spreadsheets other than Excel may require two separate charts—one for the preimage and one for the image (see Figure 4 again). You may opt to set up the spreadsheet for students or let them do it, depending on their technological sophistication. Because this lesson requires knowing how to edit the data series in a spreadsheet, it is usually easier for the teacher to create the spreadsheets so that students can focus on the mathematics and not the technology. If you don't have a computer lab, then the spreadsheet as set up in Figure 2 is also excellent for use with a single computer and an LCD projector for classroom presentation and discussion.

Making Connections

We have now created a mathematical laboratory in which students can explore and discover. When the teacher or students change numbers in "Matrix Transform," the spreadsheet will automatically "move" the image; students can immediately see the change in the image matrix (Figure 6). Students are required to work through questions designed to allow them to establish patterns and make generalizations (see Worksheets 1 and 2). The set of questions requires them to formulate hypotheses for testing. The students can now investigate how changes in "Matrix Transform" result in transformations in the product matrix and the corresponding image. They can conjecture and immediately test to see the results of their hypotheses.

One emphasis of the entire lesson is the proper use of mathematical vocabulary and connections between algebra and geometry. By watching the numbers in the preimage and image matrix change as we change numbers in the transform matrix, students can see how the x and y values are affected. Geometric mapping notation such as $(x, y) \rightarrow (x, -y)$ is then connected to the transformations and the matrices. After they complete guided activities and assignments, the students are asked to generalize about how multiplication by certain matrices results in specific transformations and to summarize what they have

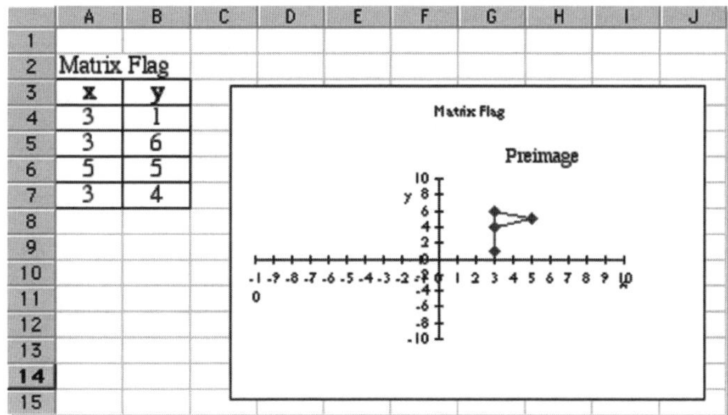

Figure 1. A numerical representation of matrix "Flag" and its corresponding graph.

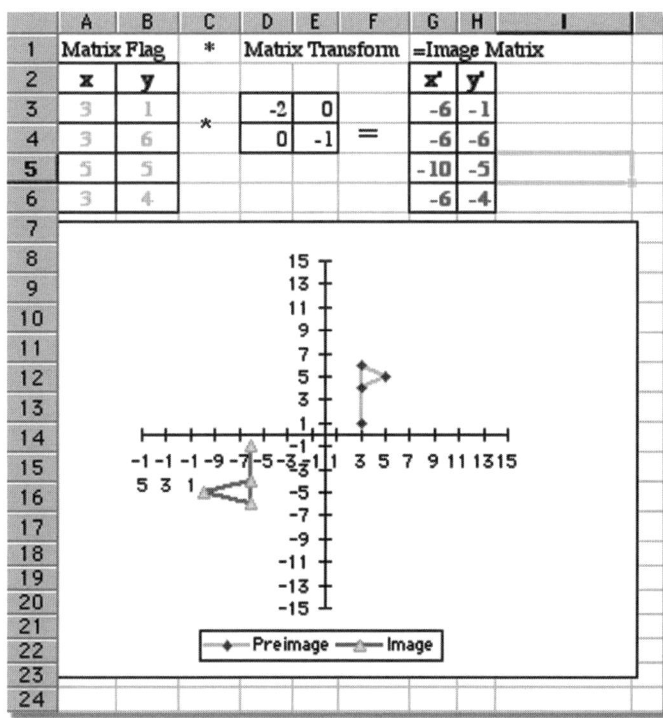

Figure 2. The spreadsheet setup as an investigative environment for students.

	A	B	C	D	E	F	G	H
1	Matı		*	Matı			=Image Matrix	
2	x	y					x'	y'
3	3	1		2	0		=MMULT(A3:B6,D3:E4)	=MMULT(A3:B6,D3:E4)
4	3	6	*	0	-2	=	=MMULT(A3:B6,D3:E4)	=MMULT(A3:B6,D3:E4)
5	5	5					=MMULT(A3:B6,D3:E4)	=MMULT(A3:B6,D3:E4)
6	3	4					=MMULT(A3:B6,D3:E4)	=MMULT(A3:B6,D3:E4)
7								

Figure 3. The formula for using the built-in matrix multiplication function of the Excel spreadsheet. Command-Return is used on a Macintosh, and Control-Shift-Enter is used in Windows to enter an array in Excel.

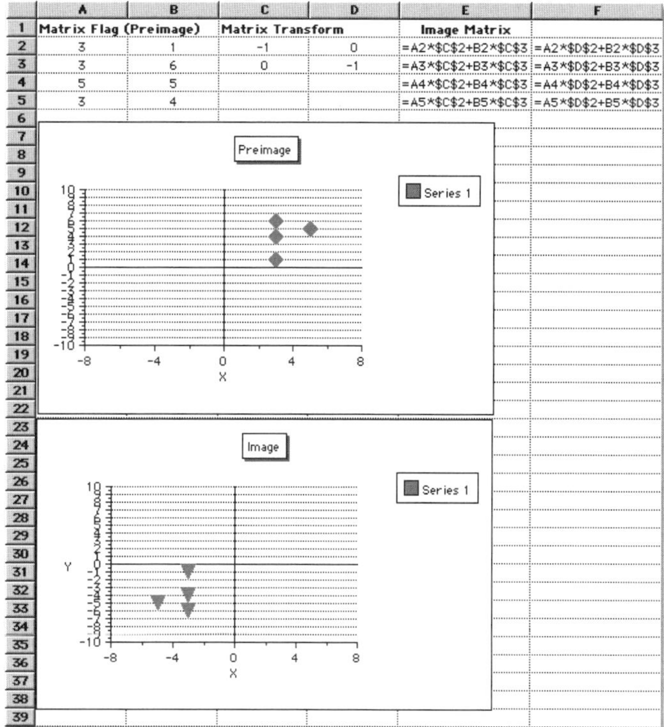

Figure 4. A numerical representation of the matrix "Flag" and its corresponding graph.

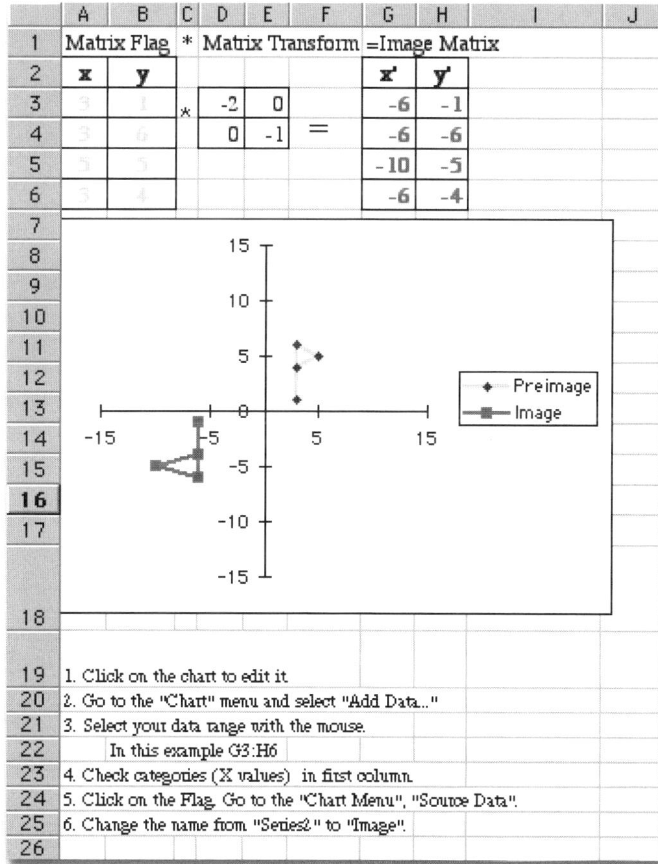

Figure 5. Directions for inserting a new data series into a chart, which puts the image on the same chart as the preimage.

learned in writing. They should state what happens when multiplying an "$n \times 2$" matrix by matrices of the form

$$\begin{bmatrix} a & 0 \\ a & 0 \end{bmatrix} \qquad \begin{bmatrix} -a & 0 \\ 0 & a \end{bmatrix},$$

and $$\begin{bmatrix} 0 & -a \\ 0 & b \end{bmatrix},$$

along with other relevant possibilities. Students can make inferences about the effect of the magnitude of the matrix elements, describing how a transformation using a number between 0 and 1 is different from a transformation using a number larger than 1.

Students complete guided group activities, assignments, and quizzes (Worksheets 3, 4, 5, and 6). They are evaluated on class participation, worksheet answers, homework assignments, quizzes, tests, and writing assignments in which they summarize what they have learned.

Through this guided investigation, my students began to understand how to produce a reflection, rotation, and dilation by using matrix multiplication. They were able to predict what happened when points in the plane are transformed by matrix multiplication of the forms

$$\begin{bmatrix} a_{11} & 0 \\ 0 & a_{22} \end{bmatrix}$$

for all values of a. They asked questions about combining and undoing transformations, as well as how a composition of two reflections can be equivalent to a rotation. I was able to help them develop the formal mathematical meanings of function composition and inverse functions. By practicing with the spreadsheet and going through this investigative lesson, my students were able to set up matrix equations describing a given geometric transformation. Conversely, they were able to describe geometric transformations given the matrix equations.

Matrix addition and geometric translations are covered in another spreadsheet but handled similarly. Translations and matrix addition can be done before or after the other transformations, depending on your preference. The spreadsheet is set up for addition with Matrix Flag (the preimage matrix), and Matrix Transform is added using typical array-adding procedures for spreadsheets (Figures 7 and 8). Matrix Transform, of course, must be the same size as the preimage matrix and is best constructed by using absolute references to an x and a y cell. These two cells are shown at the top in Figures 7 and 8. Students can change the two cells for x and y and have the spreadsheet automatically make the changes in Matrix Transform (Figure 7).

Real-World Applications

One application of this lesson that students are familiar with is computer environments and graphics. They are comfortable

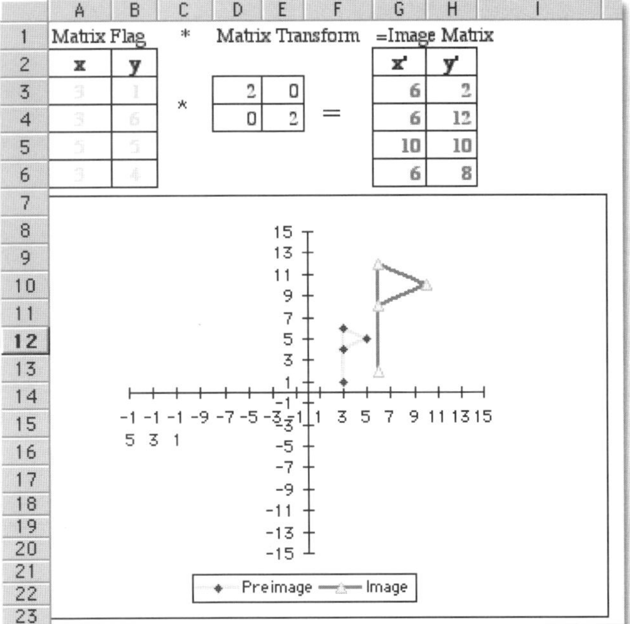

Figure 6. Notice how the change in matrix Transform shows up as a change in the image matrix and flag.

Name ______________________________

1. Multiply Matrix Flag (the preimage) by

   -1 0 0 1

   How are the numbers in the image matrix different from Matrix Flag?
   Which numbers in the image matrix did not change?
   What type of transformation took place to create the image?
   Which points in the plane remained fixed (did not move)?

2. Multiply Matrix Flag (the preimage) by

   1 0 0 -1

   How are the numbers in the image matrix different from Matrix Flag?
   Which numbers in the image matrix did not change?
   What type of transformation took place to create the image?
   Which points in the plane remained fixed (did not move)?

3. Multiply Matrix Flag (the preimage) by

   -1 0 0 -1

   How are the numbers in the image matrix different from Matrix Flag?
   Which numbers in the image matrix did not change?
   What type of transformation took place to create the image?
   Which points in the plane remained fixed (did not move)?

4. Multiply Matrix Flag (the preimage) by

   2 0 0 2

   How are the numbers in the image matrix different from Matrix Flag?
   Which numbers in the image matrix did not change?
   What type of transformation took place to create the image?
   Which points in the plane remained fixed (did not move)?

Worksheet 1.

Name ______________________________

5. Predict what type of transformation(s) will take place when Matrix Flag (the preimage) is multiplied by

   -2 0 0 2

   Do it. What type of transformation took place to create the image?

6. Predict what type of transformation(s) will take place when Matrix Flag (the preimage) is multiplied by

   2 0 0 -2

   Do it. What type of transformation took place to create the image?

7. Predict what type of transformation(s) will take place when Matrix Flag (the preimage) is multiplied by

   3 0 0 3

   Do it. What type of transformation took place to create the image?

8. Predict what type of transformation(s) will take place when Matrix Flag (the preimage) is multiplied by

   -2 0 0 1

   Do it. What type of transformation took place to create the image?

9. Predict what type of transformation(s) will take place when Matrix Flag (the preimage) is multiplied by

   2 0 0 -1

   Do it. What type of transformation took place to create the image?

10. Predict what type of transformation(s) will take place when Matrix Flag (the preimage) is multiplied by

    -2 0 0 -1

    Do it. What type of transformation took place to create the image?

Worksheet 2.

seeing an "object" move on screen, and many are familiar with copying and pasting as well as selecting and dragging objects. This method of teaching matrices and transformations provides a foundation for how computers store and manipulate information and perform graphic operations. The lesson shows how mathematics can make sense to them in their everyday lives.

Communication

The spreadsheet environment provides more opportunity for students to explore concepts of matrices than do traditional teaching methods. It stimulates reflection and abstraction, both of which enhance students' constructions of the desired concepts. The spreadsheet activities allow social interaction when

Name ______________________________

1. Write down the matrix equation that represents the following transformation.

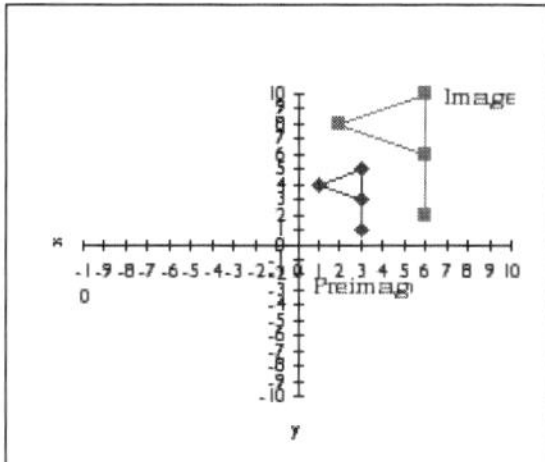

2. Write down the matrix equation that represents the following transformation.

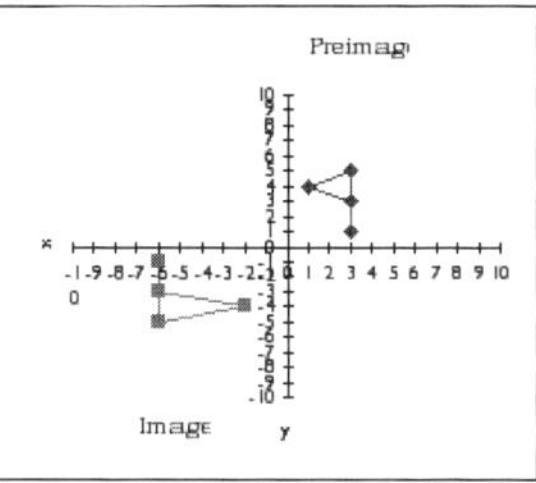

Worksheet 3.

Name______________________________

Short Answer/Essay

Write in complete sentences. Check your spelling.

1. Describe the transformation that takes place when we multiply an n*2 matrix, on the right, by a 2*2 matrix of the form

$$\begin{bmatrix} a_{1,1} & 0 \\ 0 & 1 \end{bmatrix}$$

Explain what will happen when $a_{1,1}$ is equal to 1, greater than 1, less than -1, between 0 and 1, between 1 and 0.

2. Describe the transformation that takes place when we multiply an n*2 matrix, on the right, by a 2*2 matrix of the form

$$\begin{bmatrix} 1 & 0 \\ 0 & a_{2,2} \end{bmatrix}$$

Explain what will happen when $a_{2,2}$ is equal to 1, greater than 1, less than -1, between 0 and 1, between -1 and 0.

3. Extra Credit
What happens when the elements $a_{1,2}$ and $a_{2,1}$ are changed? What type of transformation takes place? (Due one week from today.)

Worksheet 4.

Name ______________________________

1. Graph the Matrix Flag.

x	y
3	1
3	6
5	5
3	4

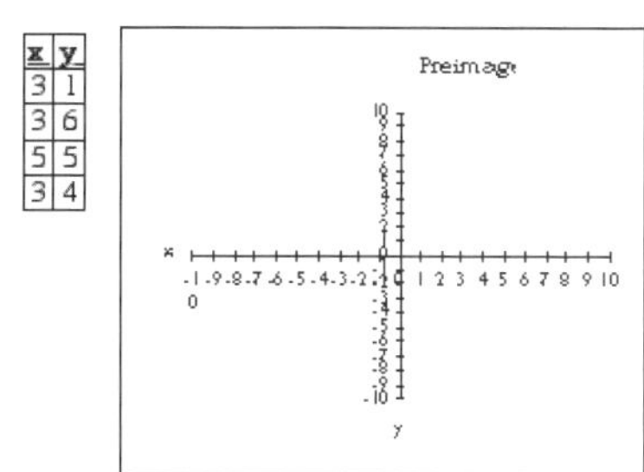

2. Fill in the correct numbers in Matrix A to produce the image matrix.

x	y
3	1
3	6
5	5
3	4

-3	1
-3	6
-5	5
-3	4

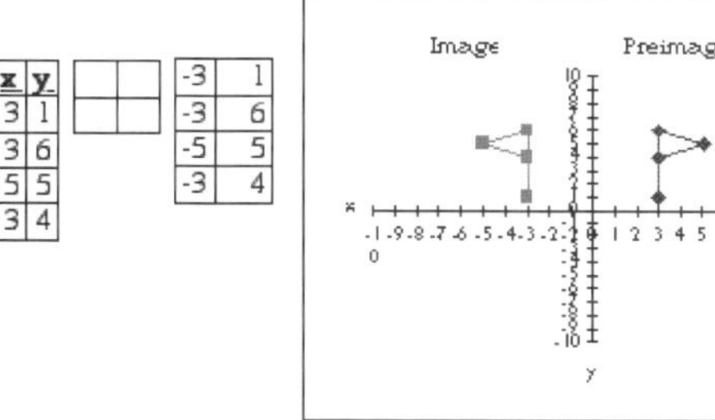

Worksheet 5.

Name ______________________________

3. Fill in the Image matrix and Matrix A that produces the image.

x	y
3	1
3	6
5	5
3	4

Preimage

Image

4. Give the correct numbers for Matrix Flag (the preimage).

-2	0
0	-2

-6	-2
-6	-10
-2	-8
-6	-6

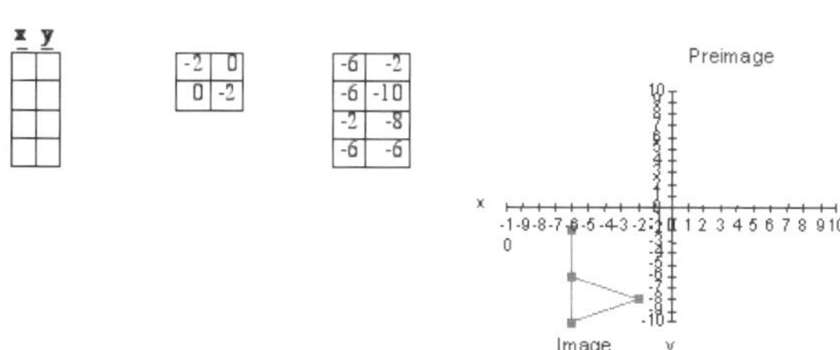

5. Create a 3X5 matrix. Use any numbers you want.

$$\begin{bmatrix} 3 & 4 & 0 \\ 2 & -1 & 3 \\ 1 & 0 & 2 \end{bmatrix} \times \begin{bmatrix} 4 \\ 2 \\ 1 \end{bmatrix}$$

6. Multiply.

Worksheet 6.

students work in groups of two, and because the activities naturally generate discussion, students' construction of concepts is further helped. If two students disagree about the outcome of a hypothesis, then they can discuss concepts and increase their understanding. The spreadsheet environment can provide a medium and language for students to express and communicate their ideas; these also help them develop a formal mathematical understanding. The environment provides a medium for students to investigate and discover mathematical concepts and systems before they have a formal mathematical background.

References

Montana Council of Teachers of Mathematics. (1996). *Integrated mathematics: A modeling approach using technology. Levels 1–6.* Systemic Initiative for Montana Mathematics and Science (SIMMS) Project. Needham Heights, MA: Simon and Schuster.

National Council of Teachers of Mathematics. (1989). *Curriculum and evaluation standards for school mathematics.* Reston, VA: Author.

National Council of Teachers of Mathematics. (1992). *Professional standards for teaching mathematics.* Reston, VA: Author.

National Council of Teachers of Mathematics. (1995). *Assessment standards for school mathematics.* Reston, VA: Author.

Resources

ClarisWorks (now AppleWorks) is available from Apple Computer, Cupertino, CA; www.apple.com/appleworks/.

Microsoft Excel is available from Microsoft Corporation, Redmond, WA; www.microsoft.com/products/prodref/447_ov.htm.

Find more resources on the Web at **www.iste.org/L&L/**.

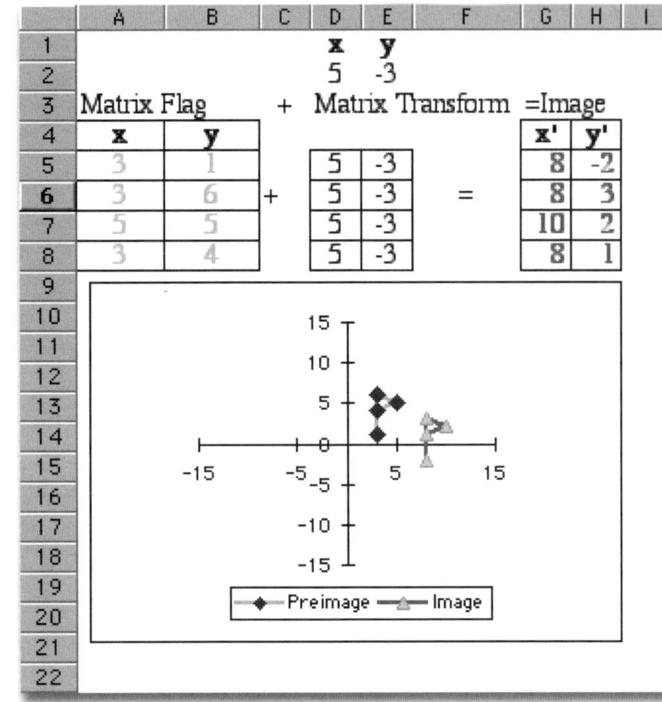

Figure 7. This spreadsheet is set up to investigate matrix addition and translations.

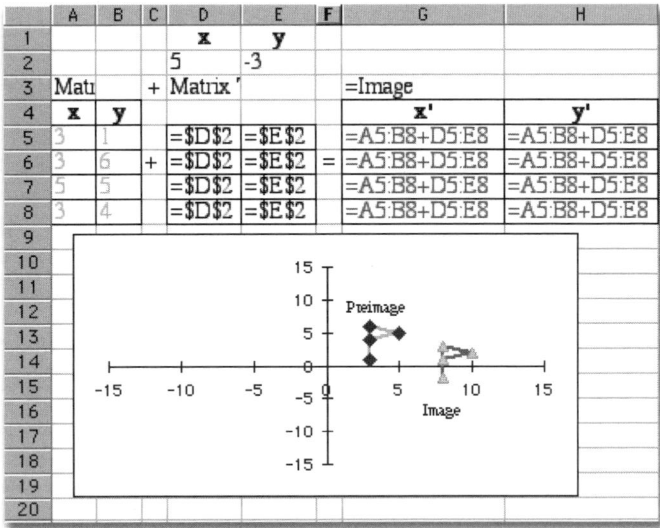

Figure 8. These are the formulas used to set up the spreadsheet for matrix addition.

Does Cincinnati Need Another Bridge?

By Deanna J. Salisbury

How do you help students in Grades 4–6 (ages 9–12) understand that it really is the shape of a bridge that creates the strength that supports it? Geometric shapes are easy to recognize, but why is it so important to know about these shapes? How do communities make decisions about the number and location of bridges that are needed in their area? At Southgate Public Elementary School, we helped students address these questions by creating a thematic unit about bridges using MicroWorlds Project Builder, a Logo-based hypermedia environment from LCSI.

Southgate is located in the rolling hills of Kentucky just below Cincinnati on the Ohio River. Students enrolled at our school see the city lights of Cincinnati reflected in the river at night before they go to bed. Their parents use bridges to cross the river every day to go to work, and the students themselves have crossed the river many times since the day they were born. When a proposal to build a new bridge across the Ohio was made, the science/math and computer teachers at Southgate wanted our students to be able to determine whether this was a good idea for our community. We also wanted students to be able to understand geometry concepts as they related to this bridge, especially the way that the bridge's geometric design contributed to its strength.

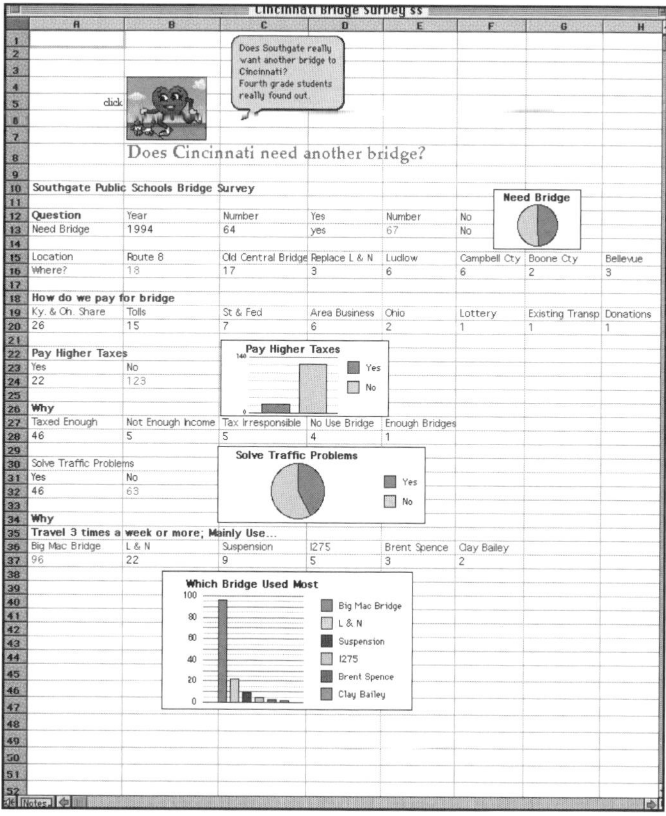

Cincinnati Bridge Survey ss

	A	B	C	D	E	F	G	H
5	click							
8		Does Cincinnati need another bridge?						
10	Southgate Public Schools Bridge Survey							
12	Question	Year	Number	Yes	Number	No		
13	Need Bridge	1994	64	yes	67	No		
15	Location	Route 8	Old Central Bridge	Replace L & N	Ludlow	Campbell Cty	Boone Cty	Bellevue
16	Where?	18	17	3	6	6	2	3
18	How do we pay for bridge							
19	Ky. & Oh. Share	Tolls	St & Fed	Area Business	Ohio	Lottery	Existing Transp	Donations
20	26	15	7	6	2	1	1	1
22	Pay Higher Taxes							
23	Yes	No						
24	22	123						
26	Why							
27	Taxed Enough	Not Enough Income	Tax Irresponsible	No Use Bridge	Enough Bridges			
28	46	5	5	4	1			
30	Solve Traffic Problems							
31	Yes	No						
32	46	63						
34	Why							
35	Travel 3 times a week or more; Mainly Use...							
36	Big Mac Bridge	L & N	Suspension	I275	Brent Spence	Clay Bailey		
37	96	22	9	5	3	2		

Figure 1. Cruncher spreadsheet with charts.

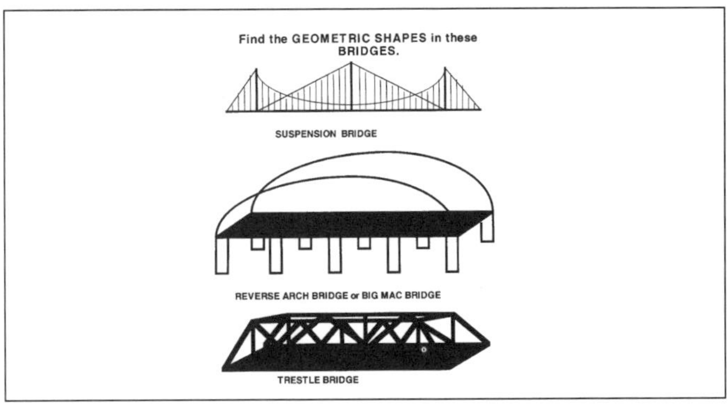

Figure 2. Bridges created with ClarisWorks.

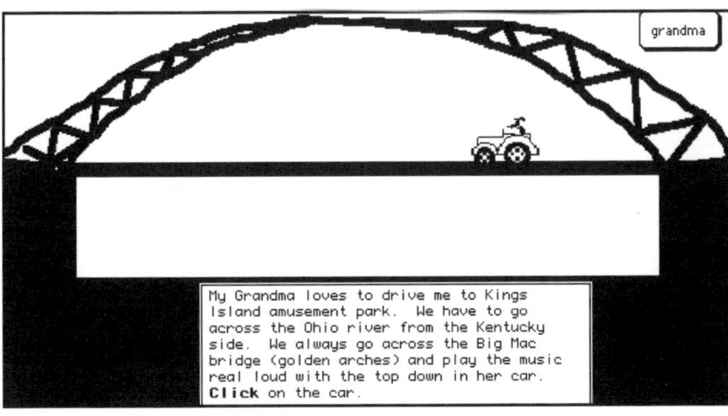

Figure 4. Student-created picture with musical component.

Figure 3. MWPB-created simulation.

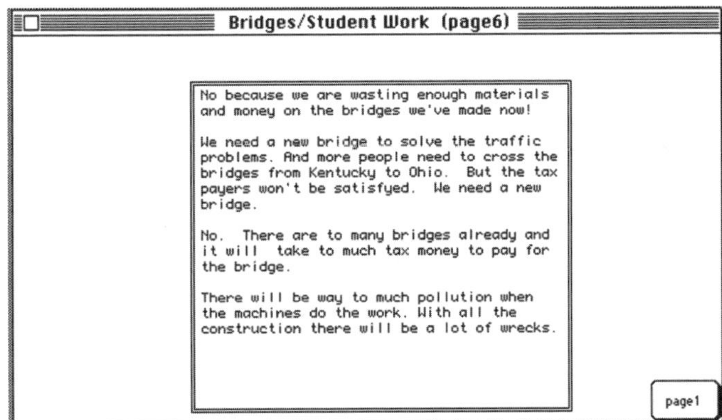

Figure 5. Student statements about the need for a new bridge.

Surveying the Community

Our project started with a community survey about attitudes toward the new bridge and the taxes required to build it. Did we need a new bridge? Why or why not? The students and teachers created the survey questions, and the results were listed in a spreadsheet format. (We used a spreadsheet because we wanted students to be able to accumulate information for a few years if they chose to expand this project as part of their portfolio for statewide assessment.) The software we used was Cruncher, a multimedia spreadsheet maker by Davidson that offers sound, animation, "speaking-text," and chart-making features. For example, after the information is entered in the spreadsheet, the numbers to be graphed are simply highlighted, and with three clicks of a mouse a full-color, 3-D spreadsheet in the format of the student's choice is produced (see Figure 1).

Integrated Studies of Bridges

The next step in our unit was to read historical documents, surveys, stories, and poems about bridges in our community. For example, we discovered that Cincinnati has a suspension bridge designed by John Roebling, the same architect who designed the Brooklyn suspension bridge. In fact, the Cincinnati suspension bridge was a model for the Brooklyn suspension bridge.

This step also required students to explore geometric shapes by using manipulatives in math. They reviewed their lesson on manipulatives in the computer lab with a ClarisWorks tangram electronic simulation from ClarisWorks in the Classroom.

Students were asked to keep records of the number of steps they had used to create a square using different tangrams. Then they were asked to repeat their trial two more times to see if they could reduce the number of steps and decrease their speed. Some of the students and I designed a ClarisWorks template in which students used the Draw tools to electronically create a historical bridge of their choice using the tangrams (see Figure 2).

In art class students replicated this activity with geometric shapes created with color paper. We believed that these activities allowed the students to gain a good understanding of how geometric shapes work together to create other geometric shapes. The colors, cooperative work groups, individual learning, and varieties of manipulation accommodated many learning styles.

The students then went on field trips to see the real bridges we were studying. They discussed the shapes they saw in the bridges and why they believed they were there. Students videotaped each of four bridges for 1 minute from 4 p.m. to 6 p.m. to monitor traffic.

Simulating Bridges and Bridge Traffic

Students used information obtained from the video to program their own electronic bridge simulation using MicroWorlds Project Builder (MWPB). MWPB allows the user to create up to 20 pages of textual information, interactive and self-running animation, drawings, simulations, and keyboard or recorded music.

Students learned to use MWPB and created their projects in eight one-hour computer classes. Elementary school students never cease to amaze me. My students had never used Logo before and had never seen what it could do. Yet they programmed this powerful simulation with only guided assistance, a manual, and some good electronic models. They were given MWPB projects that were included as models with the software. By exploring these projects, they quickly learned to manipulate the variables of angles, ratio, length, and speed in the projects. Next, students learned to use the Paint, Draw, and Stamp tools. Then they learned how to make objects move.

They began their simulation by creating and adding bridges to a teacher-created background of the Cincinnati and northern Kentucky riverfront (see Figure 3). Students created the geometric shapes they wanted for their bridges using the Shapes Center of MWPB. Then they constructed their bridges using these shapes and the Draw tools. They had to resolve problems concerning the perspective of the bridge against the background, the bridge width needed to handle traffic flow, and the proper location of the bridge.

Adding the bridge traffic to the simulation came next. Students used or modified the three car shapes MWPB provides. Next, they had to determine the rate at which the cars cross the bridge to simulate the traffic during the 4–6 p.m. time period. They added slider buttons to the car shapes so that the user could specify the speed of the car in the simulation.

Students obtained statistics from the bridge video concerning the number of cars and people crossing in one minute. This number was recorded for each of four examples of bridges on the Ohio River. (We had digitized the video, so students were able to do some careful analysis.) They were then asked to compare the size of the bridge with the number of people crossing it. Students came up with a way to relate the number of cars crossing the river to the number of people. They compared the number of cars crossing the bridge in one minute with size and type of the bridge. They had to solve these problems and explain their reasoning on a page in their MWPB project. We then introduced another bridge that students had not seen to see if they could make a good estimate of bridge traffic for the same time period by using what they had learned from the real bridge traffic they had observed and measured.

We designed another page for musical and spatial learners. This page allowed students to create a new river song on the MWPB keyboard. They were then asked to illustrate this song with a picture describing both the river and their song. One student played the notes going up and then down. His picture showed people walking up steps and then down when crossing the bridges on the Ohio River. Another song was a rain river song, and the picture showed mists and rain over the Ohio. This was the first time many of our students had made their own musical scores. It was very easy on the color-coded note and keyboard in MWPB. Figure 4 shows a student-created picture accompanied by a musical element when the user clicks on the car with the mouse.

To demonstrate to these students that geometric shape contributes to structural strength, we had them use straws to make model bridges. Each student could use from 1 to 30 straws bound together with scotch tape. Students were given 25 minutes to design a base and build their bridges. The strength of the bridges was then tested by placing a brick on them. Six bridges passed the bridge deflection test, which required that a bridge would not bend more than one-half inch when weighted down with the brick. Every bridge was completely different, and the students began to realize that shape is a very important factor in structural strength.

After completing the various activities in the unit, the students used MWPB to compose a statement about why they believed Cincinnati did or did not need another bridge (see Figure 5).

The last activity in this unit will be to walk the bridges again, but this time with a bridge architect. I expect that the questions the students ask this architect at this point in our unit will be a bit surprising. At the end of our unit, we plan to submit our findings to the city of Cincinnati for consideration.

CHAPTER 7

Measurement

The Conversation Continues

Anne: Have you ever noticed how time is relative?

Ivan: Sure. For example, when you are doing something you enjoy, time seems to fly by quickly. However, when you are doing something less appealing, or waiting for important news, time seems to creep along endlessly. This phenomenon raises the issue of when measurements are universal and when they are subjective.

Anne: Another measurement that you might think would be universal, but is quite subjective, is distance. For example, have you ever asked several different people for directions and gotten very different responses regarding distances within the directions. Further, have you noticed how directions in some cultures aren't given in distances but in minutes, or via landmarks? Again, it amazes me how much one's perspective influences how one defines or describes measurements within different contexts.

Ivan: With such technologies available as global positioning systems and mapquests on the Internet, perhaps over time measurements, as contextualized in the act of providing directions, will become more universal.

Anne: In the article titled "Measurement and Precision," the issue of how exact a measurement is will be explored. This, again, raises the question of the degree to which measurements can be universal.

THEORY INTO PRACTICE

Standard Versus Nonstandard Units

Customary Versus Metric Units

By Anne Raymond

According to the National Council of Teachers of Mathematics, students should gain both a broad understanding of what it means to measure and the specific means of determining appropriate and accurate measurements.

NCTM Measurement Standard

Measurement: Instructional programs from prekindergarten through Grade 12 should enable all students to

- Understand measurable attributes of objects and the units, systems, and processes of measurement
- Apply appropriate techniques, tools, and formulas to determine measurements

What does it mean to measure something? To measure means to determine the extent, the dimensions, the quantity, or the capacity of something. When students measure, steps they need to take include

1. determining what attribute of an object they wish to measure (length, weight, volume, etc.),
2. choosing an appropriate tool to use (such as measuring length with a paper clip rather than a ball),
3. comparing the number of units of the tool used to the item being measured.

Students usually learn how to measure using nonstandard units of measure (such as paper clips) before learning to use standard units (such as inches).

In the book *How Big Is a Foot?* (Myller, 1990) a king decides to give his wife, the queen, a bed for her birthday. Since beds have not yet been invented, the king has the queen lie down, and he measures the size needed for her bed by walking around her, determining the perimeter using his footprints. The carpenter who is to make the bed is told how many footprints it takes to make the bed. Then the carpenter makes the bed using his own footprints to measure the perimeter. Since the carpenter's feet are much smaller than the king's feet, the bed does not turn out the right size.

This story provides a wonderful setting for discussing the need for standard units of measure. In fact, legend has it that the word *ruler,* meaning a device used for measuring, comes from having used a king's foot as the primary means of measuring length.

NETS•S Productivity Standard

3. Technology productivity tools

- Students use technology tools to enhance learning, increase productivity, and promote creativity.
- Students use productivity tools to collaborate in constructing technology-enhanced models, preparing publications, and producing other creative works.

Measurement is such an interesting content area because not only do we have the opportunity to work with both standard and nonstandard units, but we also have the means of measuring with customary versus metric units of measure. In our society, we have become accustomed to working with both, depending on the setting. For example, we buy milk by the gallon, yet we buy sodas in 2-liter bottles. In the typical mathematics curriculum, students are challenged to measure items using multiple units and to compare those various measures. Students are also taught to convert measures from one unit to another. Interesting spreadsheet programs have been developed that facilitate converting customary units to metric units (for example Fahrenheit to Celsius). These conversion opportunities provide students the chance to build problem-solving skills as well as intuition about measurement.

EXAMPLES OF CUSTOMARY VERSUS METRIC UNITS

MEASURE	CUSTOMARY	METRIC
Length	inch	millimeter
	foot	centimeter
	yard	meter
	mile	kilometer
Capacity	ounce	millimeter (8 capacity ounces = 1 cup)
	cup	centiliter
	pint	deciliter
	quart	liter
	gallon	kiloliter
Temperature	Fahrenheit	Celsius
Weight	ounce	gram (16 weight ounces = 1 pound)
	pound	kilogram
	ton	metric ton

Another way to classify measurement is by considering geometric versus nongeometric measurements. Consider the following examples of each:

GEOMETRIC VERSUS NONGEOMETRIC MEASUREMENTS

GEOMETRIC	NONGEOMETRIC
length	time
perimeter	temperature
area	money
volume	capacity
weight	

When converting measurements from one unit to another, students can use spreadsheet conversion programs to assist the conversion process. For example, students are sometimes asked to convert Fahrenheit temperature measures to their equivalent Celsius temperature measures (and vice versa). Students can develop and then utilize a spreadsheet that performs such conversions.

Note: On the companion Web site for this book (http://education.bellarmine.edu/baugh/), you will find a spreadsheet file titled Temperature Conversion already set up for converting between Fahrenheit and Celsius.

No matter which way you classify or compare ways to measure, the important issues about measurement are that appropriate units are selected and that students know how to interpret those units.

Reference

Myller, R. (1990). *How big is a foot?* New York: Random House.

THEORY INTO PRACTICE

Cooperative Learning in Mathematics

By Anne Raymond

In concert with the curricular issues addressed by the National Council of Teachers of Mathematics in its *Curriculum and Evaluation Standards for School Mathematics,* concerns about the nature of mathematics pedagogy have been raised. It was suggested that in addition to curricular revision, there was a need for better mathematics instruction that would engage students in active learning (Good, Mulryan, & McCaslin, 1992). One means of engaging students in more meaningful mathematics activities is to make use of small groups, allowing opportunities for discussion of mathematics with peers.

Some mathematics educators warn that simply providing more small-group time doesn't necessarily improve the quality of instruction (Good & Biddle, 1988). However, cooperative learning, when properly implemented, can enable teachers to achieve goals such as "practicing meaningfully on mathematical topics of appropriate difficulty and interest, learning prosocial skills, taking different approaches to problem solving, verbalizing thoughts about mathematics, and enhancing social intelligence" (Noddings, 1989).

Johnson and Johnson (1985) have studied cooperative learning groups extensively and believe one key to successful cooperative learning groups is to be explicit about the task at hand and how students are to work together. When students know the task directions, the roles they are to play, and how to report what they learn, they worry less about how they are supposed to act and focus more on the mathematics.

Many activities in measurement provide opportunities for students to work cooperatively because they are tasks that individuals cannot easily complete by themselves. Two examples of such measurement activities follow.

Wicked Witch Willebrand

"Wicked Witch Willebrand" (source unknown) is a poem about a strange witch with unusual physical characteristics. The poem suggests a variety of measurements related to the witch. Students are to work in cooperative learning groups to determine relevant measurement information within the poem in order to plan and create an appropriate drawing of Wicked Witch Willebrand. As students plan, they have to agree on how to start, how to make sure they can get the entire picture on the paper provided, and how to measure accurately.

> *Wicked Witch Willebrand*
>
> Wicked Witch Willebrand was 4 feet tall and very scary indeed;
> A nose like a banana, eyes like seeds.
> Her mouth was 2" wide, it dimpled her cheeks,
> And made it difficult for her to speak.

Willebrand's ears were large and very furry,
Her fair was bright red and always quite curly.
On top of her head was a hat 14" tall,
If she didn't stand up straight, most likely she'd fall.

Two-inch long fingers were on her broad hands,
She wore rings of many colors: purple, red, green, and tan.
Her nails were 2" long and pointed like a steeple,
Wow, what a scary sight she must have been to people.

Her black dress was long, usually dragging on the street,
And the funniest parts were her tiny 3" feet.
Her five toes curled up, as she walked to and fro,
But where can such a funny lady really go?

Students have to deal with both issues of estimation and precision, all the while working as a team to discuss and obtain their results. Since students are dealing with measurements in different units, it is appropriate to use a calculator to compute necessary calculations so that they can work with the measuring tools they have and check their estimations when required.

Bungee Barbie and Kamikaze Ken

A second activity that engages students in cooperative learning and measurement is Bungee Barbie and Kamikaze Ken. In this task students engage in a simulation of bungee jumping using rubber bands and a meter stick. Barbie or Ken is attached to a rubber band at the ankle. Then, the doll is "bungeed" from the top of a meter stick, whereupon students have to determine how far he or she falls by reading the point on the meter stick that the doll reaches. Three trials are completed, and then an average is taken of the measurements. Then a second rubber band is added and three trials are performed. Rubber bands are continually added until students have been able to conduct trials for bungees with six rubber bands.

Students then take the trial data and project how many rubber bands it would take for Barbie or Ken to bungee from a high distance without hitting her or his head on the ground. For some experiments, this distance may be over a stairwell balcony, out a multistory window, or perhaps from a rooftop. Students may use Excel or a graphing program to project the number of rubber bands needed for the final bungee jump. Note that in this activity, students must work together because it is virtually impossible to gather the data alone.

Resources

Good, T., & Biddle, B. (1988). Research and the improvement of mathematics instruction: The need for observational resources. In D. Grouws & T. Cooney (Eds.), *Perspectives on research on effective mathematics teaching* (pp. 114–142). Reston, VA: NCTM.

Good, T.L., Mulryan, C., & McCaslin, M. (1992). Grouping for instruction in mathematics: A call for programmatic research on small-group processes. In D. A. Grouws (Ed.), *Handbook of research on mathematics teaching and learning* (pp. 165–196). New York: Macmillan.

Johnson, D., & Johnson, R. (1985). The internal dynamics of cooperative learning groups. In R. Slavin, S. Sharan, S. Kagan, R. Hertz-Lazarowitz, C. Webb, & R. Schmuck (Eds.), *Learning to cooperate, cooperating to learn* (pp. 103–124). New York: Plenum.

Noddings, N. (1989). Theoretical and practical concerns about small groups in mathematics. *Elementary School Journal, 89,* 607–623.

INSIGHT

Animal Measurements

By Anne Raymond

We often hear people refer to the ages of their canines in dog years versus human years. Generally, people believe that 1 dog year equals 7 human years, but others would say that the comparison is not a perfectly linear relationship. Nevertheless, given the fact that humans and dogs age at different rates, this allows us to compare the measures of years in multiple ways. Students can develop tables or charts to project a comparison of dog years to human years. Further, they can develop a graphical representation of the relationship between the measures.

Considering the notion that 1 dog year equals 7 human years, students might develop a spreadsheet that can project the ages of dogs in human years at different dog years or even at a dog's age of 6 months.

CHART COMPARISON

DOG YEARS	1	2	3	4	5	6...
HUMAN YEARS	7	14	21	28	35	42...

GRAPHICAL REPRESENTATION

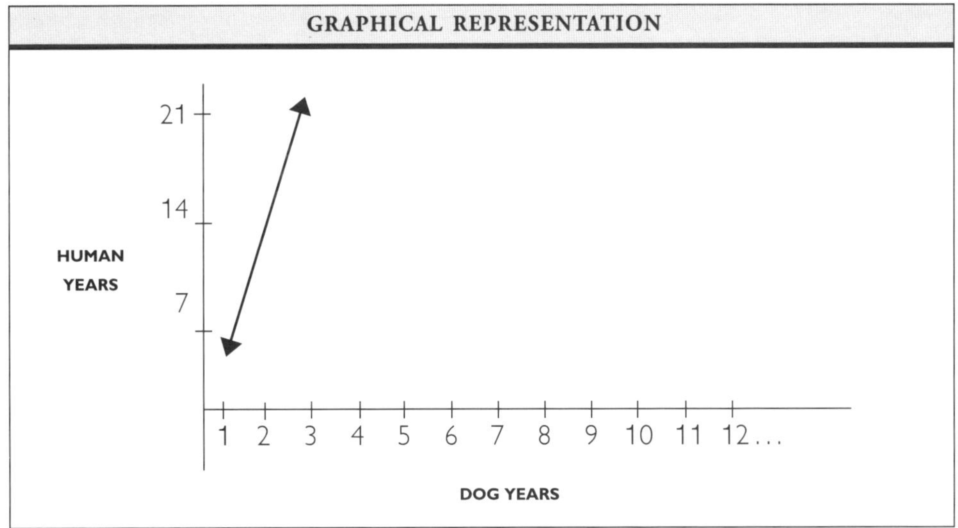

In the 2000 newsletter developed to promote the World's Largest Math Event, the National Council of Teachers of Mathematics suggests the following table of nonlinear data is a more accurate expression of the relationship between dog years and human years.

CHART COMPARISON												
DOG YEARS	3 mo	6 mo	1 yr	2 yr	4 yr	6 yr	8 yr	10 yr	14 yr	18 yr	20 yr	21 yr
HUMAN YEARS	5 yr	10 yr	15 yr	24 yr	32 yr	40 yr	48 yr	56 yr	72 yr	90 yr	94 yr	101 yr

Students can use graphing software to plot the points and try to determine a means of describing the nonlinear relationship. Further, using either the linear or nonlinear image of the relationship between dog years and human years, students can answer such questions as, "How old would you be in dog years?" or "How many dog years would it take for your dog to be considered a teenager in human years?"

Horses also offer interesting means of comparing measurements. First, a horse year is supposedly equivalent to 3 human years. So, much like comparing dog years to human years, students can compare horse years to human years and ultimately discover the linear relationship $y = 3x$, where x represents the number of horse years and y represents human years.

Another interesting measurement related to horses is in the measurement called a hand. Historically, people used different parts of the body as measuring instruments. The hand was the width of an adult hand and was used to measure the height of a horse from the ground by one of the front hooves to the withers (shoulder area). Thus, when describing a horse, one would say it was so many hands high. The notation used was hh. If a horse was more than a certain number hands—for example, 16 hands and 2 inches—the horse would be referred to as 16.2 hh.

Now, with the confusion of the width of different human hands, some agreement had to be reached to be able to universally know how big a hand was. In the horse trade, the measurement of a hand has come to be known as 4 inches. In fact, horse measuring sticks have been developed that are essentially wooden rulers marked off in hands. This measurement provides yet another opportunity for students to develop charts, graphs, or formulas to compare measures. They can also develop spreadsheets for converting horse measures to human measures.

Resource

Web Site

Equine World UK, About Horses, Measurement: www.equine-world.co.uk/about_horses/height.htm

INSIGHT

How Tall Will You Be in a Scale?

By Ivan W. Baugh

Oftentimes, when presenting a plan for a building or a tract of land, the designer will prepare a scale model to accompany the presentation. What is involved in creating a scale model?

First, you must complete your design to meet the specifications expressed by your client. Next, you prepare a layout of the area where the design will be built. Then you plan a scale model of the completed structure. Scale models help the client visualize how the completed project will look. Computer-aided drafting (CAD) software facilitates drawing plans to scale. When the client accepts the plan, the same program can generate the blueprints from which to build the structure.

Landscape architects use specialized CAD software to plan plantings around houses. Interior decorators also have specialized CAD software that enables them to draw a room to scale, draw the furniture to scale, and then virtually prepare various arrangements before actually moving a piece of furniture.

Specialized drafting programs such as 3rd PlanIt or CadRail enable model railroaders to draw a layout to scale for the room in which they will build the model railroad. Some of this specialized software will also create three-dimensional representations and allow the user to operate virtual trains on the layout. This helps the modeler avoid operational problems on the completed layout.

Another use of scale modeling occurs in hobbies. Boy Scouts may earn a merit badge for railroading. One of the activities they may complete involves scale modeling. While working to prepare materials to place online to help Scouts complete their railroading merit badge, I developed the following activity to help them grasp the concepts of scale.

If the user is not an even number of feet in height, he enters the feet in one cell, and in a second cell he enters the inches. A formula converts the feet to inches and adds the two numbers to express his height in total inches. Another formula takes his height and divides it by the scale ratio to get his height in each scale. The Scout will then use a sheet of paper, a ruler, and pen or pencil to draw a line showing his height in each scale. He will label each line with the scale name and create marks on the Y axis to show the height. Your students may complete this activity using the file Scale-Converter on the companion Web site to this book (http://education.bellarmine.edu/baugh/).

Students often build dioramas or draw floor plans of their rooms at home. With measurements in hand, they can use a spreadsheet to determine the ratio of actual size to the scale size they want to use to create their diorama or draw the floor plan. By using a spreadsheet, they can determine the largest scale they can use to fit the sheet of paper on which they will draw. Then they will draw the floor plan. On the paper they will indicate the scale they used to complete the floor plan.

Invite an architect to speak to your class. Ask the architect to bring some drawings to show the class what a person uses to build a structure. The architect may bring a scale model he or she has created. If a scale model is brought, it would be helpful if the plans that produced the scale model could be included.

Model railroaders often host open houses or have displays at malls. If their schedules permit, they would welcome the opportunity to visit your class to share how they use scale in modeling their miniature of the real world. Some modelers may have a module or a portable layout they can bring to show the class.

You can also visit Web sites that feature layout tours (www.midcentral-region-nmra.org) as well as scale drawings of layouts. Your students can see how the modeler has created a miniature of the world in the scale of choice.

Immediately, I can hear some of you saying, "Girls don't want to do that!" Did you know that we have women who are model railroaders? You may have one in your area. I know a husband and wife who work as a team on a model railroad layout. The wife creates the buildings that appear on the layout in the chosen scale. She has created scale models of houses that were important in her life—where she lived as a child, houses of relatives, etc. They appear on the layout in a city scene. At shows, her models attract considerable attention because of the quality of her models and the special significance they have for her.

I know another husband and wife who work together on model railroading. In fact, that is how they met. They often present model railroading clinics at regional and national conventions to help other modelers improve or acquire new skills.

Girls could make a scale model of their home to use as a dollhouse. I can remember my younger sister as a child using cardboard boxes to make rooms and dollhouses. She decorated the rooms and houses, created simple scale-appropriate furniture, and played for hours in her miniature world.

Lead your students to investigate how scale is used in cartography. This could become another way to help them grasp the use of scale. They could draw a scale model of the school grounds. Students could write a description of the school grounds and exchange it with classes at other area schools. As a culminating activity, the class could go on a mystery field trip. They would not know to which school they were going. The challenge for them would be to recognize which school they are visiting as they travel to the destination.

Scale drawing plays an important role in daily life. Use these ideas as springboards to create other meaningful activities to help your students grasp the value of scale drawing. After using the scale-converter.xls spreadsheet on the companion Web site, adapt it to the scale you want to use in the model you plan to build.

Measurement & Precision

"People are integrating math education. Other people are integrating science education. Alas, the math integrators and the science integrators aren't talking to each other and creating courses in which math and science work together to help students survive and thrive in an integrated, intertwingled world."— Laran Stardrake

By Bob Albrecht

Well, here we are again, beginning our third year of Power Tools for Math and Science. This year, Power Tools for Math and Science will describe our progress as we help develop a new course and two curriculum construction sets.

Integrated Math, Science, and Technology is a new course at The Center for Technology, Environment, and Communication (C-TEC), which is a highly restructured learning community at Piner High School in Santa Rosa, California. For information about C-TEC, see this column in TCT Feb. '93, Mar '93, Apr. '93, and May '94. Sixty incoming freshmen and sophomores will take this course instead of the existing Integrated Science course. This new course will also replace algebra or geometry for these students. This will be a learn-by-doing course, designed around hands-on activities, experiments, investigations, and projects. It will be team taught with very flexible scheduling. Students will do much writing, but also report their work in new media ways, such as video and computer animation. We'll expect students to quickly become proficient in power tools used routinely by all C-TEC students, such as Microsoft Word and Excel. Students can elect to learn how to use other power tools, such as QBasic, Visual Basic, AutoCAD, and animation software.

Curriculum Construction Sets contain lots of parts, plus lots of information on how you can use the parts to build your own customized curriculum. You can browse a Curriculum Construction Set for things that you can do in one or two class periods in a traditional math or science course. You can also find more time-consuming activities, investigations, and projects. You can use a Curriculum Construction Set to help build your own integrated science course, or integrated math course, or integrated math/science course. We've been developing and testing parts for many years, and will continue indefinitely. Ultimately, we'll publish Curriculum Construction Sets in CD-ROM, video, networks, and perhaps other media— anything but paper! In the meantime, we'll share some stuff here in TCT and figure out other ways to distribute units during the 1994–1995 school year.

By the time you read this, we expect to have a C-TEC bulletin board set up to communicate information on 1) the Integrated Math, Science, and Technology course, and 2) Curriculum Construction Sets. For more information,

	A	B	C	D
1	Integrated Math, Science, and Technology Course, 1994-1995.			
2	C-TEC, 1700 Fulton Road, Santa Rosa, CA 95403			
3				
4	May 15, 1994. Filename: METERSTK.XLS			
5				
6	Measure length and width of a table using four meter sticks. Enter your estimates.			
7	Also enter lower bound = one estimation unit less than your estimate and			
8	upper bound = one estimation unit greater than your estimate.			
9	Calculate areas using your estimate, lower bound, and upper bound.			
10	Calculate area differences: your estimate - lower bound & your estimate - upper bound.			
11				
12		Lower	Your	Upper
13		bound	estimate	bound
14	Uncalibrated meter stick			
15	Estimate to 0.1 meter			
16	Length (L)	1.4	1.5	1.6
17	Width (W)	0.6	0.7	0.8
18	Calculate area. A = L * W	0.84	1.05	1.28
19	Area: your estimate - lower bound		0.21	
20	Area: your estimate - upper bound		-0.23	
21				
22	Decimeter-calibrated meter stick			
23	Estimate to 0.01 meter (0.1 decimeter).			
24	Length (L)	1.51	1.52	1.53
25	Width (W)	0.75	0.76	0.77
26	Calculate area. A = L * W	1.1325	1.1552	1.1781
27	Area: your estimate - lower bound		0.0227	
28	Area: your estimate - upper bound		-0.0229	
29				
30	Centimeter-calibrated meter stick			
31	Estimate to 0.001 meter (0.1 centimeter).			
32	Length (L)	1.515	1.516	1.517
33	Width (W)	0.756	0.757	0.758
34	Calculate area. A = L * W	1.14534	1.147612	1.149886
35	Area: your estimate - lower bound		0.002272	
36	Area: your estimate - upper bound		-0.002274	
37				
38	Millimeter-calibrated meter stick			
39	Read to 0.001 meter (1 mm).			
40	Length (L)		1.517	
41	Width (W)		0.758	
42	Calculate area. A = L * W		1.149886	

Figure 1. Four meter sticks: data and area calculations.

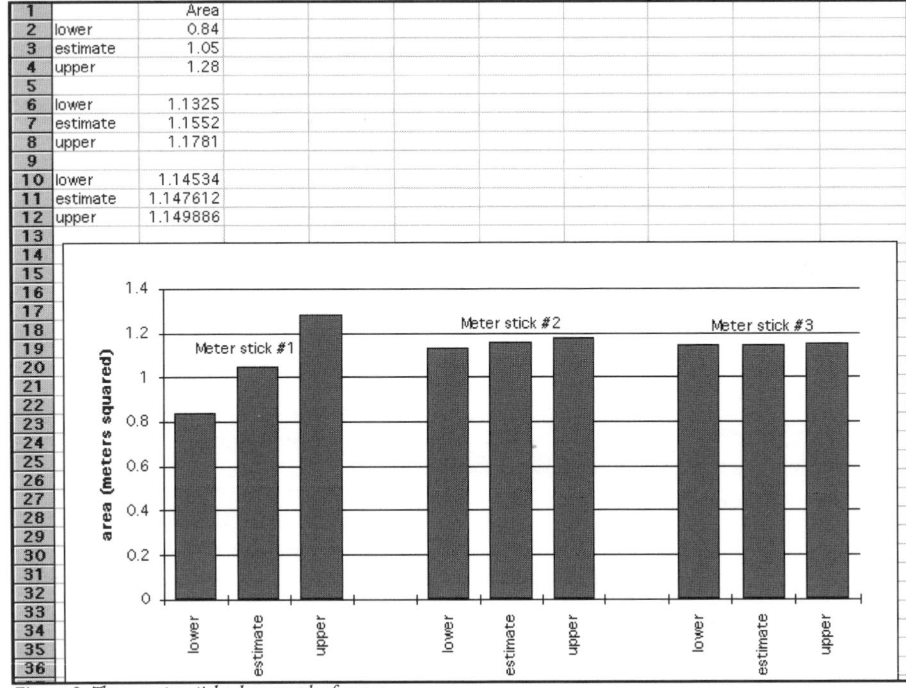

		Area
2	lower	0.84
3	estimate	1.05
4	upper	1.28
5		
6	lower	1.1325
7	estimate	1.1552
8	upper	1.1781
9		
10	lower	1.14534
11	estimate	1.147612
12	upper	1.149886

Figure 2. Three meter sticks: bar graph of areas.

send a self-addressed, stamped envelope to James Gonzalez, Chair, C-TEC, 1700 Fulton Road, Santa Rosa, CA 95403.

A Significant Investigation: Four Meter Sticks

Measurement will be a major strand in the new integrated course. This includes preicision of measuring instruments, measurement errors, propagation of errors in calculations based on measured values, estimates of experimental error, significant digits, units, and dimensional analysis. There will be many activities in which students make the same measurements using two or more measuring instruments of different precisions. Here is an example of this type of activity.

A standard meter stick, hopefully found in all math and science classrooms, is usually calibrated in centimeters (cm), and millimeters (mm). Our standard stick has big numbers that label tens of centimeters (10, 20, 30,...), smaller numbers that label in-between centimeters (21, 22, 23,...), and unlabelled ticks for millimeters. A well-made meter stick is a precision instrument that you can use to measure length to an accuracy of 1 millimeter (mm) = 0.001 meter (m).

Aha! A splendid opportunity. Let's make less precise meter sticks, then use sticks of different precisions to measure the same objects, make calculations based on these measurements, and compare the results. We decided to use sticks in four precisions.

- Meter stick #1 is uncalibrated. It is one meter long with no markings. You can use it to measure multiples of one meter and estimate length to 0.1 meter (1 decimeter).
- Meter stick #2 is calibrated in decimeters. There are 10 calibration marks, 1 at each decimeter. You can use this stick to measure multiples of 1 decimeter and estimate length to 0.01 meter (1 centimeter).

- Meter stick #3 is calibrated in centimeters. There are 100 calibration marks, one at each centimeter. You can use this stick to measure multiples of 1 centimeter and estimate length to 0.001 meter (1 millimeter).
- Meter stick #4 is a standard meter stick calibrated in millimeters. You can use it to measure to 1 millimeter (0.001 meter) with an error of, say, 0.5 mm. We'll use this stick as a standard for comparing measurements made with the less precise sticks.

Greg Ellisen runs the woodshop at Piner High School. He quickly made 30 bare sticks in beautiful hardwood. Students carefully calibrated the sticks. We are ready to run meter stick investigations early Fall '94.

In Paul Davis's classroom, students work in teams at eight tables. Each team will use each meter stick to measure the length and width of their table in meters (m). They will then use Excel to calculate the area of the tables in meters squared (m^2) for each set of measurements. They will also enter a lower bound and upper bound for each measurement and calculate lower bound and upper bound areas. It might go like this:

Meter stick #1 is uncalibrated. We use it to estimate the length and width to 0.1 meter. We get 1.5 m for the length and 0.7 m for the width. We are quite sure that the length is between 1.4 m and 1.6 m, so we assign 1.4 m as the lower bound and 1.6 m as the upper bound of our length estimate. We are quite sure that the width is between 0.6 m and 0.8 m, so we assign 0.6 m as the lower bound and 0.8 m as the upper bound of our width estimate.

We use meter stick #2 to make the same measurements and get 1.52 m for the estimated length and 0.76 m for the estimated width. This gives us lower and upper bounds as shown here:

Length: lower bound = 1.51 m, upper bound = 1.53 m
Width: lower bound = 0.75 m, upper bound = 0.77 m

We use meter stick #3 to get estimates, lower bounds, and upper bounds for the length and width. Then we use meter stick #4 (the standard stick) to measure the length and width to the nearest millimeter. We'll use these as our almost exact measurements for comparisons with measurements made with the less precise instruments.

Figure 1 is an Excel worksheet showing all the measurements, appropriate lower and upper bounds, and calculations based on the data. Our Excel template is protected, except for the cells in which students enter data and make calculations. Students can change information only in cell ranges B16: D18, C19: C20, B24: D26, C27: C28, B32: D34, C35: C36, and C40: C42.

Let's look at the data for meter stick #1. Here are the area calculations:

Area from lower bound data = 0.84 m^2
Area from measurements = 1.05 m^2
Area from upper bound data = 1.28 m^2

We are quite sure that the area is in the range 0.84 m^2 to 1.28 m^2. Using the measurements gives 1.05 m^2, but we really can't claim that all three digits are significant! How should we report this value in a scientific journal? How should a student report it in the required investigation report? After students gain experience with several investigations of this type, we can home in on the mysterious realms of measurement errors and significant digits, essential in real-world science. Also essential in college science courses. We hope that the student will conclude that the area calculated from meter stick #1 measurements should be reported as 1 m^2.

The Excel worksheet also calculates the difference between the area calculated from the measured (estimated) values and the areas calculated from the lower and upper bounds. You might want to have students calculate these values as percents of the estimated area, about 20% for meter stick #1.

We encourage you to replicate the above analysis for the data and calculations using meter sticks #2 and #3. What values should be reported for the calculated areas? Also compare results using meter sticks #1 to #3 with the data and calculations using meter stick #4.

Whenever possible, we like to present data visually. So we quickly transferred the calculated areas to a new Excel worksheet and made the chart shown in Figure 2—about five minutes work and, we think, well worth the effort. We encourage you to make visual comparisons of the areas using sticks #1 through #3 with the area using stick #4.

This summer we'll document this investigation more completely, and develop similar activities. For example, 2-meter tape measures and 30-meter or 50-meter measuring ropes with various calibrations. For the ropes, nylon won't do—it stretches. We'll look for non-stretchy rope that isn't too expensive.

For measuring small objects, we'll make some 30-centimeter sticks, uncalibrated, decimeter-calibrated, and centimeter-calibrated. We think that millimeter-calibrated 30-centimeter sticks are available to use as standards. Help build your own Curriculum Construction Set ideas!

Eating Right? Fat Chance!

Teaching Math and Nutrition With Spreadsheets

Editor's Note: The National Council of Teachers of Mathematics Curriculum Standards call for enhancing the mathematics curriculum by increasing attention to connecting mathematics to other subjects, to the world outside the classroom, and, in particular, to the students' world. The attention to this area is considered so important that one standard is labeled Mathematical Connections. Perhaps you have noticed that students in middle school and high school have little knowledge about nutrition. They eat food that gives them energy. They eat lots of fast food. Many of them worry about eating—some to bulk up, and others to lose weight. The concern is whether they are eating right. And, of course, literally no one can tell them what is good for them to eat. The activity described in this article uses spreadsheet technology to connect the mathematics students are learning to their own eating habits. Are they eating right? Encourage them to reflect on the results!

By Brenda Justice

One of my focus areas this past year in mathematics was to teach students to use mathematical thinking and computers to solve problems in other curricular areas. To accomplish this goal, I designed a lesson titled Eating Right? Fat Chance! that combines the study of nutrition, weights, and measurements. In this lesson, students use basic math skills to generate formulas in a computer spreadsheet, as well as their understanding of weights and measurements to determine the number of servings in a portion of food. After they enter this information into a computer, the spreadsheet calculates the number of servings in each food group, the total grams of fat consumed, the total caloric intake, and the percentage of fat eaten by students in a typical day.

Teaching nutrition in this way enables students to concentrate on the analytical process and nutritional concepts rather than the mathematical computations. Students use the knowledge acquired to design their own balanced menu that supplies the correct number of servings from each food group and meets the caloric and fat requirements for an average student of their own age and activity level.

The lesson enhances the study of nutrition, reinforces mathematical concepts, encourages communication, and teaches students new computer skills. Students will want to

The Food Pyramid

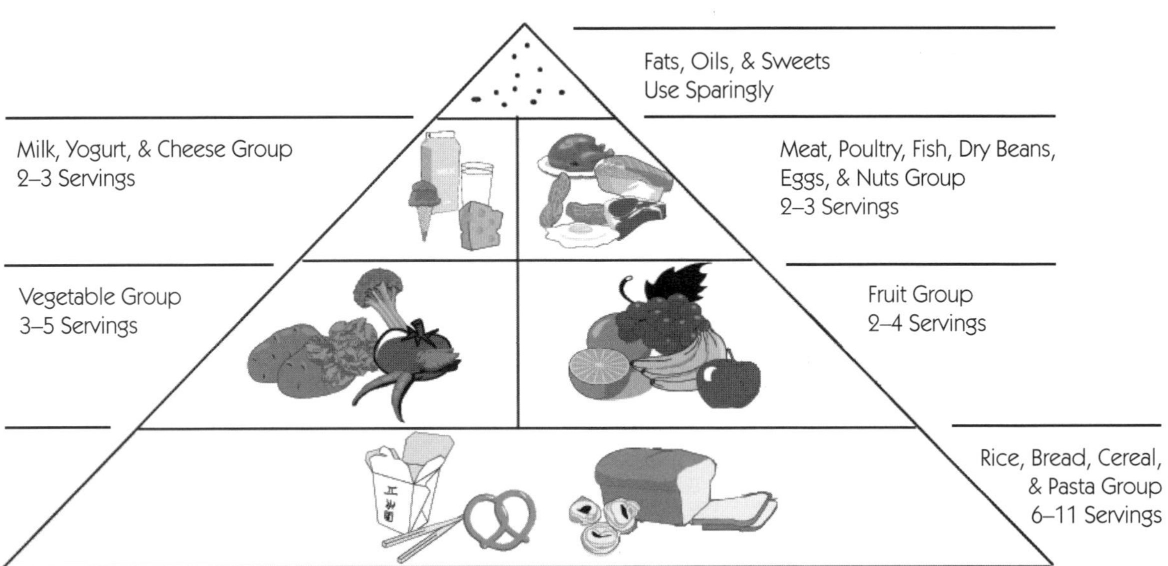

What's in a Serving?

Rice, Bread, Cereal, & Pasta Group

1 slice of enriched or whole grain bread

125 milliliters (1/2 cup) of cooked cereal, rice, pasta, or grits

250 milliliters (1 cup) of ready-to-eat cereal

1 enriched or whole grain muffin, biscuit, or roll

Vegetable Group or Fruit Group

1 medium fruit

a bowl of tossed salad

a lettuce wedge

1 medium potato

125 milliliters (1/2 cup) of sliced fruit or vegetables

125 milliliters (1/2 cup) of juice

1/2 medium grapefruit

Milk, Yogurt, & Cheese Group

250 milliliters (1 cup) milk or yogurt

45 grams (1 1/2 slices) of cheddar cheese

500 milliliters (2 cups) of cottage cheese

250 milliliters (1 cup) of pudding

375 milliliters (1 1/2 cups) of ice cream

Meat, Poultry, Fish, Dry Beans, Eggs, & Nuts Group

1 oz meat or chicken or 2 eggs

60 milliliters (1/4 cup) of peanut butter

125 milliliters (1/2 cup) of nuts

250 milliliters (1 cup) of cooked dry legumes

Fats, Oils, & Sweets

Use very small amounts, like 1 tablespoon, etc.

Sweets: candy, jam, soft drinks, syrup

Fats: butter, margarine, salad dressing

look beyond the assignment to discover ways a spreadsheet can be used in other subject areas.

Teachers can use Eating Right? Fat Chance! in their eighth- and ninth-grade health classes and in their sixth- and seventh-grade math classes. The lesson takes approximately four to five 45-minute class periods, depending on the students' math and computer skills and their prior experience with keyboarding. This includes class time as well as computer lab time.

Students in my classes respond to this lesson with energy and excitement. They leave class not only understanding what it means to eat right but also with a skill they can use in real life! I encourage you to try this activity with your students, and help them learn how to start eating right.

Objectives

1. Teach students to identify the food groups, number of servings, and serving sizes as outlined in the Food Pyramid.
2. Help students reinforce their knowledge of weights and measurements to determine serving sizes.
3. Teach students to create a simple spreadsheet (see Figure 1) on the computer and use it to analyze the fat content in their diet.
4. Help students understand the mathematical power of a spreadsheet to manipulate information and produce a balanced menu.

Activity 1

Time: One 45-minute class period

Display a large poster of the Food Pyramid and distribute a smaller version so that students can refer to it. Discuss the Food Pyramid's structure. Explain that it is in the form of a pyramid to make it easier to remember to eat fewer of the foods in each group as we approach the top. Have students bring samples of the different foods to class. Let them determine the actual size of a single serving of each food by using measuring cups and food weights. Have them determine how much cereal, pasta, rice, bread, milk, juice, fruit, vegetables, cheese, margarine, peanut butter, chicken, and meat is in a recommended single serving. Ask them to dish out the amount of food they normally eat and calculate the number of single servings it represents. (They will probably be surprised at how many servings they actually eat.) The lesson should provide students with background information on the different foods and food groups, the number of suggested servings per group, and the appropriate serving size for typical foods in each group. The measuring cups and food weights help students learn to visualize the amount of food in a single serving. (Contrary to popular belief, a plate of spaghetti is not one serving!)

Fat Analysis Template1

	A	B	C	D	E	F
1						
2	First and Last Name					
3	**Food Group**	**#Gms Fat/srv**	**#Cal/srv**	**# Servings**	**Total Fat**	**Total Calories**
4	Bread/Rice Pasta					
5					=B5*D5	=C5*D5
6					=B6*D6	=C6*D6
7					=B7*D7	=C7*D7
8					=B8*D8	=C8*D8
9					=B9*D9	=C9*D9
10	Total for Bread				=SUM(E5:E9)	=SUM(F5:F9)
11	Fruits					
12					=B12*D12	=C12*D12
13					=B13*D13	=C13*D13
14					=B14*D14	=C14*D14
15					=B15*D15	=C15*D15
16					=B16*D16	=C16*D16
17	Total for Fruit Group				=SUM(E12:E16)	=SUM(F12:F16)
18	Vegetable					
19					=B19*D19	=C19*D19
20					=B20*D20	=C20*D20
21					=B21*D21	=C21*D21
22					=B22*D22	=C22*D22
23					=B23*D23	=C23*D23
24	Total for Vegetable Group				=SUM(E19:E23)	=SUM(F19:F23)
25	Proteins					
26					=B26*D26	=C26*D26
27					=B27*D27	=C27*D27
28					=B28*D28	=C28*D28
29					=B29*D29	=C29*D29
30					=B30*D30	=C30*D30
31	Total for Protein Group				=SUM(E26:E30)	=SUM(F26:F30)
32	Milk/Yogurt/Cheese					
33					=B33*D33	=C33*D33
34					=B34*D34	=C34*D34
35					=B35*D35	=C35*D35
36					=B36*D36	=C36*D36
37					=B37*D37	=C37*D37
38	Total for Milk Group				=SUM(E33:E37)	=SUM(F33:F37)
39	Fats					
40					=B40*D40	=C40*D40
41					=B41*D41	=C41*D41
42					=B42*D42	=C42*D42
43					=B43*D43	=C43*D43
44					=B44*D44	=C44*D44
45	Total for Fat Group				=SUM(E40:E44)	=SUM(F40:F44)
46	Miscellaneous Group					
47					=B47*D47	=C47*D47
48					=B48*D48	=C48*D48
49					=B49*D49	=C49*D49
50					=B50*D50	=C50*D50
51					=B51*D51	=C51*D51
52	Total for Miscellaneous Group				=SUM(E47:E51)	=SUM(F47:F51)
53	Totals for the Menu				=E10+E17+E24+E31+E38+E45+E52	=F10+F17+F24+F31+F38+F45+F52
54	Percentage Fat in Meal	=E53/F53*900				

Fat Analysis Template

Figure 1. Fat analysis spreedsheet template.

Activity 2

Time: Homework done outside of class

Ask students to pick a typical day and record all food, including snacks, they consume during a 24-hour period. They can use a worksheet to record this information. (See the Typical Daily Menu worksheet shown in Figure 2.) Their listings should be specific and include the number of servings for each food. Remind them to record condiments they add to their food, including mayonnaise, sugar, ketchup, and butter.

Activity 3

Time: One or two 45-minute class periods, depending on the students' computer skills

Have students use a computer spreadsheet program to create a Fat Analysis Template similar to the one shown in Figure 1. (The spreadsheet shown was created with Microsoft Excel; however, any spreadsheet program can be used.) Have students save this spreadsheet as a template with the formulas and labels already listed. They will use it in the next activity to enter food data from their Typical Daily Menu worksheets.

Activity 4

Time: One 45-minute class period

Hand out the instruction sheet shown in Figure 3 for entering data from the Typical Daily Menu worksheet. Ask students to type in the information on each food and the number of servings they recorded in their Typical Daily Menu under the appropriate categories. They can use reference sources to determine the grams of fat and calories per serving. Weight measurements should not be entered in the cell for the number of servings; instead, the number of servings should be entered. For example, a half-cup of a 1-cup serving should be entered as .5.

Activity 5

Time: Two 45-minute class periods

Ask students to evaluate and discuss their results. They should analyze the number of servings consumed for each food group as well as the number of calories and the percentage of fat in their daily diets. After they have discussed the results, ask them to design a balanced menu and enter the information into the Fat Analysis Template. Their menu should include at least the minimum recommended number of servings from each food group and satisfy the caloric requirements for a student of their own age and activity level. The total fat content should not exceed 30% of the total caloric intake.

Typical Daily Menu

Keep an accurate record of all foods and the amount of each that you consumed within a 24-hour period. Be sure to include all snacks and drinks.

Meal	Food	Number of Servings
Breakfast		
Lunch		
Dinner		
Snacks		

Figure 2. Typical daily menu worksheet.

Grading

Base the grade for each student on the satisfactory completion of the following components:

- An accurate spreadsheet template
- A typical daily menu
- A balanced menu
- Participation and effort shown during the computer labs and discussion

Instructions for Entering Foods From the Typical Daily Menu Into Your Fat Analysis Template

1. Open *Excel.*
2. Go to **File.** Open the Fat Analysis Template.
3. Notice that the title is Fat Analysis Template1.
4. Look at your menu sheet and type each food eaten into its correct food group. You will need to break down some foods to place them correctly. For example, a sandwich may be included in the bread, meat, vegetable, cheese, and fat groups.
5. Enter the number of grams of fat per serving in column B, the number of calories per serving in column C, and the number of servings in column D. Enter only numbers in those columns; do not include any measurement units.

   (Do not type anything in columns E and F. The spreadsheet will automatically calculate those totals for you. Watch the totals change as you enter the numbers. That is the beauty of a spreadsheet! You can change any number and it will automatically recalculate the totals. Try it! Enter all foods consumed.
6. Choose **Save As** from the **File** menu.
7. Change the file name to Typical Daily Menu.
8. Select **Print** from the **File** menu.
9. Close the file or quit *Excel.*

To enter foods from the Balanced Menu into your Fat Analysis Template, repeat the above steps, except this time save the file as Balanced Menu.

Figure 3. Instruction sheet.

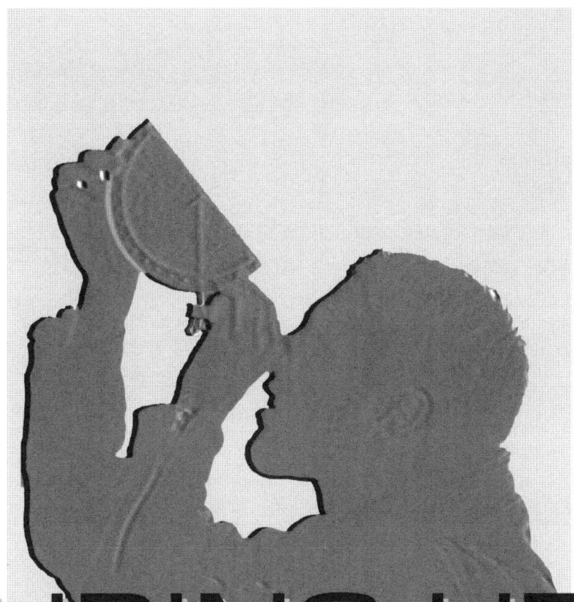

MEASURING HEIGHTS

OR WHAT TRIGONOMETRY TABLES ARE ALL ABOUT

By Ihor Charischak

What comes to mind when you hear the word "trigonometry"? Unless you are currently teaching the subject or are well versed in mathematics, you probably don't remember very much about it. All that may remain are vague recollections of sines, cosines, and trig tables. You probably don't remember what these tables are all about and why they are important.

In their effort to be mathematically precise, mathematicians use symbols in ways that do not necessarily facilitate understanding. Take Einstein's famous formula $E = mc^2$. You may know what the letters mean and how to get a value for E, but do you really understand this formula and how it is derived? Like most of us, students can "pretend" to know mathematics by applying formulas to their homework and exam problems. These students miss an opportunity to genuinely understand the meaning behind the symbols and to experience the power of real problem solving. This article describes an activity that sheds a little conceptual light on the teaching of a small segment of trigonometry—a branch of mathematics that is often used but not fully understood.

Learning Without Meaning

Here is a scenario of a typical trigonometry lesson.

Students were asked to measure the height of a pole that was not accessible to direct measurement. They began by making a device known as a clinometer, which is really a protractor with a piece of weighted string attached to it. Their specific instructions for the activity were as follows:

1. Mark off a distance of 10 meters from the pole. This distance is called the baseline.
2. Get an angle-reading with the clinometer. Subtract this reading from 90. This is your angular distance.
3. From a table of tangents (see Table 1 at the end of the article), find the number that corresponds to your angle.
4. Multiply this number (the tangent of your angle) by the baseline. (Example: If your angle of elevation is 67°, the tangent is 2.35. The baseline is 10. Therefore, the height of the pole is 2.35 x 10, or 23.5 meters.)

The students successfully plugged their values into the formula, got the right answer, and then moved on to the next problem. No one bothered to ask why the product of the tangent of the angular distance and the baseline results in an estimate for the height of the pole!

So what is this tangent? Does it have anything to do with what students already know? The secret lies in knowing what the tangent tables are all about. It's time to add some meaning to the activity.

Pole Measuring Revisited

Here's a different scenario for another trigonometry activity.

In a science class a group of students were building rockets they would launch later in the week. They wanted to be able to determine which rocket went highest. The teacher suggested that they use a clinometer.

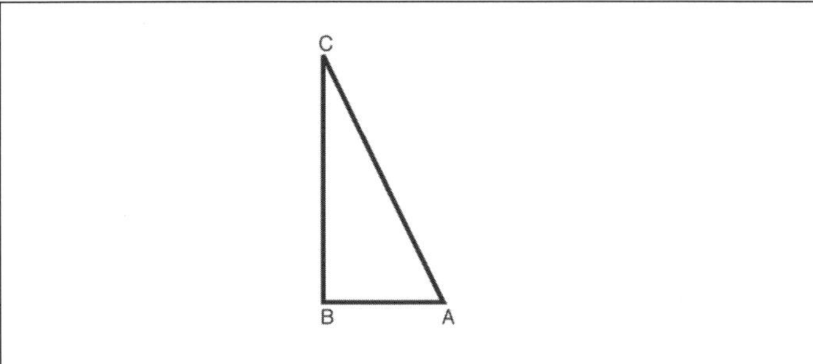

Figure 1. Students' scale drawing.

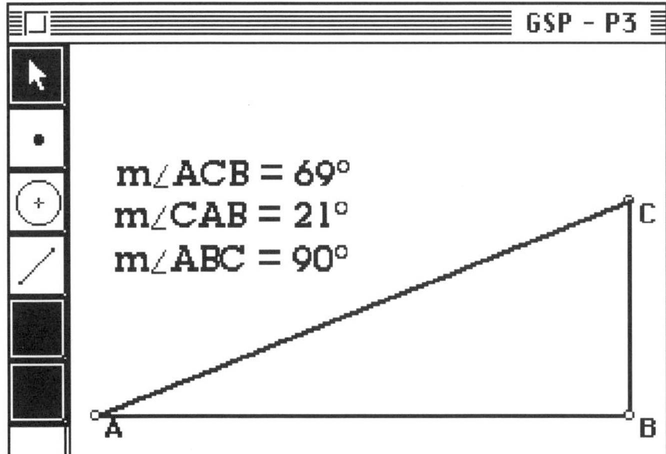

Figure 2. Dynamic right triangle.

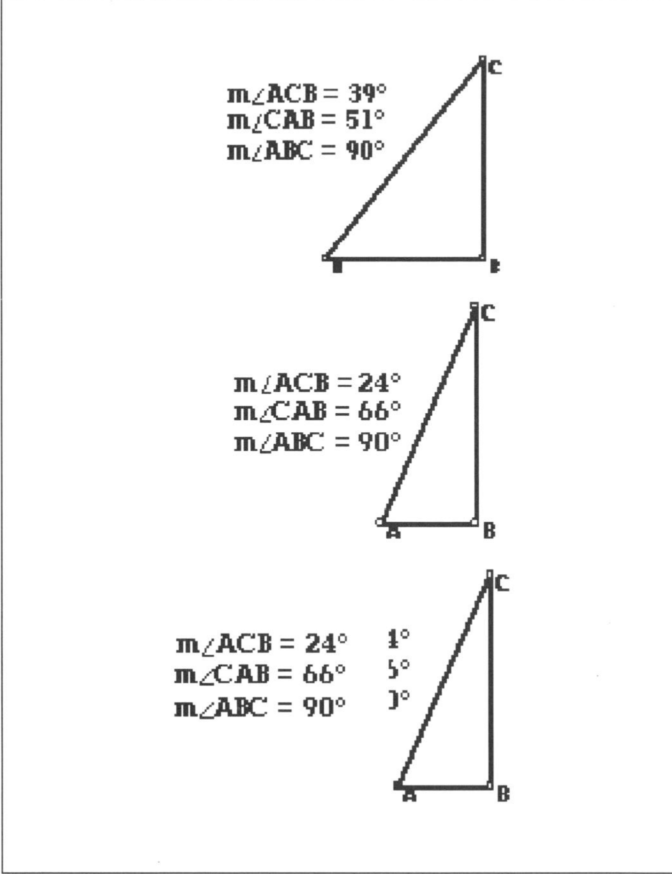

Figure 3. Manipulating triangles.

They first learned how to use this instrument by measuring the height of a telephone pole across the street. The teacher demonstrated the use of the clinometer by taking the group outside, measuring off 10 meters from the pole, and using the clinometer to measure the angle of elevation. Working in pairs, with one student measuring and the other reading, the students took two readings, recorded their results, and came back to class the next day with the measurements.

The students agreed that 21° was a reasonable clinometer reading. After some discussion they decided to make a scale drawing of the scene (see Figure 1). A few minutes later, one of the students realized that 21° could not be the angle of elevation because it didn't look right in relation to their scale drawing. Then they realized that the clinometer reading gave them the measure of angle ACB, not CAB! But how could they find angle CAB?

Noting that her students were stumped, the teacher directed their attention to a "dynamic" right triangle she had created using The Geometer's Sketchpad (see Figure 2). What made the triangle dynamic was that the teacher could change the triangle by dragging a vertex to a different position, which changed the measure of the acute angles.

She first established that this triangle was indeed a right triangle by noting that angle ABC was 90°. She challenged the students to see if they could "mess up" the triangle's "rightness" by moving the vertices to different positions. After some effort,

Figure 4. The Geometer's Sketchpad scene.

the students realized they could change the lengths of the sides and the measures of angles BAC and ACB, but not angle ABC, which in all cases remained 90°. Next she made angle BCA equal 21°. The students noticed that CAB became 69°. Finally, the triangle looked the way it was supposed to. The teacher focused their attention on how to figure out angle CAB if they knew the measurement for angle ACB, which is the angle the clinometer measured. By dragging ACB and watching the measures of ACB and CAB change, they realized that the sum always remained 90° (see Figure 3). Thus, after experimenting with the triangles, the students had a major insight: To determine the angle of elevation, you take the clinometer reading and subtract it from 90°.

Returning to their scale drawing, they were able to come up with the unknown height. They solved the problem, but were not happy. They did not relish the thought of having to make scale drawings for every rocket that went up. They wanted to devise a better way.

The teacher then showed them a digital image of the pole they had measured. The image was taken by a QuickTake 150 camera and copied as a PICT file into The Geometer's Sketchpad (see Figure 4).

The students discussed the relationship of this image with its real counterpart. They realized that the digital snapshot was really a small-scale drawing of the real triangle and could be measured just like their own small-scale version by using The Geometer's Sketchpad's measuring tools. They calculated the following measurements for triangle ABC: baseline = 1.4", height = 3.6", angle of elevation = 69°.

Next, the teacher asked the students to determine how much bigger the real triangle was than its The Geometer's Sketchpad counterpart. She described how the smaller triangle could be stretched to match the real triangle. The students

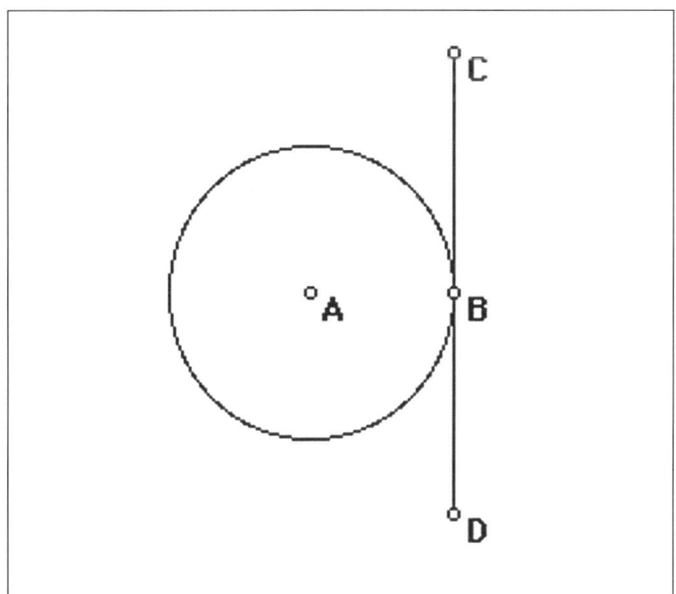

Figure 5. Illustration of a tangent.

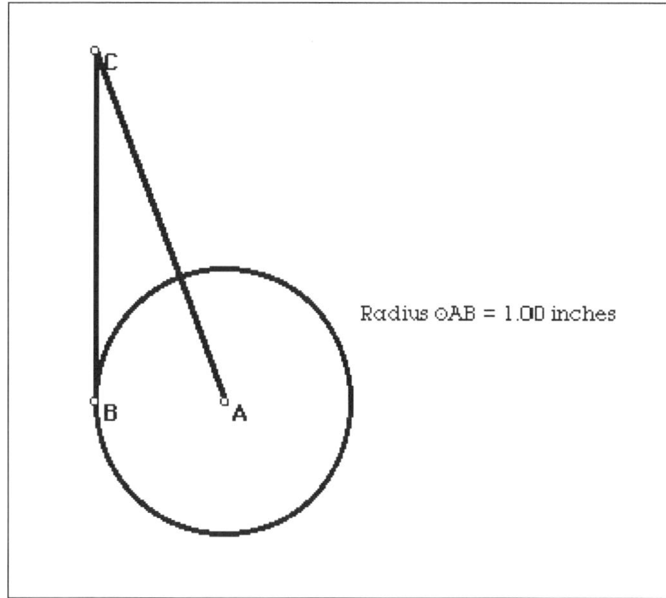

Figure 6. Tangent drawn using The Geometer's Sketchpad.

came to realize that whatever it took to stretch 1.4" (the base of the onscreen triangle) to equal 10 meters (the base of the real triangle) would also be needed to stretch the longer leg (3.6") of the small triangle to become the unknown height.

The teacher explained that this "stretch factor" was the number of times that the real baseline was longer than its respective leg on the smaller triangle. Thus, the students divided 10 by 1.4 to get the ratio, but they all agreed that this process was a pretty "ugly" way to arrive at an answer. Once again, they wanted to find an easier way.

As it turns out, there is an easier way. And here is where the trigonometry comes in.

The Tangent to the Rescue

The problem would be easier if the enlargement factor was a whole number and if there was a quick way to get the measurement of the side corresponding to the pole for any angle of elevation. In other words, we need a "library" of measurements for every possible angle and an enlargement factor that is easy to determine and apply. The teacher told the students that a library of tangents would help them. The teacher began by defining a tangent: A tangent is a line that touches a circle in one and only one place (see Figure 5).

The students were then shown how to draw a tangent using The Geometer's Sketchpad. The result is shown in Figure 6. Here are the steps they used.

1. Draw a unit circle. (The diameter is 1 unit.)
2. Draw a radius AB.
3. Highlight line AB and point B.
3. From the Construct menu, choose Perpendicular Line.
4. Place a point C on the line.

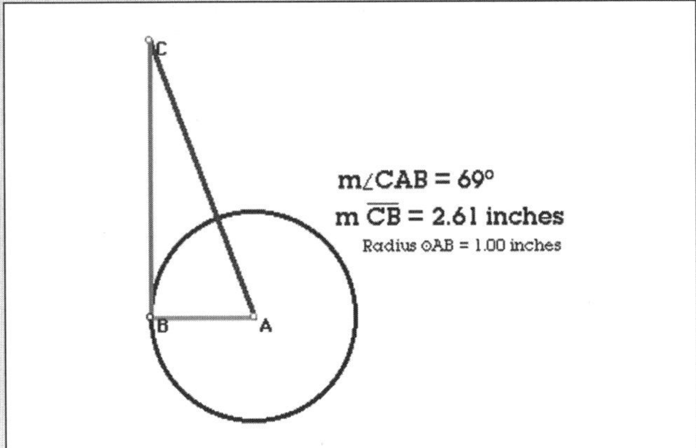

Figure 7. Completed calculations for student problem.

5. Hide the line.
6. Draw line segments AC and BC.
7. Measure angle BAC and segment BC.

It turns out that the measure of BC is the tangent of angle CAB. But how does this help us? First, this triangle is similar to the one we are trying to solve in our problem. Second, the radius BA equals 1, so the length of the baseline is the stretch factor. Third, we can find any angle measure we need by dragging vertex ACB up or down until it becomes the desired angle. Then the measure of BC becomes the length multiplied by the stretch factor, which gives us the length of the missing side.

Now the students could compare the measurements this triangle generates with the values in the tangent table (see the table on the right). The table of tangents can be thought of as a library of triangles with the measurement of the tangent line for the given angle. (But what happens when angle BAC gets closer to 90 degrees? That is the reason it is undefined.)

Notice that the enlargement factor is now easy to determine because the side of the small triangle corresponds to the baseline and is always 1. This means that the baseline is always equal to the enlargement factor! After the students got their tangent length to correspond to the angle of elevation, all they needed to do was multiply the tangent times the baseline length, and they had their answer. Figure 7 shows their completed calculations. And now, as Paul Harvey likes to say, you know the rest of the story.

Conclusion

Some teachers might question my decision to define a tangent of an angle as a measurement rather than as a ratio, which is how it is usually defined. (The tangent of angle A is defined as the opposite side divided by the adjacent side.)

A "Library of Tangents" Table

Angle A	Tangent of A	Angle A	Tangent of A	Angle A	Tangent of A	Angle A	Tangent of A
1	0.017	26	0.487	51	1.234	76	3.999
2	0.035	27	0.509	52	1.279	77	4.318
3	0.052	28	0.531	53	1.326	78	4.689
4	0.070	29	0.554	54	1.375	79	5.125
5	0.087	30	0.577	55	1.427	80	5.648
6	0.105	31	0.600	56	1.481	81	6.285
7	0.123	32	0.624	57	1.538	82	7.078
8	0.140	33	0.649	58	1.599	83	8.095
9	0.158	34	0.674	59	1.662	84	9.447
10	0.176	35	0.700	60	1.730	85	11.332
11	0.194	36	0.726	61	1.802	86	14.146
12	0.212	37	0.753	62	1.878	87	18.804
13	0.231	38	0.781	63	1.960	88	28.011
14	0.249	39	0.809	64	2.047	89	54.816
15	0.268	40	0.838	65	2.141	90	#####
16	0.287	41	0.869	66	2.243		
17	0.306	42	0.900	67	2.352		
18	0.325	43	0.932	68	2.471		
19	0.344	44	0.965	69	2.600		
20	0.364	45	0.999	70	2.742		
21	0.384	46	1.035	71	2.898		
22	0.404	47	1.071	72	3.071		
23	0.424	48	1.110	73	3.263		
24	0.445	49	1.149	74	3.479		
25	0.466	50	1.191	75	3.722		

This is a spreadsheet version of a table of tangents. The tangent function requires that the angles be in radian measure. What is needed is the length of the arc intercepted by a central angle of 45° in a unit circle. Because 45° is 1/8 of the complete circle and the circumference of the circle is $2\pi r$, the radian measure of 45° is $2\pi r/8$, or $\pi/4$, when $r = 1$.

Table 1.

But since the adjacent side is 1, the tangent of the angle equals the length of the side. I find that thinking about tangents (and sines and cosines) as measurements is an easier way for students to understand what is going on. Besides, the Greeks thought of sines, cosines, and tangents as lengths. The ratio notion didn't come into play until the 18th century!

Software

The Geometer's Sketchpad. Key Curriculum Press, 1150 65th St., Emeryville, CA 94608; 800/955-MATH; Fax 800/541-2442.

Author's note: For more information and updates about this article, visit www.ciese.org/mathprojects/articles/measure.html.

Measurement Long Ago and Today

By Bob Albrecht

Different measurement units have been used through the years. In this month's Power Tools column, the authors use ancient and modern units to measure length, time, mass, and weight. The activities they describe are fun for students at all levels, whether they are elementary students learning basic concepts of measurement or high school students studying Newtonian physics.

Two major systems of measurement prevail in the world today: the United States Customary System (USCS, formerly called the British system of units), used in the United States of America and in Burma, and the Système International (SI, known also as the international system and the metric system), used everywhere else. Each system has its own standards of length, mass, and time. The units of length, mass, and time are sometimes called the fundamental units because, once they are selected, other quantities can be measured in terms of them.

— *Paul Hewitt (1994)*

Zounds! It's August already as we write this episode, but another school year will be well along by the time you read it. Our Power Tools theme this year is a new project called "Hands-On and Far-Out Physics" at C-TEC, a project-based learning community at Piner High School in Santa Rosa,

California. In this project, our remote mentors will be Galileo Galilei, Sir Isaac Newton, and Carl Sagan. In the October issue of L&L we shared ideas about the project and suggested that you visit Rice University's Galileo Project; to do so, send your Web browser to **http://es.rice.edu/ES/humsoc/Galileo/index.html**.

The Rice site is constructed by history students who intertwingle history, hands-on science, and math. Beautiful.

Length: How Long? Distance: How Far?

Length is a fundamental physical quantity. In Galileo's time, the *cubit* was one unit for measuring length. The cubit, in fact, goes back to ancient Egypt and Babylon. It is defined as the length from the point of the elbow to the tip of the middle finger on the same arm—Bob's cubit is 0.475 meters, but George's is somewhat longer.

What does the Internet say about cubits? We used a search engine—AltaVista—to look for information. We entered the search key "cubit," and AltaVista replied, "About 972 documents match your query."

Oops, that was too many. We refined the search by requiring both "cubit" and "length" to appear in the search target. To do so in AltaVista, we put a plus sign (+) in front of each word.

- Search key: **+cubit +length**
- AltaVista: "About 370 documents match your query."

Still too many. We added "+measure*." The asterisk (*) told AltaVista to require words that began with the word *measure,* such as *measurement.*

- Search key: **+cubit +length +measure***
- AltaVista: "About 135 documents match your query."

We browsed a bit more and discovered several biblical references. We then told AltaVista to exclude documents that contained the words *bible* and *biblical* and other words that began with the prefix *bibl.* To exclude a word, put a minus sign (-) in front of it.

- Search key: **+cubit +length +measure* -bibl***
- AltaVista: "About 65 documents match your query."

That helped a lot! We noticed that several references contained the word *ark,* so we excluded it.

- Search key: **+cubit +length +measure* -bibl* -ark**
- AltaVista: "About 53 documents match your query."

We finally had brief descriptions of 53 documents. We found several documents related to our quest and some pleasant yet unrelated surprises. Serendipity! The picture at the Origin of Measurement, a Japanese site, is worth *sen no kotoba* (1,000 words). Send your Web browser to **www.yokogawa.co.jp/Information/Museum/MUSEUM_E.HTM.**

The Number 28 site makes connections between Egyptian measures, cubit and *digit* (the width of a finger), and the number 28, which is both a perfect number and a triangular number. To see for yourself, go to **http://acorn.educ.nottingham.ac.uk/ShellCent/Number/Num28.html.**

In Babylon, a cubit was 30 fingers, or so says the Physics 109: Babylon site. On the Internet, it's less than a cubit away at **www.phys.virginia.edu/classes/109N/lectures/babylon.html/.**

As Gaia turns, the Internet changes. If you try our search keys, you may see different sites and find documents that were not on the Internet when we first searched. May dragons of good fortune dance on your keyboard.

Using What We've Found

With your handy cubit (that is, your forearm), make some measurements. Measure the length and width of a table in cubits and compute the table's area in square cubits. Different people may get different values for the same measurement. The table Bob measured is 2.5 cubits long and 1.6 cubits wide. George claims the table is shorter and narrower than Bob's measurements. Is George's forearm longer or shorter than Bob's?

Measure other people's heights using your cubit (forearm). Awkward, isn't it? We'd love to watch you measure your own height in cubits using your forearm. And measuring the length and width of your classroom also might be a bit cumbersome. Hmm. Let's videotape people doing these and other measurements.

You may get different measurement values between today and tomorrow and between tomorrow and next week. To avoid this problem, make a measuring stick that is the length of your own "standard" cubit. You can now repeat measurements with some expectation of getting the same result. Make a physical bar graph by putting students' measuring sticks side by side, perhaps in order from shortest to longest. This provides a visual display of the range and the median. Measure each cubit stick with a meter stick, construct a spreadsheet and bar graph, and calculate and display the mean and deviation from the mean for each cubit.

The *pace,* or the *stride* or the *step,* is another linear measure that has been around since antiquity. Some ancient cultures defined the pace as one step, while others defined it as two. We will use one step or stride for this measure. Your step may be a better instrument than your forearm for measuring the length of your classroom. Measure several things by walking and counting steps; measure them today, tomorrow, and next week. Do you get the same answers? Make a step stick of your "standard" step. How many cubits equals one step for you? Elementary school students: Will your cubit and step sticks be accurate a few months from now when you have grown a little taller?

Different students may have cubits and steps of different lengths. Perhaps they might agree to use a step stick that is the average of their individual sticks; cubit sticks, too. A classroom culture might thus adopt a standard measure.

Today the meter is the worldwide standard measure of length. After playing with measures from long ago, we will replicate our experiments using the meter (abbreviated *m*) as our standard measure. To encourage students to learn about such things as errors of measurement and significant digits, we will ask them to measure objects using the following four types of meter sticks:

- ***uncalibrated***, or estimating length to the nearest 0.1 m
- ***decimeter calibrated***, or estimating length to the nearest 0.1 decimeter (0.01 m)
- ***centimeter calibrated***, or estimating length to the nearest 0.1 centimeter (0.001 m)
- ***millimeter calibrated***, or measuring length to the nearest millimeter (0.001 m)

This same meter stick investigation is described in this column in the September 1994 issue of *The Computing Teacher*.

Time: How Long?

Galileo did his own hands-on and far-out physics—with rudimentary equipment. How do you measure time without a stopwatch or a clock with a second hand? Galileo used his resting pulse, musical rhythm and tempo, and a *clepsydra* (water clock).

We'll ask students to use their pulses to measure time. How many pulses between two musical tones? How long—that is, how many pulses—can George hold his breath? How long does it take Laran to read aloud a page of her haiku? How long does it take to download a file from the Internet?

Students with different pulse rates will get different values for the same measurement. Aha! This means we need a device to measure time that does not vary from person to person. The Galileo Project describes in meticulous detail how to make a clepsydra, and it presents spreadsheet data from students' work. The project site also describes how Galileo discovered that the pendulum can be used to time events with good precision, at least for his own time period. According to the Galileo Project, Galileo's physician friend Santorio Santorio used the pendulum principle to invent a *pulsilogium,* a handheld pendulum for measuring pulse rate. Other sources say that Galileo invented the pulsilogium himself and presented it to the medical faculty at the University of Pisa, who then took credit for the invention.

Using your own pulse, measure the frequency of a pendulum (swings per pulse) and then measure your pulse with a pendulum (pulses per swing). Which is more constant over time: the frequency of the pendulum or your pulse? Yes, we will replicate Galileo's pendulum experiments and use pendulums to measure time during other experiments.

Speed: How Fast?

Length and time are fundamental physical quantities. Speed is a *derived* quantity: the distance (d) traveled by a moving object divided by the time (t) it takes for the object to travel that distance. In the metric system, we might use meters per second (m/s) for a person's walking speed, kilometers per hour (km/h) for an automobile's freeway speed, kilometers per second (km/s) for planets traveling around the sun, and lots of kilometers per second (about 3.0×10^5 km/s) for light, the fastest known thing in the universe.

Using What We've Found (Again)

Go outside on a nice day and mark off a distance of, say, 30 meters using a metric tape measure or trundle wheel. You can find both in school science and math catalogs. (We found a 50-meter tape for approximately $50 and a trundle wheel for $28 or so at the Sargent Welch Internet site, **www.sargentwelch.com**. The NASCO paper catalog has a 10-meter tape for $15.30 and a trundle wheel for $21.45. For a copy of the catalog, call 800/558-9595 or fill out a form on the NASCO site, **www.nascofa.com**.)

As George walks the distance at constant speed (well, as constant as he can), use your pulse (that is, count pulses) to measure the time it takes him to go from start to finish. Divide the distance by the number of pulses to get his average speed in meters per pulse. Yup, people with different pulse rates will get different average speeds. Other instruments for measuring time are independent of a person's physiological and emotional state. Let's try a few.

- Make a clepsydra and use it as a timer. Calculate average walking speed in meters per drop, per unit volume, or per unit mass of water that dripped into a container. How might Galileo relate drops, volume, or mass to time?
- Agree on a pendulum of standard length and use it as a timer. Calculate average speed in meters per pendulum swing. How might Galileo relate pendulum swings to time?
- Let's jump ahead to the present day and use a stopwatch. How does an old-fashioned mechanical stopwatch measure time? How does an electronic stopwatch measure time? (Hint: Something oscillates.)
- For distances of 10 meters or less, use a Calculator-Based Lab (CBL) or microcomputer-based lab (MBL) motion sensor. For information about these amazing measurement marvels, see this column in the October 1997 *L&L.*

As did Galileo, we'll soon roll balls down an inclined plane, grab data, fit mathematical functions to the data, and discover relationships among: time, distance, velocity, and *acceleration*. Look for these explorations in future issues.

Acceleration is a derived quantity. It

is the rate of change in speed, expressed as meters (or even cubits) per second per second, or in meters per second *squared* (m/s^2 or $m \cdot s^{-2}$). First we'll do it Galileo's way and then do it again using stop watches, CBL and MBL motion sensors, and photogates (gadgets that record the time when a moving object breaks a light beam). Wouldn't Galileo have loved these devices?

If all goes well, we'll learn a lot about distance, speed, and acceleration of an object moving in Earth's gravity. Later, when we begin to explore Isaac Newton's work, we'll learn why objects move as they do.

Mass: How Much Stuff? Force: What Does It Weigh?

Much confusion! Mass and weight are frequently confused with each other, especially in the United States, which clings to the backward, cumbersome, hard-to-learn, and hard-to-teach USCS system. The rest of the world uses the elegant, simple, easy-to-learn, and easy-to-teach Système International (SI), also known as the metric system. Scientists worldwide use this system, and up-to-date science and math books use SI units.

Mass is a fundamental physical quantity. The mass of an object is the amount of "stuff" in it—that is, the quantity of *matter* in the object. If you look up the words *mass* and *matter* in a dictionary, you may find that they are defined in terms of each other, thus completing a circular definition. Mass in physics is like a postulate in geometry. Scientists accept mass as a fundamental physical quantity and use its concept to explore the universe. The SI unit for mass is the *kilogram* (kg).

Mass is a measure of an object's *inertia*, or its resistance to a force applied to the mass in an effort to change its state of motion. Galileo knew about inertia and acceleration, but he did not construct a mathematical model that related mass, acceleration, and force to one another; Newton did that. Newton's second law of motion is a great conceptual leap; it states that the acceleration of an object (a) is proportional to the net force applied to the object, (b) is in the direction of the applied force, and (c) is inversely proportional to the mass of the object. This law is expressed in physics or math textbooks by equations such as $a = F/m$ or $F = ma$, where F is force, m is mass, and a is acceleration.

Weight is the force of gravity on an object. Because weight is a force, it can be expressed in terms of mass and acceleration, as stated in Newton's second law. You may see this written in physics or math textbooks as $w = mg$, where w is the weight, m is the mass, and g is the acceleration of the object produced by the gravitational force. The SI unit for force or weight is the newton (N), and the unit for acceleration is meters per second squared (m/s^2 or $m \cdot s^{-2}$). Because force = mass × acceleration, the newton is equal to kilograms × meters per second squared ($kg \cdot m/s^2$). The USCS unit for force is the pound (lb), the unit for mass is the slug, and the unit for acceleration is feet per second squared (ft/s^2 or $ft \cdot s^{-2}$). We prefer metric!

On July 4, 1997, the Mars Pathfinder spacecraft landed on the Red Planet and deployed a rambunctious rover named Sojourner. Sojourner's mass is—oops, we forgot. No problem. While Sojourner roves Mars, we'll rove the Internet.

- AltaVista search key: **+sojourner +kilogram***
- AltaVista says: "36 documents match your query."

We quickly found the desired data at the Mars Rover Fact Sheet site: **http://robotics.jpl.nasa.gov/tasks/rovertech/factsheet/homepage.html**. This site has just the right stuff we need to know about Sojourner and Rocky 7, a prototype of future rovers that is scheduled to go to Mars in 2001.

Sojourner's mass is 11.5 kilograms (kg). So, what is Sojourner's weight? Well, it depends on where Sojourner is roving. If she is meandering over Earth's surface at the equator, then she weighs 112 newtons (25.3 pounds). If Sojourner is cruising Earth's surface at the North Pole, then it weighs slightly more than 113 newtons (25.4 pounds). As you can see, at the surface, Earth's gravitational force is slightly more at the poles than at the equator. At C-TEC (north latitude 38.45 degrees), Sojourner weighs slightly less than 113 newtons (25.4 pounds). Wherever she is on Earth, her mass is 11.5 kilograms, but her weight differs with location because the force of gravity differs slightly in different locations. On Mars, Sojourner's mass is still 11.5 kilograms. Her weight, though, is only 42.4 newtons (9.54 pounds). You can use Newton's second law of motion in the form $w = mg$ to crunch out Sojourner's weight in the places mentioned throughout this paragraph.

At the equator, $g = 9.78\ m/s^2$, so

$$w = (11.5\ kg)(9.78\ m/s^2) = 112\ kg \cdot m/s^2 = 112\ N.$$

At the poles, $g = 9.83\ m/s^2$, so

$$w = (11.5\ kg)(9.83\ m/s^2) = 113\ kg \cdot m/s^2 = 113\ N.$$

At C-TEC, $g = 9.80\ m/s^2$, so

$$w = (11.5\ kg)(9.80\ m/s^2) = 113\ kg \cdot m/s^2 = 113\ N.$$

On Mars, $g = 3.69\ m/s^2$, so

$$w = (11.5\ kg)(3.69\ m/s^2) = 42.4\ kg \cdot m/s^2 = 42.4\ N.$$

To convert the above values from newtons (N) to pounds (lb), multiply each value by 0.2248 pounds per newton (lb/N).

Reference

Hewitt, P. G. (1994). *Conceptual physical science.* New York: HarperCollins.

To find out how the Mars Pathfinder explorer vehicle got her name, go to the July 14, 1995, press release at **http:// esther.la.asu.eduasu_TES_Editor/ PATHFINDER/sojourner_7_95.html**.

> The name Sojourner was chosen for the Mars Pathfinder Rover after a year-long worldwide competition in which students up to 18 years old were invited to select a heroine and submit an essay about her historical accomplishments. The students were asked to address in their essays how a rover named for their heroine would translate these accomplishments to the Mars environment.
>
> Valerie Ambrose, 12, of Bridgeport, Conn., submitted the winning essay about Sojourner Truth, an African-American reformist who lived during the tumultuous era of the U.S. Civil War. An abolitionist and champion of women's rights, Sojourner Truth, whose legal name was Isabella Van Wagener, made it her mission to "travel up and down the land," advocating the rights of all people to be free and the rights of women to participate fully in society. The name Sojourner was selected because it means "traveler."

During our search for information about Sojourner, we found several serendipitous sites, including Live from Mars—Teachers' Guide and Mars Pathfinder Welcome to Mars (the latest information from the Carl Sagan Memorial Station). We later searched for "Sojourner Truth," but that's another story.

CHAPTER 8

Algebra

The Conversation Continues

Ivan: A significant body of research has developed regarding women and mathematics. In my experience, more women than men teach mathematics until you look at the baccalaureate level and beyond. Yet there exists a longtime view that women cannot "do" mathematics as well as men. What gives?

Anne: Well, women who have pursued mathematics beyond the high school level have generally been perceived as *really* smart. Many women feel men are afraid of smart women, so they have sometimes tried to hide how smart they really are. That may explain why so many women did not pursue higher mathematics in the past. Perhaps nowadays, the perception of women as leaders and as smart is more publicly acknowledged.

Ivan: Isn't it odd that society has long accepted women as teachers of mathematics even with the doubts about their ability to do mathematics?

Anne: Teaching at the K–12 levels has traditionally been a low-paying job. Consequently, this drove many men out of the teaching workforce into higher-paying jobs. Thus, women have been the sustaining force in the teaching profession. So whether or not society has had confidence in women's abilities in mathematics, people have accepted women as teachers of mathematics.

Ivan: We frequently hear that female students start to lose interest in mathematics when they reach the study of algebra. This concerns me, particularly because we also hear about how important it is for students to learn algebra. Why is algebra important to learn?

Anne: There's a discussion of the importance of algebra in the next commentary, "Algebra as the Gateway." We'll also see this issue addressed in the article "Julio's Run: Studying Graphs and Functions."

INSIGHT

Algebra as the Gateway

By Anne Raymond

More and more these days, businesses need employees with higher-level mathematics skills. When interviewing candidates, employers say they need workers who have solid mathematics skills and are able to solve problems involving mathematics. Today's workforce must not only be able to engage in critical thinking skills, but must also be able to apply them using technology. Algebra is one of those basic mathematical content areas that helps broaden one's reasoning and problem-solving skills.

Algebra is generally viewed as "the branch of mathematics that deals with symbolizing general numerical relationships and mathematical structures and with operating on those structures" (Kieran, 1992, p. 391). Although algebra is formally taught at the middle and high school levels, informal algebra and conceptual understanding of structures that lead to algebraic thinking can be introduced at the elementary level. The National Council of Teachers of Mathematics' standard on algebra describes the algebraic concepts and structures that students should be addressing as they progress through formal schooling.

NCTM Algebra Standard

Algebra: Instructional programs from prekindergarten through Grade 12 should enable all students to

- Understand patterns, relations, and functions
- Represent and analyze mathematical situations and structures using algebraic symbols
- Use mathematical models to represent and understand quantitative relationships
- Analyze change in various contexts

The content of algebra has changed very little over the years. However, the ways we can teach algebra have broadened. A number of algebraic manipulatives and related programs are increasingly being infused into the algebra classroom. For example, Algebra Tiles and Hands-On Equations are two algebra programs that have been used to provide a more concrete introduction to algebraic thinking.

In a study of the effectiveness of the Hands-On Equations algebra system, Raymond and Leinenbach (2000) found that eighth-grade students' confidence and ability in algebra increased beyond initial expectations with the use of hands-on equations. The study also found that students were able to maintain their learning and, as they moved to the high school setting, to transfer it to the more traditional approach to algebra.

NETS•S Problem-solving Standard

6. Technology problem-solving and decision-making tools

- Students use technology resources for solving problems and making informed decisions.
- Students employ technology in the development of strategies for solving problems in the real world.

For many reasons, algebra is often referred to as the gateway to the study of advanced mathematics and, consequently, to career choices. Students often lose interest in mathematics once they reach middle school and experience mathematics at a more abstract level. Many students do not take algebra at the middle school level and many do not go beyond algebra at the high school level. However, most college entrance standards require students to take algebra by the end of high school, and in many cases, they require students to take courses beyond algebra. Thus, in this sense algebra may serve as a gateway to a college education.

In addition, algebra serves as the foundation for continued mathematics study in precalculus and calculus. In fact, many mathematicians agree there is a direct correlation between student success in algebra and student success in calculus. Thus, algebra serves as a gateway to further success in advanced mathematics courses.

Further, when we examine the many college majors available to students, most of the majors within the broad areas of science, mathematics, and technology require the study of mathematics beyond algebra. Thus, students who do not master algebra are severely limited in their choice of a college major.

Finally, career choices are often directly related to what was studied in college. Consequently, those students who were not able to master algebra at some point in their studies are excluded from many career opportunities.

Since educators realize they often lose mathematics students because of students' experiences in algebra in the middle and high schools, increasing efforts are being made to help young students see algebra as a problem-solving endeavor and to help them see connections between algebra and future career options. One effort to help students become better problem solvers in middle school is the national Mathcounts contest, which is specifically designed for middle school students (teachers can contact their state's department of education for information on Mathcounts contests in their area). Students compete in individual and team competitions at the regional, state, and national levels. The focus is on problem solving.

A growing effort to help students find connections between mathematics and future career opportunities has resulted in middle and high school career days, during which people who have careers that involve mathematics come to the schools and talk about how they use mathematics in their jobs. In addition, regarding the study of algebra, middle and high school teachers are becoming better equipped to answer the question "When are we ever going to use this?" The following list summarizes some examples of connections between school mathematics and related careers. Some of the ideas were taken from Saunders (1988).

A SAMPLING OF CAREERS THAT REQUIRE THE USE OF ALGEBRAIC FORMULAS AND LINEAR EQUATIONS	
Accountant	Income tax specialist
Airline passenger service agent	Industrial engineer
Auditor	Insurance claims supervisor
Auto mechanic	Landscape architect
Carpenter	Machinist
Carpet layer	Medical lab technician
Cartographer	Navigator
Civil engineer	Oceanographer
Electrical engineer	Pharmacist
Electrician	Photographer
Environmental analyst	Political campaign manager
Firefighter	Purchasing agent
Farm advisor	Technical researcher
Forestry land management planner	Television repair technician
Heating and air conditioning specialist	

Beyond having more concrete introductions to algebra using manipulatives, teachers are trying to help students contextualize algebraic understanding through meaningful problem solving. They are also having students approach algebra from multiple perspectives, for example, by viewing algebraic equations formulaically and graphically. (*Note:* The idea of examining algebra from multiple perspectives is addressed in a featured article in this chapter titled "Julio's Run: Studying Graphs and Functions.")

Word problems in algebra can be divided into three categories: traditional word problems, problems approached from a functional perspective, and open-ended generalization problems (Kieran, 1992). An example of a traditional word problem is:

> *Katie has 5 more marbles than Ben. The sum of the marbles the two have altogether is 49. How many marbles does each person have?*

In traditional problems, student are generally encouraged to solve the problem by forming an equation (or equations) involving unknowns and operations. An example of a functional-type problem is:

> *One long-distance telephone calling plan charges $1 for the first 10 minutes and $.05 per minute after that. Another calling plan charges $2 for the first 10 minutes and $.03 per minute after that. For how many minutes will the two plans cost exactly the same?*

Students can solve this problem in a variety of ways. They might be encouraged to use a computer program to test possibilities rather than use the traditional algebraic formula approach. Here the emphasis would be to separate the attempt to merely answer a specific question from the general function of how a change in one variable affects the outcome.

Finally, an example of a more open-ended problem is:

> *A girl multiplies a number by 5 and then adds 12. She then subtracts the original number and divides the result by 4. She notices that the answer she gets is 3 more than the number she started with. She says, "I think that would*

> *happen, whatever number I started with." Using algebra, show that she is right (Kieran, 1992, p. 407).*

This problem-solving question guides students to examine many examples and then try to make sense of why this happens in general. They are required to explain their reasoning. This is also a perfect moment to allow students to develop and test their theories using a computer technology such as computer-driven spreadsheets. For more algebraic problem-solving ideas, refer to the works of Schadler (1992), Koirala and Goodwin (2000), and Martinez (2002).

References

Kieran, C. (1992). The learning and teaching of school algebra. In D. A. Grouws (Ed.), *Handbook of research on mathematics teaching and learning* (pp. 390–490). Reston, VA: National Council of Teachers of Mathematics.

Koirala, H. P., & Goodwin, P. M. (2000). Teaching algebra in the middle grades using mathmagic. *Mathematics teaching in the middle school, 5*(9), 562–566.

Martinez, J. G. R. (2002). Building conceptual bridges from arithmetic to algebra. *Mathematics teaching in the middle school, 7*(6), 326–331.

Raymond, A. M., & Leinenbach, M. (2000). Collaborative action on the learning and teaching of algebra: A story of one mathematics teacher's development. *Educational Studies in Mathematics, 41*, 283–307.

Saunders, H. (1988). *When are we ever gonna have to use this?* Palo Alto, CA: Dale Seymour.

Schadler, R. (1992). *Algebra problems: One step beyond.* Palo Alto, CA: Dale Seymour.

THEORY INTO PRACTICE

Helpful Spreadsheet Features

Functions and Conditional Formatting

By Ivan W. Baugh

Spreadsheets offer a number of functions. They enable the user to complete an involved calculation by simply telling the software to perform the selected function. One example would be to find the standard deviation of a group of numbers. Doing this manually (without a computer) involves multiple steps. On the computer, you enter the function name in the formula, identify the cells containing the numbers for which you want the standard deviation, and enter the formula. Immediately you have the results.

Microsoft Excel groups approximately 235 functions in these categories: Database, Date and Time, External, Engineering, Financial, Information, Logical, Lookup and Reference, Mathematic and Trigonometry, Statistical, and Text and Data. Lotus 1-2-3 offers a similar number of functions. AppleWorks, which offers more than 100 functions, groups them in these categories: Business and Financial, Date and Time, Information, Logical, Numeric, Statistical, Text, and Trigonometric. Other spreadsheets may organize them in a similar fashion. Look at help screens to learn the available functions and acquire information to facilitate their use.

These functions enable users to complete complex calculations. All functions require an argument that appears in the parentheses following the name of the function; some functions (e.g., Pi, Random) have an empty argument, but you must include the space for the argument: pi() or rand(). By selecting a range of adjacent cells rather than listing each of them individually, the user can include a greater number of cells in the formula where needed. In an argument, you may include a range of cells. For example, to select all the cells between C1 and D5 one would enter (for Microsoft Excel and Microsoft Works) C1:D5 or (for Lotus 1-2-3 and AppleWorks) C1..D5. Individual cells may be listed separately, but they need to be separated by a comma (C1, D5, H12, Q15, Z100). This allows you to include nonadjacent cells in the argument. You may also combine in the argument a range (cells that are adjacent) and individual cells, for example, C1:D5,H12,Z100.

Microsoft Excel offers a very useful feature called Conditional Formatting. The user selects a cell or range of cells and specifies the condition and the action taken when the data meet the condition. For example, in a grade book, you can (1) set the condition for the lowest grade a person may make to earn a C in a course, and (2) set the formatting to bold and red when the condition does not exist. This enables you to see at a glance which students in a class need preventive help. Figure 8.1 shows the window that appears in Excel when you choose conditional formatting.

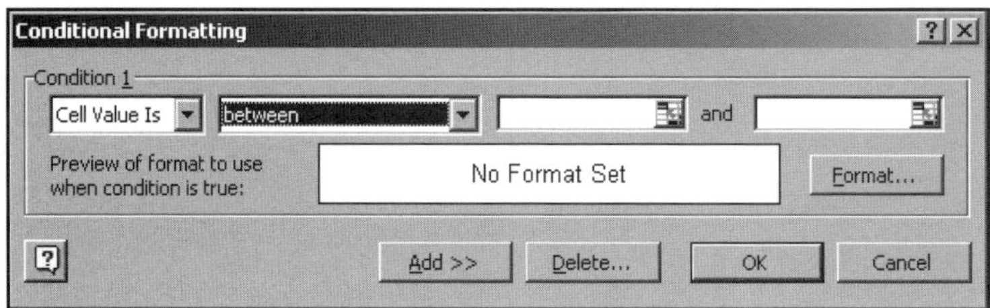

Figure 8.1.

The first box allows you to specify Cell Value Is or Formula Is. The second box offers eight choices for the condition to be met. The choice in the adjacent box determines whether you will have one or two boxes in which to enter criteria. The third and fourth boxes provide space for entering the criteria. The Format button accesses a window such as that shown in Figure 8.2.

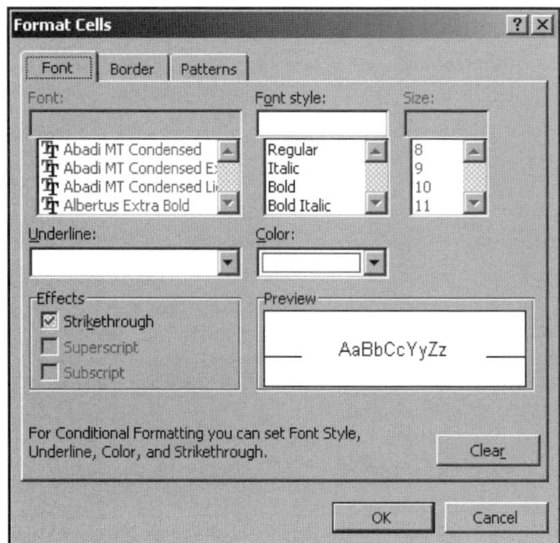

Figure 8.2

In this window you choose Font, Border, Patterns, Font Style, Size, Underline, Color, and Effects.

The companion Web site for this book (http://education.belarmine.edu.baugh/) provides a spreadsheet to accompany the article "Using Computer Spreadsheets to Solve Equations." Conditional formatting enables the user to see the precise location where the values for X and Y are equal. This identifies the location where the condition is true, facilitating the cells that provide an answer based on the data the user has entered. Click in a cell, from the Format menu choose Conditional Formatting. Look at the manner in which I instructed the computer to activate the condition. The screen shot in Figure 8.3 shows how I specified the condition and the action to occur when the condition is met.

Spreadsheets use two types of cell references when you copy and paste formulas: relative and absolute. Relative cells maintain the same relationship in a copied formula. For example, I enter a formula in cell C5 that reads A5+B5. I want to clone that formula in other cells. I copy the formula, select the cells to which I want this formula to also apply, and paste. If I went to cell C10 to paste the formula, it would read A10+B10—a relative reference. Absolute references do not maintain the physical relationship. In Figure 8.3, Cell E5 is added to each cell in row 7, the value of X. By

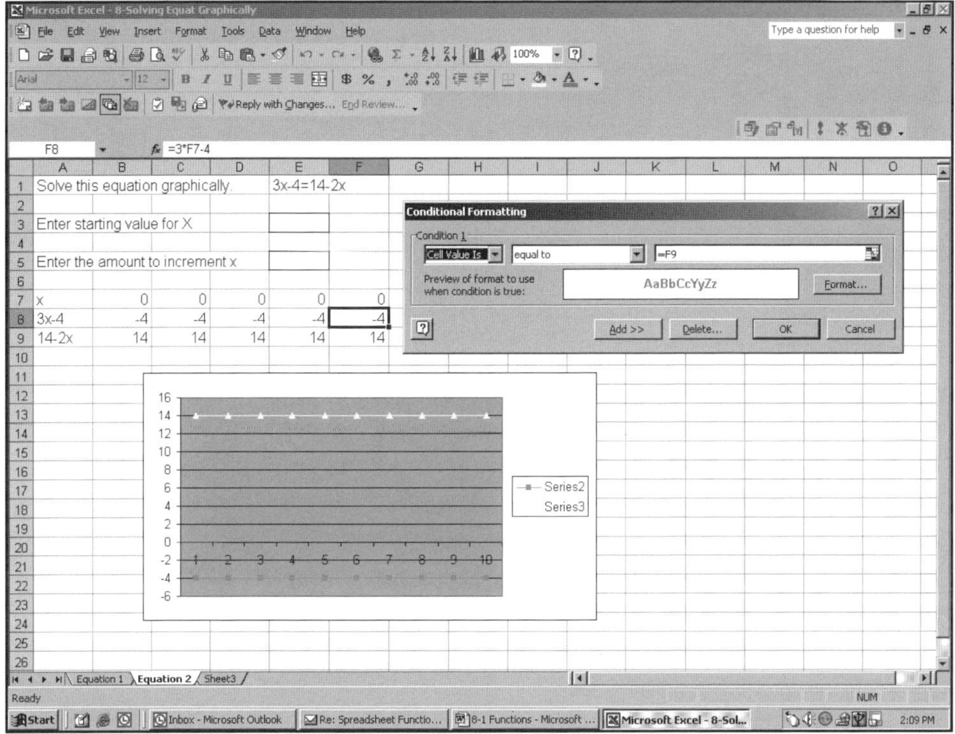

Figure 8.3

using an absolute reference in my formula, =B7+$E$5, I can copy and paste the formula without having to type in the value in E5 each time. The $ before E locks it into column E; the $ before 5 locks it into row 5. In the article "Using Computer Spreadsheets to Solve Equations," the author described completing the task using relative references. On the companion Web site I include a file, Solving Equations Graphically, that makes use of absolute cell references. Absolute cell references may involve a column reference, a row reference, or a cell reference. The Help screens of your software will explain how to instruct your software to use absolute cell references in formulas.

You will want to make certain your students completely understand how the functions work before starting to use them on the computer. With this thorough understanding of the results a function produces, spreadsheets enable students to spend their time analyzing the data rather than doing involved calculations. It helps prepare them for the world of work, where businesspeople use functions to analyze data as part of the decision-making process.

THEORY INTO PRACTICE

Gender Issues

By Anne Raymond

Historically, research has probed the question of whether, in the field of mathematics, there are measurable intellectual or affective differences related to gender. Some findings suggest that intellectually, males naturally have a stronger spatial sense and thus may be more successful in mathematics (e.g., Fennema & Leder, 1990). Other studies find that the level of confidence as it relates to mathematics is higher in males than in females, even when there are no differences in achievement (e.g., Fennema & Sherman, 1977).

The issue of gender differences in mathematics continues to be studied from a variety of perspectives. Some changes in approach have accompanied changes in pedagogical practices. For example, studies used to focus on determining what was wrong with girls when it came to mathematics and on how to change girls to be stronger and more confident in mathematics. Mathematics educators have come to realize that perhaps the "problem" is not with the girls themselves, rather with the ways in which mathematics has been taught. Thus, research on gender and mathematics has taken a new direction: how to adapt the teaching of mathematics to meet the needs of both the male and female learners (Campbell, 1995). Further, many efforts have been implemented to change the underrepresentation of girls and women in mathematics and math-related careers. For example, there have been increased efforts to expose girls to career opportunities related to mathematics early on and throughout the middle and high school years.

Some possible causes of gender inequity in the mathematics classroom fall into two categories: teacher interactions and belief systems related to gender (Van de Walle, 2001). Some possible links to teacher interactions and gender include the notions that boys get more attention during mathematics instruction than girls do, boys receive more criticism and praise for answers to mathematics questions, teachers wait longer for responses from males than from females, and girls receive more time for low-level question responses and are less likely to be called on to answer high-level questions. Some issues concerning how belief systems related to gender have an impact on inequity are (1) that in middle school, girls are often afraid to act too smart in front of boys and (2) that when teachers and parents believe girls can't do math, the wrong messages are sent to girls.

Efforts to meet the needs of females in the mathematics classroom have included implementing feminist pedagogy. According to Jacobs and Rossi (1997), feminist pedagogy in the mathematics classroom includes such practices as (a) using students' own experiences to build knowledge, (b) writing in the mathematics classroom, (c) cooperative learning in the mathematics classroom, and (d) developing a community of learners. Certainly, one can argue that many male learners would also benefit from these types of teaching practices.

In a study of students' images of mathematicians and mathematics, Picker and Berry (2001) examined seventh graders' work related to the assignment of drawing their perceptions of a mathematician. Students were also to list characteristics of mathematicians on their drawings. The results of the study indicate that today's students have some of the same limited and negative images of mathematicians that have persisted for decades. For example, 95% of the male students drew pictures of male mathematicians, and 58% of the female students drew pictures of male mathematicians. Also, the pictures of the mathematicians typically illustrated men with unkempt hair and clothing, with pencils and other tools in their pockets, and with glasses. Further, when students stated characteristics of mathematicians, they included such items as: They have no friends, they are not married, they are unlucky, and they are very unstylish. These perspectives are quite negative and suggest that teachers need to be working to change students' images of mathematicians.

Two possible ways to address students' perceptions of mathematicians are to introduce them to the human side of mathematicians and to expose them to more female mathematicians. Many great stories about mathematicians' lives tell of their childhood experiences and how they became interested in mathematics. Other sources focus specifically on women in mathematics and how they often had to struggle against society's image of male mathematicians in order to be able to study and enjoy mathematics themselves (e.g., Cooney, 1996; Reimer & Reimer, 1990; Reimer &

NOTED FEMALE MATHEMATICIANS
Hypatia
Sophie Germain
Emmy Noether
Mary Somerville
Ada Lovelace
Sonya Kovalevsky

Reimer, 1995). Here's a beginning list of interesting female mathematicians worth reading about:

A third suggestion for addressing gender inequity issues in mathematics is to incorporate more use of technology in the classroom. When students have technological tools available for individual or group exploration, the mathematics can be explored in a more gender-neutral and more student-directed manner. As students explore solving equations with technology in the article "Using Computer Spreadsheets to Solve Equations," students can use the spreadsheets to make and test hypotheses in a nonthreatening environment. The trial and error endeavors along with the visual representations in this article, as well as in the article "Julio's Run: Studying Graphs and Functions," provide students the privacy to explore an often difficult topic for students without their having to make known some of their initial misunderstandings in a whole-class setting.

References

Campbell, P. C. (1995). Redefining the girl problem in mathematics. In W. G. Secada, E. Fennema, & L. B. Adajian (Eds.), *New directions for equity in mathematics education* (pp. 225–241). New York: Cambridge University Press.

Cooney, M. P. (Ed.). (1996). *Celebrating women in mathematics and science.* Reston, VA: National Council of Teachers of Mathematics.

Fennema, E., & Leder, G. C. (1990). *Mathematics and gender.* New York: Teachers College Press.

Fennema, E., & Sherman, J. (1977). Sex-related differences in mathematics achievement, spatial visualization and affective factors. *American Educational Research Journal, 14*(1), 51–71.

Jacobs, J. E., & Becker, J. R. (1997). Creating a gender-equitable multicultural classroom using feminist pedagogy. In J. Trentacosta & M. J. Kenney (Eds.), *Multicultural and gender equity in the mathematics classroom: The gift of diversity* (pp. 107–114). Reston, VA: National Council of Teachers of Mathematics.

Picker, S. H., & Berry, J. S. (2001). Your students' images of mathematicians and mathematics. *Mathematics teaching in the middle school, 7*(4), 202–208.

Reimer, L., & Reimer, W. (1990). *Mathematicians are people, too: Stories from the lives of great mathematicians* (Vol. 1). Palo Alto, CA: Dale Seymour.

Reimer, L., & Reimer, W. (1995). *Mathematicians are people, too: Stories from the lives of great mathematicians* (Vol. 2). Palo Alto, CA: Dale Seymour.

Van de Walle, J.A. (2000). *Elementary and middle school mathematics: Teaching developmentally* (4th ed.). Upper Saddle River, N.J.: Pearson Addison Wesley.

Find the Formula

Using a Spreadsheet to Solve a Pattern

By Louis Feicht

Subject: Algebra

Grade Level: 6–12 (Ages 12–18)

Technology: Spreadsheet (e.g., Microsoft Excel), graphing calculators

Standards: *NETS•S 3.* (Find out more about the NETS Project at www.iste.org—choose Standards Projects.) *NCTM 2.* (Read the math standards online at www.nctm.org.)

One of the many powerful teaching applications of computer spreadsheets is to use them to help students investigate and solve problems that involve finding a pattern and formula. A familiar problem I've seen in many eighth-grade classrooms is:

> to figure out how many total bricks there would be if the pyramid was 50 bricks high with 50 bricks on the bottom row. We can see by counting that there are 10 total bricks if the stack has four bricks on the bottom (Figure 1).

One obvious method of attacking this problem is to create a table (Figure 2). With a table it may be possible to establish a pattern. The problem can be reworded as *The Total Number of Bricks* = $1 + 2 + 3 + 4 + 5 + 6 + \ldots + 50$. So, we are asking students to add consecutive integers from 1 to 50. Most mathematics teachers have the experience and tools necessary to apply strategies to solve this problem, as they've seen some variation of it before. But even the brightest of middle school students—and probably high school students, too—will solve the problem

Figure 1. How many total bricks would be in this pyramid if the bottom row is 50 bricks wide and the pyramid is 50 bricks high?

Spreadsheets are powerful tools that help students understand and communicate about math.

by brute force, if at all. Most students would not bother to find the sum of the first 50 integers manually, and I don't blame them. With a spreadsheet, we can give the majority of our students a tool to solve the problem. The computer can help even more by developing abstract thinking and generalizations through the language of spreadsheet formulas and regression.

The spreadsheet also brings back "guess and check" as a viable strategy, which should be used and encouraged. My experience, regardless of what I'd like to believe about my teaching, is that many students will still use guess and check to solve algebraic problems, no matter how much algebra we try to instill in them. A spreadsheet helps bridge the gap between concrete and abstract thinking. In many cases, it allows students to guess and check every single possibility in a given problem, quickly and easily, while still using a level of abstraction in the application of the spreadsheet formulas.

	A	B
1	Row	Total Bricks
2	1	1
3	2	3
4	3	6
5	4	10
6	5	15
7	6	21
8	7	28
9	8	36
10	9	45
11	10	55
12	11	66
13	12	78
14	13	91

Figure 2. Making a table is a good start.

Had I not seen this problem before, the part of me seeking an elegant solution would discover a pattern that I would then attempt to generalize into a formula. However, if I needed a solution quickly—a more realistic situation in the workplace—I would use a spreadsheet to find the correct answer.

Solving the Brick Problem with a Spreadsheet

I am of course aware of many varied strategies for solving this problem, such as regrouping (the famous Gauss story), using the area of a triangle of bricks, or solving a system of equations or matrices. The following methods can also be done on a graphing calculator. However, it has been my experience in schools that a far greater number of students have access to computers than graphing calculators.

One of the reasons I prefer teaching with a spreadsheet to solve problems such as the brick problem is that the user interface is "friendlier" than the calculator's. The user can have all representations of a problem on the same screen. And when students are working in pairs or groups on this type of problem, the spreadsheet promotes communication among them about abstract solutions to the problem. And it provides an environment for exploration and quick testing of student hypotheses.

A little basic knowledge of a computer spreadsheet will provide us with a quick answer to how many bricks are needed and even generalize the problem by providing us with a function (formula). Figures 3 and 4 show how students with a little knowledge of spreadsheets can quickly arrive at a correct solution. Students will need to have some knowledge of spreadsheet formulas before attempting this problem on

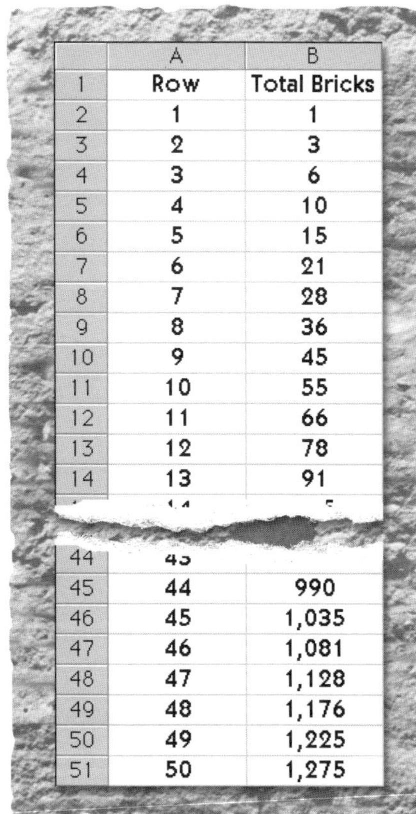

	A	B
1	Row	Total Bricks
2	1	1
3	2	3
4	3	6
5	4	10
6	5	15
7	6	21
8	7	28
9	8	36
10	9	45
11	10	55
12	11	66
13	12	78
14	13	91
[illegible]	[illegible]	[illegible]
44	[illegible]	[illegible]
45	44	990
46	45	1,035
47	46	1,081
48	47	1,128
49	48	1,176
50	49	1,225
51	50	1,275

Figure 3. Using a spreadsheet to count the integers.

	A	B
1	Row	Total Bricks
2	1	1
3	2	=A3+B2
4	3	=A4+B3
5	4	=A5+B4
6	5	=A6+B5
7	6	=A7+B6
8	7	=A8+B7
9	8	=A9+B8
10	9	=A10+B9
11	10	=A11+B10
12	11	=A12+B11
13	12	=A13+B12
14	13	=A14+B13
15	14	=A15+B14
16	15	=A16+B15
17	16	=A17+B16
18	17	=A18+B17
19	18	=A19+B18
20	19	=A20+B19
21	20	=A21+B20
44	43	=A44+[illegible]
45	44	=A45+B44
46	45	=A46+B45
47	46	=A47+B46
48	47	=A48+B47
49	48	=A49+B48
50	49	=A50+B49
51	50	=A51+B50

Figure 4. The formulas underlying the spreadsheet.

	A	B	C	D
1	Row	Total Bricks	First Differences	Second Differences
2	1	1		
3	2	3	2	
4	3	6	3	1
5	4	10	4	1
6	5	15	5	1
7	6	21	6	1
8	7	28	7	1
9	8	36	8	1
10	9	45	9	1
11	10	55	10	1
12	11	66	11	1
13	12	78	12	1
14	13	91	13	1
15	14	105	14	1
16	15	120	15	1
17	16	136	16	1
18	17	153	17	1
44	43	946	43	
45	44	990	44	1
46	45	1,035	45	1
47	46	1,081	46	1
48	47	1,128	47	1
49	48	1,176	48	1
50	49	1,225	49	1
51	50	1,275	50	1

Figure 5. The first and second differences calculated using a spreadsheet. Because the second differences are constant, a second-degree polynomial is implied.

their own. Students can start to use formulas in a spreadsheet at least as early as sixth grade. By developing and practicing the use of spreadsheet formulas, students will begin to communicate and thus think naturally in the language of the spreadsheet, talking in terms of "A2 + A3." This is communication using variables and a level of abstraction and thus lays the groundwork for more formal mathematics.

For many problems, including this one, I still have students develop the problem using paper and pencil. And students must decide which variable is to be independent and which is to be dependent. In the brick problem, the row of the pyramid would be the dependent variable, and the total number of bricks *depends* on how many rows we have in the pyramid. This is a subtle yet important concept if students are to be expected to set up these types of problems on a spreadsheet by themselves. As a generalization for simple problems, the spreadsheet cells that contain the independent variable will not have to contain a formula that depends on cells in other columns. The cells representing the dependent variable, however, will contain a spreadsheet formula that depends on other cells in the spreadsheet. I place heavy emphasis on correct labeling of the spreadsheet columns, as this may aid students in understanding and finding solutions.

The next step is for students to find a formula to find the total number of bricks. Often students will not immediately be able to find the correct formula, but they will be able to verbalize how to get the total bricks for the next row. When asking students how they can arrive at the total number of bricks for the next row, you may often hear responses such as, "I took this number (pointing to their paper or the computer screen) plus this number." Here is where we take students from adding two numbers to adding the variables that represent those numbers. Instead of, "I added six plus four," we want to lead students to adding A5 + B4. Once the correct pattern is established, students can use the Fill Down features of the spreadsheet to solve the problem. You may encourage students to check the problem at more than one spreadsheet row to add confidence to their solution.

Extending the Problem

We have found a solution to our brick problem—1,275 bricks. But we can go much further. We are now ready to extend the problem and find a general solution for any number of bricks. We use the first and second differences to give us a clue about the formula (Figure 5). The constant *second* differences suggest a *second-degree* polynomial. We now take advantage of the regression capabilities of the spreadsheet. A chart of the data series must first be made. In Microsoft Excel we make a *scatterplot* of the data, then click directly on a data point on the chart. All the data points will be selected. Next go to the Chart Menu and choose Add Trendline. Choose a Polynomial and degree of 2. Then choose the Options Tab and check the box that says Display equation on chart.

The regression equation of $y = .5x^2 + .5x + 2E{-}12$ is displayed on the screen (Figure 6). The term of "*2E–12*" should be recognized by students as being very close to zero. Students should also be required to explain what the x and y represent in terms of the original problem. In fact, I like to have students reword the equation, replacing the variables with words such as *bricks, row,* and *total number of bricks.* Making students explain what the spreadsheet has generated keeps them from just handing in some "equation the computer generated."

Another way to check student understanding is to ask them to find an answer to a similar problem without using technology or use numbers so large they are beyond the limitations of the spreadsheet. Two similar problems follow. The spreadsheet also supports students in exploring these problems.

Similar Problem, Different Words

An ice cream store has 31 flavors of ice cream. How many different two-scoop cones are possible? If we do not allow two scoops of the same flavor, we have a standard combination problem of 31 things taken 2 at a time:
$C\binom{31}{2}$.

If we do allow two scoops of the same flavor, as is more realistic, then it will soon become clear that the problem is the same as the sum of the first 31 consecutive integers (Figure 7).

If there is only one flavor—chocolate—we have only one possible combination. If we add one flavor—vanilla—we have three possible combinations. Adding one more—strawberry—gives us six combinations. Four flavors—add marshmallow—gives us 10 combinations.

The Old Handshake Problem

Each person in a room shakes hands with every other person. If there are "n" people in the room, how many handshakes take place? If there is one person in the room, zero handshakes take place. If two people are in the room, one handshake takes place. If 3 people are in the room, 3 handshakes will occur; for 4 people, 6 handshakes; and for 5 people, 10 handshakes. Again a familiar pattern, but slightly different because we start at zero instead of one.

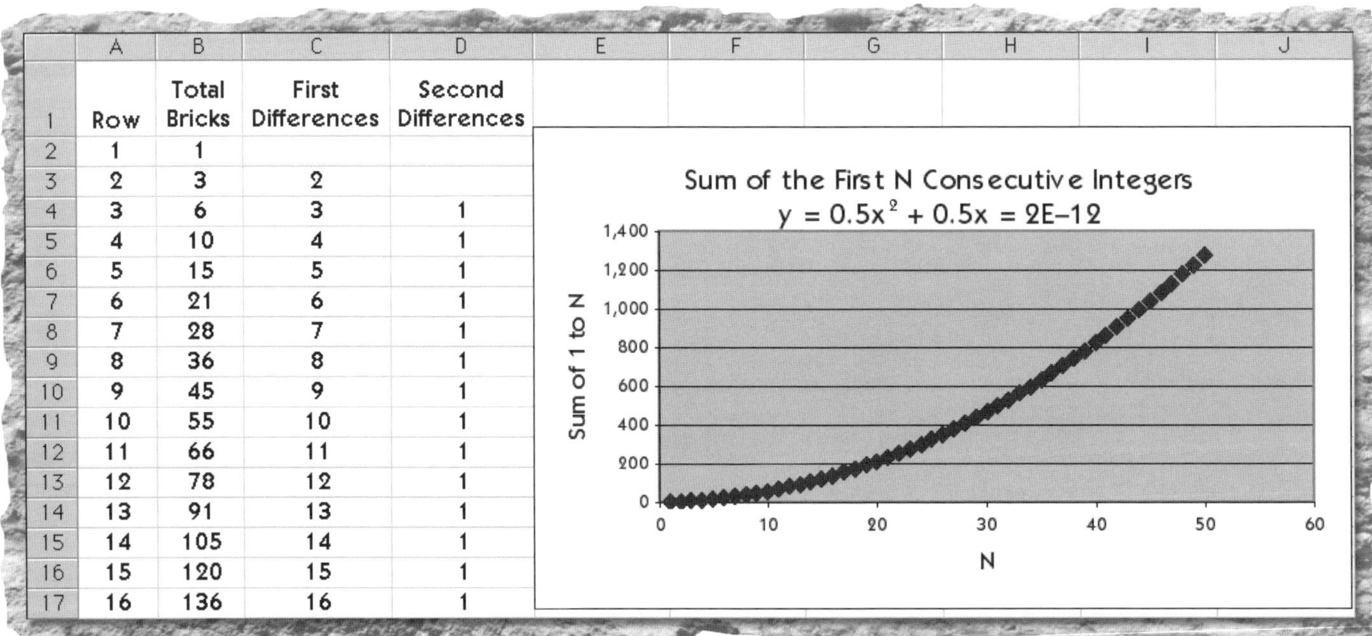

	A	B	C	D
1	Row	Total Bricks	First Differences	Second Differences
2	1	1		
3	2	3	2	
4	3	6	3	1
5	4	10	4	1
6	5	15	5	1
7	6	21	6	1
8	7	28	7	1
9	8	36	8	1
10	9	45	9	1
11	10	55	10	1
12	11	66	11	1
13	12	78	12	1
14	13	91	13	1
15	14	105	14	1
16	15	120	15	1
17	16	136	16	1

Figure 6. Using a spreadsheet, students can see numerical, symbolic, and graphical representations of the same problem at once.

Summary

The spreadsheet puts a powerful problem-solving tool in students' hands. And on a single computer screen, we have a numerical, symbolic, and graphical representation (Figure 6). If you require students to label their problems correctly and write summaries, you will also have a linguistic representation of the problem and solution. Once you begin this type of problem solving with students using real-world technology, you will begin to see endless possibilities for applications of technology in the mathematics classroom. Your students may begin to see the computer as a natural problem-solving tool, and thus improve their chances of succeeding in math.

Resource

Excel is available alone or as part of Microsoft's office suite at your local software reseller or online at www.microsoft.com.

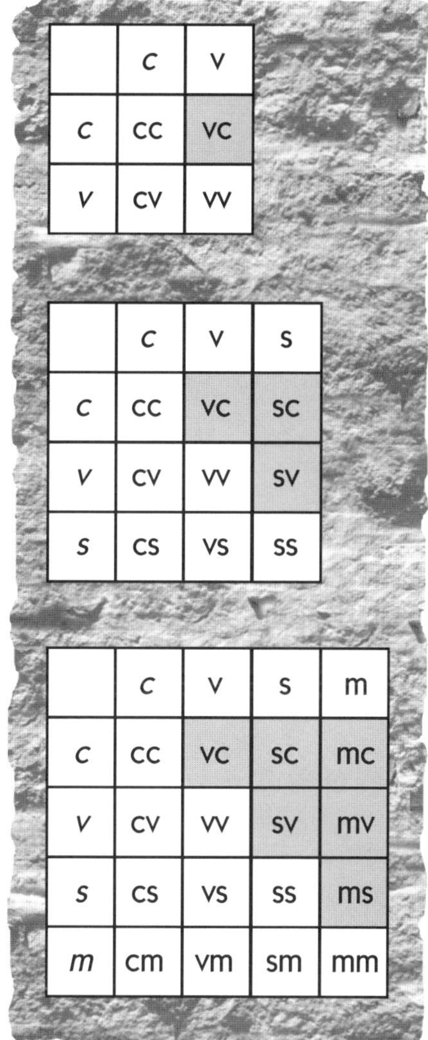

	c	v
c	cc	vc
v	cv	vv

	c	v	s
c	cc	vc	sc
v	cv	vv	sv
s	cs	vs	ss

	c	v	s	m
c	cc	vc	sc	mc
v	cv	vv	sv	mv
s	cs	vs	ss	ms
m	cm	vm	sm	mm

Figure 7. A matrix of possible two-scoop cones for two, three, and four flavors. The repeated cones are grayed out.

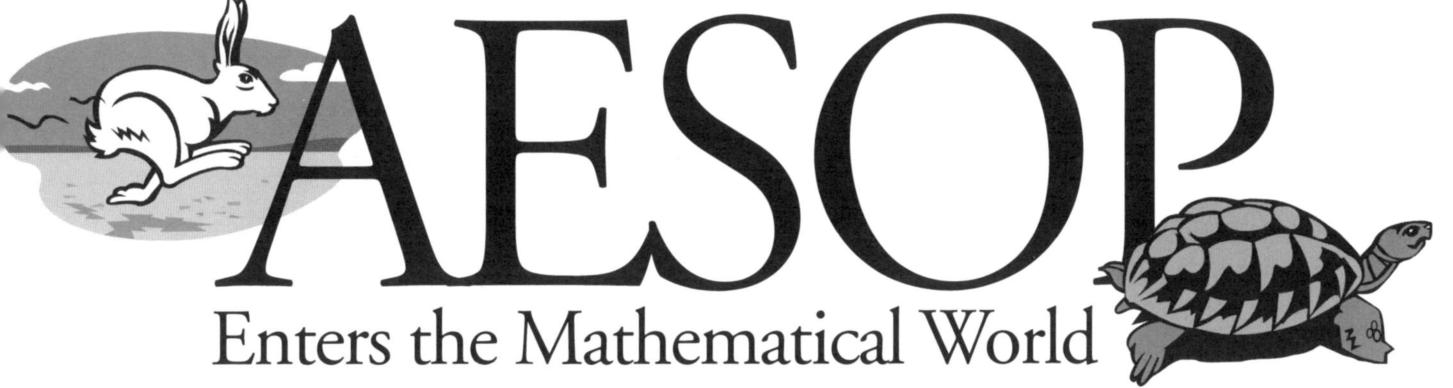

AESOP Enters the Mathematical World

One day a hare was making fun of a tortoise. "You are a slowpoke," he said. "You couldn't run if you tried."

"Don't laugh at me," said the tortoise. "I bet that I could beat you in a race."

"Couldn't," replied the hare.

"Could," said the tortoise.

"All right," said the hare. "I'll race you, but I'll win, even with my eyes shut."

They asked a passing fox to set them off. "Ready, set, go!" said the fox.

The hare went off at a great pace. He got so far ahead he decided he might as well stop for a rest. Soon he fell asleep. The tortoise came plodding along, never stopping for a moment.

When the hare woke up, he ran as fast as he could to the finish line . . . but he was too late. The tortoise had already won the race!

By Janet M. Walker

Subject: Prealgebra to Precalculus

Grade Level: 7–12 (Ages 12–18)

Technology: TI-83 graphing calculator and TI-Graph Link (Texas Instruments)

Standards: *NETS* 3 and 5. (See http://cnets.iste.org for more information about the NETS project.) *NCTM* 1, 2, and 8. (See www.nctm.org for more information on current NCTM standards.)

This old fable, "The Tortoise and the Hare," (Clark & Voake, 1990) provides an excellent situation for a mathematical investigation for 7th through 12th graders or any class that studies patterns and functions. Using the recommendations from the National Council of Teachers of Mathematics (NCTM, 1989) for worthwhile mathematical tasks for students, a problem relating to this fable can be used to implement four aspects of the NCTM's standards. First, students must use technology to model and solve the problem. This includes using a graphing calculator and a spreadsheet to investigate graphs and the relationships between them. Second, students are exposed to mathematics that involve different types of functions and graphs. Third, students are expected to communicate their findings in a clear and concise manner and in the form of a creative and accurate commentary. Finally, students are encouraged to work cooperatively on the project.

Establishing the Problem Situation

Although the story of the race between the hare and the tortoise may not model a real-world problem, the tale provides a fun and intriguing investigation for students. A third animal (the beaver) is added to make the problem more challenging for students. We state the problem as follows.

> Three animals—Tortoise, Hare, and Beaver—run a 100-meter dash. Each animal runs according to these rules:
>
> Tortoise $d = 12 + 3.75t$
> Hare $d = 0.24t(t - 5.1)$
> Beaver $d = [t(t - 20)(t - 9.1)]/10$
>
> where distance d is in meters and time t is in seconds (adapted from Lovitt, 1991).

The obvious question for this problem would be: Who wins the race? Although this is an important question that must be answered, it's not the main focus of this investigation. Using the above models, students are asked to predict mathematically what will be happening at any point in the race. Their task is to write and perform a commentary that gives listeners an idea of what is happening from beginning to end.

Simply answering the question "Who won the race?" is of no real value in learning mathematics. Most investigations require an analysis of the mathematics involved and the ability of students to communicate that analysis. Specifically, investigations are best presented in the form of a report or a story about the problem. Communicating the results of a problem is the most important aspect of problem solving and should be stressed with students (NCTM, 1989). This problem involves students in a collaborative and creative investigation that uses problem solving and communication.

Mathematical Model

Several aspects of this problem are set for students to help them begin their investigation. Students may work as individuals or in small groups to complete the problem. They can explore each animal's path by graphing each equation on the same set of axes. Students will have to decide the "best" window for their graph to view each animal's path. The graph of each animal models linear, quadratic, and cubic functions, respectively. Students will need to analyze each graph individually and discuss the possible reasons that each animal progressed as it did. Figure 1 shows a graph of the animals on their 100-meter trek.

The animals appear to finish at the same time, so determining which animal actually wins is a bit difficult. Students will need to analyze each animal's progress toward the finish line to determine the winner. By using the zoom feature on the graphing calculator and focusing closely on the finish line, they will be able to transform the graphs to determine which animal wins. Students may want to add an additional equation of $y = 100$ to represent the finish line. This will help them determine the race winner. If they perform all the steps correctly, they'll see that the beaver wins the race (see Figure 2).

Another nice way to have students determine the winner of the race is to use a spreadsheet. The students will have to figure out how to show the information they need. Figure 3 shows a spreadsheet made by students.

Students next write a commentary that gives listeners an idea of what's happening at each stage of the race—for example, at every second or at every meter. This could include where the animals are, the direction they're going, and their acceleration and speed. Students also should include special moments in the race, such as when animals pass one another or change directions. To support their commentary, students should produce computer- or calculator-generated graphs.

Finally, have students present the commentary and act out the race for the class. They can bring in or make props to use in the presentation (see "Sample Commentaries" on pp. 24 and 25).

Assessment

Teachers will find it easy to assess students on this investigation. Obviously, the students must participate and complete the activity. Teachers who use it should already have established guidelines for their students. Considering the commentaries in the accompanying sidebar, three points should be noted.

First, did the students interpret each graph correctly? Although they were certainly creative, the students in the second commentary did not fully comprehend the movement of the animals and thus did not understand the graphs. These students had Tortoise starting the race fast and then slowing down. In fact, Tortoise's speed was constant. They also had Beaver returning to the starting line but not beyond. Beaver actually traveled more than 44 meters beyond the starting line.

Second, did the students support their commentary with accurate and precise information? Students who complete this activity are asked to turn in their commentaries, along with the graphs they generated by computer or calculator to help them interpret the models. These graphs can be used to determine *how* the students analyzed the graphs.

Third, the students can be assessed on their cooperation within their groups. They are expected to work collaboratively on the investigation, and

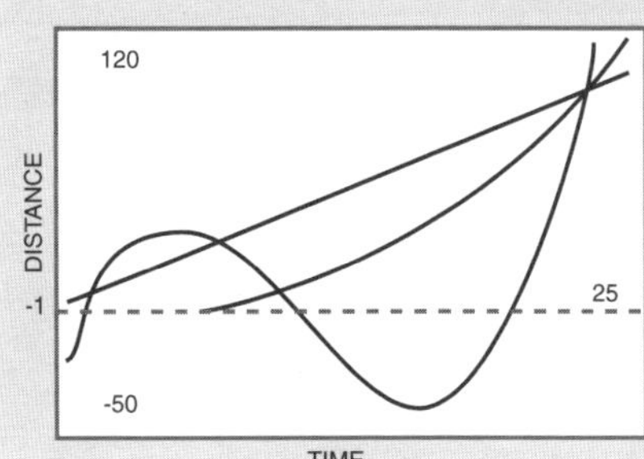

Figure 1.

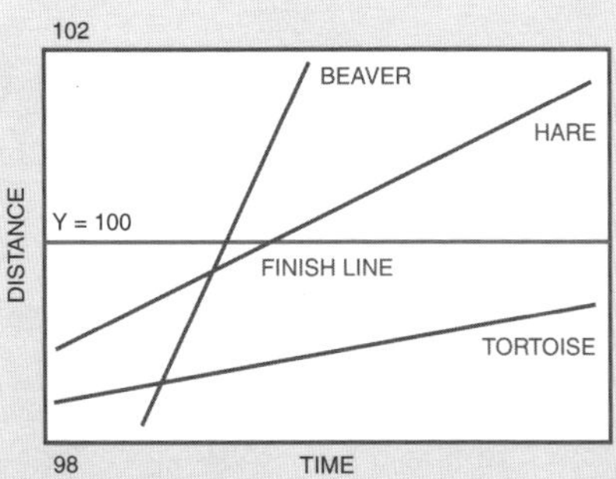

Figure 2.

APPENDIX A: Spreadsheet ---------- Distances ----------			
Time	Tortoise	Hare	Beaver
1	15.75	-0.984	15.39
2	19.5	-1.488	25.56
3	23.25	-1.512	31.11
4	27	-1.056	32.64
5	30.75	-0.12	30.75
6	34.5	1.296	26.04
7	38.25	3.192	19.11
8	42	5.568	10.56
9	45.75	8.424	.99
10	49.5	11.76	-9
11	53.25	15.576	-18.81
12	57	19.872	-27.84
13	60.75	24.648	-35.49
14	64.5	29.904	-41.16
15	68.25	35.64	-44.25
16	72	41.856	-44.16
17	75.75	48.552	-40.29
18	79.5	55.728	-32.04
19	83.25	63.384	-18.81
20	87	71.52	0
21	90.75	80.136	24.99
22	94.5	89.232	56.76
23	98.25	98.808	95.91
23.1	98.625	99.792	100.254
23.2	99	100.7808	104.6784
23.3	99.375	101.7744	109.1838
23.4	99.75	102.7728	113.7708
23.5	100.125	103.776	118.44
23.6	100.5	104.784	123.192
23.7	100.875	105.7968	128.0274
23.8	101.25	106.8144	132.9468
23.9	101.625	107.8368	137.9508
24	102	108.864	143.04
24	102	108.864	143.04
25	105.75	119.4	198.75

Figure 3.

Sample Commentaries

I assigned the tortoise, hare, and beaver problem to my elementary preservice teachers who were studying patterns and functions. The students really enjoyed this investigation. As their teacher, I enjoyed listening to their thought processes and final commentaries. With their permission, here are two commentaries taken from students in my class.

Jason (J): Welcome to the First Annual Great Animal Race. Competing in today's race are Tortoise, Hare, and Beaver.

Nancy (N): Tortoise, from Pittsburgh, Pennsylvania, is weighing in at 4 pounds 2 ounces and approximately 4 inches in height. Because of a prior injury, Tortoise is running the race with only 3 legs. Hare traveled all the way from Los Angeles and is weighing in at 3 pounds 7 ounces and is 4.5 inches in height. Last is Beaver from Lexington, Kentucky, weighing 3 pounds 9 ounces and 5 inches in height.

J: With these three over-qualified contenders, today's race should be a good one. It appears that the three are now lining up to begin the race. Because of Tortoise's handicap, the officials have awarded him a 12-meter head start. It appears that Hare is taking a little longer to reach the starting line. He better hurry up; the gun is about to sound.

N: Ladies and Gentlemen, the race has begun. Hare is about a meter back, and it appears that he is going in the wrong direction.

J: Beaver has gotten off to a fabulous start! He just blew right past Tortoise at 16 meters.

N: I can't believe it! Beaver's speed is amazing. He is closing in on 32.64 meters at only 4 seconds. But wait! What is he doing? It looks as though he is turning around.

J: Maybe he is hurt.

N: He could have pulled a muscle.

N: Looks like Hare is back on track. He appears to have turned around.

J: Beaver is still headed back to the starting line and has just crossed Tortoise's path, who is now in the lead traveling at a constant speed.

N: Just as Beaver makes his way to the starting line, Hare also reaches the starting line, headed in the right direction.

J: Tortoise is continuing at his same speed with a commanding lead.

N: It has been confirmed that Beaver has pulled his hind leg muscle and is continuing on to the locker room to have it taped up.

J: Hare is beginning to pick up speed and is quickly approaching Tortoise.

N: Looks like Beaver is ready to get back on the track. At 15.5 seconds, he is headed toward the starting line.

J: Whatever they did in that locker room worked wonders for Beaver because he is coming out like a bandit. He is now going his fastest speed trying to catch up to Tortoise and Hare.

N: It is coming to the moment we have all been waiting for: the end of the race. Looks like Tortoise is going to walk away with the gold today.

J: But WAIT! Hare and Beaver have just made a tremendous comeback and caught up. It looks like a tie.

J: We will have to check with the officials on this one.

N: The results are in. Beaver gets the gold finishing at 23.094 seconds, Hare comes in close behind at 23.125 seconds. Tortoise finishes third at 23.5 seconds.

J: What a race! Stay tuned. We will be broadcasting live interviews with the animals themselves after this commercial break.

Shannon (S): Ladies and Gentlemen, welcome to the Second Annual Great Animal Race. This year we have three competitors: Tortoise, Hare, and Beaver. The winner of the race will be the first animal to travel 100 meters in the least amount of time. So without further ado, let the race begin! Animals are you ready? On your mark, get set, GO!

Tricia (T): Well the race has begun and Beaver is in the lead. He has almost traveled a complete four meters, but for all you Tortoise fans, don't worry because he also got off to a great start. Hare, on the other hand, looks as though he is just not ready to start the race.

S: Beaver looks like he is determined to win this year's race just like he did last year. But what about Tortoise? It looks as though he has slowed down just a bit because he is not moving as fast as he was before. And then there is Hare, still saving his energy.

T: We are now approaching four seconds and Beaver has really risen to a peak, he has traveled more than 32 full meters. He is really moving fast today, must have eaten a great breakfast. Tortoise seems to be just moving along at a nice easy pace, he has traveled 27 meters. And Hare, I am starting to worry about him, maybe he just doesn't want to participate in the race today.

S: OH, MY GOODNESS! *What* is going on? Beaver looks as though he is going in the wrong direction, and yes, he certainly is. Within four seconds, Beaver has backtracked himself and has almost lost half of his distance.

T: What is happening now? It looks as though Hare has finally decided to come out of his shell. After a total of six seconds since the race started, Hare has moved a little more than a meter. What an exciting moment for all you Hare fans. Beaver is *still* going the wrong way. He is now back to the starting line, and I'm not sure what he is waiting for, but let's hope that it is something worth his while. And Tortoise just keeps going! He has traveled almost 46 meters. He is almost there. Can he do it?

S: Well, we are now more than halfway through the race and Tortoise is still in the lead. Hare definitely deserves some kind of praise from his fans for making somewhat of a comeback. And Beaver is still resting at the starting line. I'm not sure, but maybe Beaver and Hare had worked out some kind of plan before the race, they both seem to have accomplished the resting thing, but at different times.

T: Oh, my goodness! Beaver has just taken off like a bolt of lightning, and boy, oh, boy, is he going fast. Looks as though he is determined to win the race.

S: This is the closest race I have ever seen. You will never guess who has just crossed the finish line. Ladies and Gentlemen, *Beaver has won the race!* He has reached the finish line at a total of 23.094 seconds. Yes, everyone—I said Beaver. The only animal who went the wrong direction and decided to take a short nap in the middle of the race. Right behind him was Hare in second place with a finishing time of 23.125 seconds. And the animal who I thought was definitely going to win the race came in third place, or what they call *last*. Tortoise finished the race in 23.5 seconds.

T: What an amazing race. Well, ladies and gentlemen, that concludes this year's race, and I thank you all for coming out and supporting these three fine animals. I hope that everyone had fun, I know I did. Hope to see you next year at the Third Annual Great Animal Race!

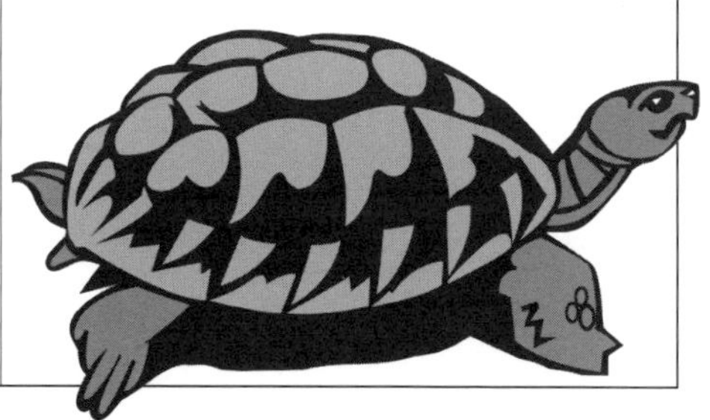

just how much or how little this occurs can be assessed through individual, peer, and teacher evaluations.

Modifications

Several modifications of this activity can be made depending on the level of mathematics a teacher wants to address. With algebra students, the teacher may want to use just the tortoise and the hare and make the equations a bit easier. For example, use $y = 4x + 8$ for Tortoise and $y = 4.4x$ for Hare.

When working with advanced algebra or precalculus students, a teacher could have the students convert each equation into a parametric equation. The students could then graph the equations on the calculator using the parametric mode and graphing the equations simultaneously. This will show the actual progress of each animal at specific times in the race. Using parametric equations, students can see a more accurate representation of the animals' actual race.

Teachers also could modify the activity by giving the students a commentary and having them draw the graphs and complete the model. Students can also create their own graphs or commentaries or both to share with the class.

Students could use the original tortoise and hare story and create an accurate mathematical model or graph for the story. This would include using piecewise functions to show that students understood range and domain.

Conclusion

This problem's use of higher order thinking skills, the ability to chart a pathway to a solution, its required interpretation of mathematics, and the necessary communication of results are important goals for students to meet. This investigation lends itself to each goal, plus the students enjoy the investigation and work hard to complete it.

References

Clark, M., & Voake, C. (1990). *The best of Aesop's fables.* Boston: Little, Brown & Co.

Lovitt, C. (1991). *Math problem solving and modeling.* Melbourne: Thomas Nelson Australia Publishing Co.

National Council of Teachers of Mathematics. (1989). *Curriculum and evaluation standards for school mathematics.* Reston, VA: Author.

Acknowledgments

The author wishes to thank Jason Groom, Nancy Malone, Tricia Vacchiano, and Shannon Commesso for allowing their commentaries to be used for this publication.

Understanding Algebra through Graphing Calculators

The activity described explains how graphing applications help students understand slope and y-intercept.

Traditional algebra focuses heavily on refining students' manipulation skills through the use of paper and pencil. However, algebra is not a collection of procedures and formulas to memorize and then practice through 20 to 30 mundane homework problems. Algebra is a way of thinking, communicating, and reasoning. As teachers, we want our students to be able to understand algebra and use it appropriately. The use of technology during instruction changes the focus from manipulation to understanding algebra as a language (Demana & Waits, 1990). The exploratory and visual nature of graphing calculators and computer graphing applications allows students to investigate algebraic properties; make, test, and refine conjectures; and find counterexamples (Hirschhorn & Thompson, 1996). The graphing calculator can help students "develop understanding about variables, basic concepts of algebra, and explore mathematical topics" (Demana, 2000, p.1).

Introducing technology such as the graphing calculator and computer graphing applications in middle school and providing learning experiences that use such technologies allows students a deeper understanding of the content they are learning. The high school mathematics teacher can then build on the foundation laid by the middle school mathematics teacher. (*Editor's note:* See the Resources section at the end of this article for a list of further readings on this topic.)

By Janie M. Cates

Subject: Algebra

Audience: Teachers, teacher educators

Grade Level: 6–8 (Ages 11–13)

Technology: Graphing calculator

Standards: *NETS•S III, VI; NETS•T II, III, IV* (www.iste.org/standards); *NCTM* Grades 6–8 Algebra (http://standards.nctm.org/document)

The Activity

By using technology such as the graphing calculator or computer graphing applications to explore linear equations, students are not bogged down with trying to create an accurate graph to reach a correct conclusion. Instead, the technology enables students to create graphs instantly, explore similarities and differences between graphs, and more easily make, test, and refine their conjectures.

In this activity, students explore the graphs of various linear equations to learn how m and b affect the graph of $y = x$ in the equation $y = mx + b$. Using graphing calculators, students compare graphs of equations with different slopes and y-intercepts, first as separate concepts and then combined in slope-intercept form, to find how changes affected the equation $y = x$. They also predict what the graphs of specific equations will look like and then use the graphing calculator to check their predictions. In my classroom, the students had little background knowledge of the graphing capabilities of graphing calculators, requiring me to model how to graph and view equations. When creating graphs, students first create the graph on their graphing calculators, then draw them on their assignment sheets to turn them in. Figure 1 provides examples of student work with the y-intercept from the activity. Figure 2 shows examples of student work with slope.

These explorations can lead to a deeper conceptual understanding of slope and y-intercept. As a class discussion during each lesson, student volunteers share the equations of lines that reflected the focus of the lesson, such as slope and y-intercept. They also explain

"Technology can help students think more deeply about mathematics, facilitate generalization, empower students to solve difficult problems and furnish concrete links between geometry and algebra, algebra and statistics, and real problem situations and associated mathematical models" (Demana & Waits, 1990, p. 28).

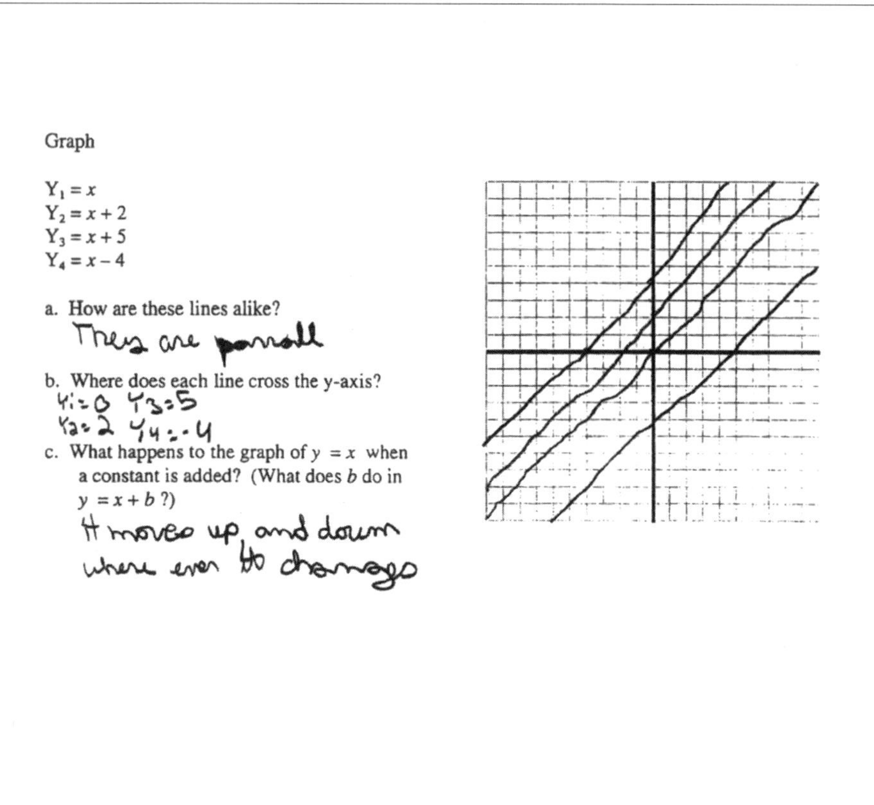

Figure 1. Student work with the y-intercept. The graph is copied from the student's graphing calculator.

how the slope and y-intercept affected the graph of the line. Not only are students able to discuss what the slope and y-intercept are and graph the lines of such equations, they can also provide the equations for lines that match specific criteria (e.g., slopes to the right, crosses the y-axis above the x-axis, etc.).

Assessment

Using technology to enrich lessons becomes wasteful when students are assessed using only traditional paper and pencil methods. Mercer (1995) states, "It is very important for us, as teachers, to realize that graphing calculators [and other technologies] change not only the way we present material but also the types of questions we should use on our homework, quizzes, and tests. For example, it is better to ask for equations of graphs than to ask for graphs of equations" (p. 273). We must ask the students to do both. At the end of the unit, the teacher gives students a separate

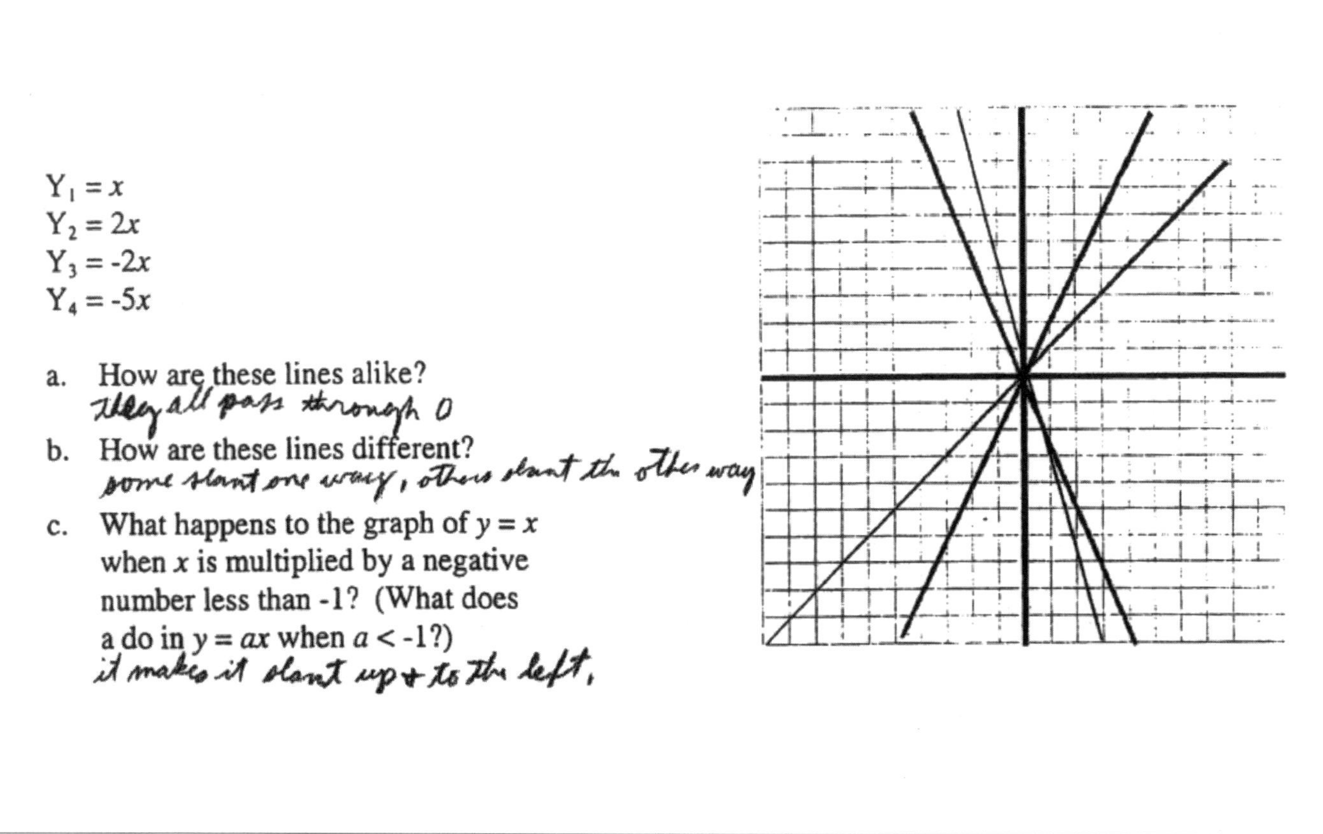

Figure 2. Student work with slope. The graph is copied from the student's graphing calculator.

Table 1. Percentage of Students with Correct Responses on Conceptual Understanding

Assessment Item	*Graphing Calculator (18 students)*	*Paper and Pencil (15 students)*
Students wrote the equation of a line that:		
Slopes to the right	100%	60%
Slopes to the left	59%	53%
Has a more horizontal slope	33%	13%
Intercepts the y-axis above the x-axis	72%	33%
Intercepts the y-axis below the x-axis	67%	27%
Students identified the slope as negative in the line drawn	56%	73%
Students identified the y-intercept as 4 in the line drawn	28%	7%
Students knew that m is the slope and tells the steepness and direction of the line	50%	33%
Students knew that b is the y-intercept and tells where the line crosses the y-axis	44%	27%

evaluation and asks them to write equations that fit specific criteria. The students must write the equation of a line that:

1. intercepts the y-axis above the x-axis,
2. intercepts the y-axis below the x-axis,
3. slopes to the left,
4. slopes to the right,
5. has a steep slope, and
6. has a more horizontal slope.

The students also describe the graph of a provided linear equation with a positive slope and a positive y-intercept. The students do not have to find the exact equation for the line, just identify that the slope and y-intercept are positive. They must explain how the coefficients, m and b, affect the graph of $y = x$. I compared the group of students taught using the graphing calculator to a group of students taught using paper and pencil methods. I evaluated both groups on their procedural and conceptual knowledge. The students using the graphing calculator scored slightly better on the chapter test testing procedural knowledge than the students learning with the traditional method of instruction—an average of 75% correct compared to an average of 71.2% correct, respectively. Table 1 shows the percentages of students who provided correct responses on the conceptual understanding portion of the unit evaluation. Students who were taught with the graphing calculator performed better on most of the conceptual items.

Conclusion

Technology is a valuable asset when introducing new concepts in mathematics. With technology, students can explore these new mathematical concepts and construct their own understanding. Assessment also takes on new meaning as a guide for instruction. Furthermore, assessment becomes a more appropriate evaluation of students' conceptual understanding. Technology is an integral part of our society, and the mathematics classroom offers a wonderful avenue to use it meaningfully.

Resources

Dessart, D. J., DeRidder, C. M., & Ellington, A. J. (1999). The research backs calculators. *Mathematics Education Dialogues, 2*(3), 6.

Dunham, P. H., & Dick, T. P. (1994). Research on graphing calculators. *The Mathematics Teacher, 87*(6), 440–445.

Heid, M. K. (1990). Uses of technology in prealgebra and beginning algebra. *The Mathematics Teacher, 83*, 194–198.

Mercer, J. (1992). What is left to teach if students can use calculators? *The Mathematics Teacher, 85*(6), 415–417.

National Council of Teachers of Mathematics. (1989). *Curriculum and evaluation standards for school mathematics.* Reston, VA: Author.

National Council of Teachers of Mathematics. (2000). *Principles and standards for school mathematics.* Reston, VA: Author.

References

Demana, F. (2000, August). *Using technology to prepare all students for success in algebra.* Paper presented at the T3 World-Wide Conference, Tokyo, Japan.

Demana, F., & Waits, B. K. (1990). The role of technology in teaching mathematics. *Mathematics Teacher, 83*, 27–31.

Hirschhorn, D. B., & Thompson, D. R. (1996). Technology and reasoning in algebra and geometry. *The Mathematics Teacher, 89*(2), 138–142.

Mercer, J. (1995). Teaching graphing concepts with graphing calculators. *The Mathematics Teacher, 88*(4), 268–273.

Using Computer Spreadsheets to Solve Equations

Algebra is the language through which most of mathematics is communicated. . . . Algebra as a means of representation is most readily seen in the translation of quantitative relations to equations or graphs. . . . Computing technology enables schools to provide a richer set of algebra experiences for all students (National Council of Teachers of Mathematics, 1989, p. 150).

Consider the algebra curriculum. What is the focus? Yes, we want students to develop a conceptual understanding of the subject matter, but when repetition and rules are emphasized, the focus shifts in students' minds from the concept to the algorithm.

By Margaret L. Niess

Subject: Algebra

Grade Level: 8–12 (Ages 13–18)

Technology: spreadsheet software (e.g., ClarisWorks, Microsoft Excel or Works)

Consider the problem of helping students solve for x, an unknown, in the equation $6x + 2 = 3x - 4$. By tradition, we focus on solving this equation *algorithmically.* Of course, we first try to have the students understand the concept. Assigned sets of problems, however, ask students to solve for x, and teachers often resort to a set of "rules" to guide students toward solutions. Unfortunately, when using such rules students tend to focus on algorithmic rather than conceptual understandings of solutions; as a result, they become limited in their thinking.

Using a Spreadsheet

A computer spreadsheet's graphing capabilities constitute a tool that can conceptually engage students and help them see equations and their solutions in new ways. The examples in this article have been done using a ClarisWorks spreadsheet, but the process can be easily adapted to any standard spreadsheet program.

Finding x in $6x + 2 = 3x - 4$. Begin by setting up the spreadsheet and placing the following formula in cell C3:

```
=B3+1
```

The formula instructs the spreadsheet to add 1 to the value in cell B3.

Copy this formula to cells D3 through K3. The result is that the program will add 1 to the value in the cell to the left of each cell D3 through K3. The values displayed in the cells thus become 2, 3, 4, 5, 6, 7, 8, 9, and 10.

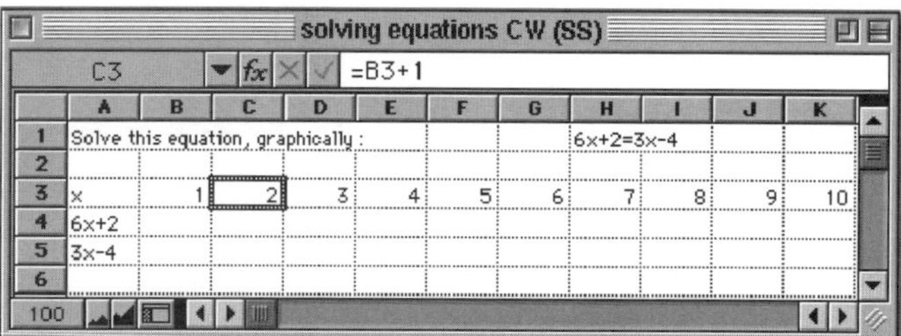

C3 | =B3+1

	A	B	C	D	E	F	G	H	I	J	K
1	Solve this equation, graphically:							6x+2=3x-4			
2											
3	x	1	2	3	4	5	6	7	8	9	10
4	6x+2										
5	3x-4										
6											

Instruct the spreadsheet to calculate the values for these expressions: $6x + 2$ and $3x - 4$. In cell B4, enter the following formula:

=6*B3+2

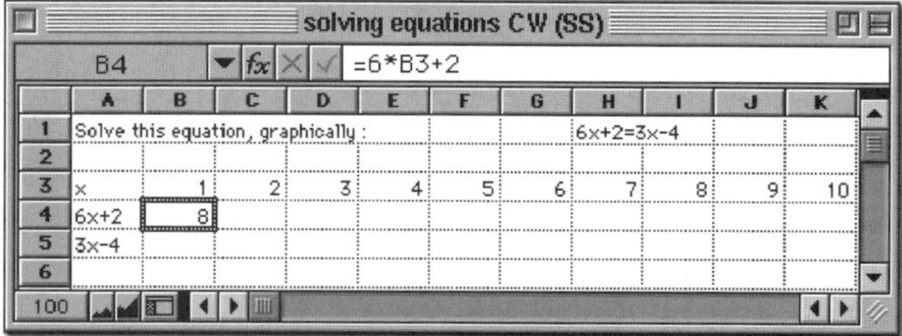

B4 | =6*B3+2

	A	B	C	D	E	F	G	H	I	J	K
1	Solve this equation, graphically:							6x+2=3x-4			
2											
3	x	1	2	3	4	5	6	7	8	9	10
4	6x+2	8									
5	3x-4										
6											

In cell B5, enter this formula:

=3*B3–4

B5 | =3*B3-4

	A	B	C	D	E	F	G	H	I	J	K
1	Solve this equation, graphically:							6x+2=3x-4			
2											
3	x	1	2	3	4	5	6	7	8	9	10
4	6x+2	8									
5	3x-4	-1									
6											

Copy the formulas across columns C through K.

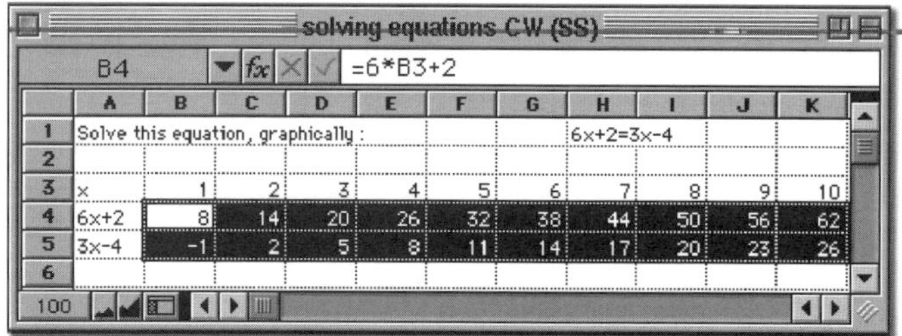

B4 | =6*B3+2

	A	B	C	D	E	F	G	H	I	J	K
1	Solve this equation, graphically:							6x+2=3x-4			
2											
3	x	1	2	3	4	5	6	7	8	9	10
4	6x+2	8	14	20	26	32	38	44	50	56	62
5	3x-4	-1	2	5	8	11	14	17	20	23	26
6											

Create a graphical display of the data. The resulting graph shows:

- The expression $6x + 2$ is greater than the expression $3x - 4$ for the values of x from 1 through 10.
- The expressions are moving to a point of intersection for values of x less than 1. (They are converging.)
- The expressions are diverging for values of x greater than 1.

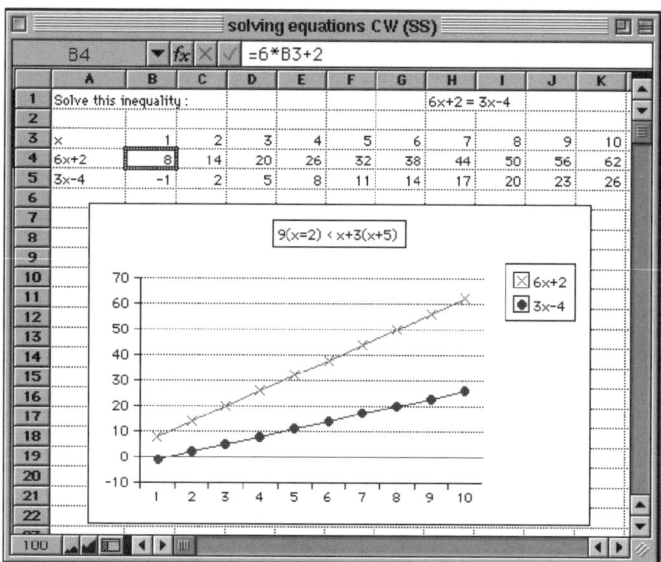

B4 | =6*B3+2

	A	B	C	D	E	F	G	H	I	J	K
1	Solve this inequality:							6x+2 = 3x-4			
2											
3	x	1	2	3	4	5	6	7	8	9	10
4	6x+2	8	14	20	26	32	38	44	50	56	62
5	3x-4	-1	2	5	8	11	14	17	20	23	26

The intersection is clearly less than 1, but the spreadsheet is set up to begin with 1. To adjust it, simply change the initial value in cell B3. Enter –5 in cell B3, and watch the values and the graph be instantly updated.

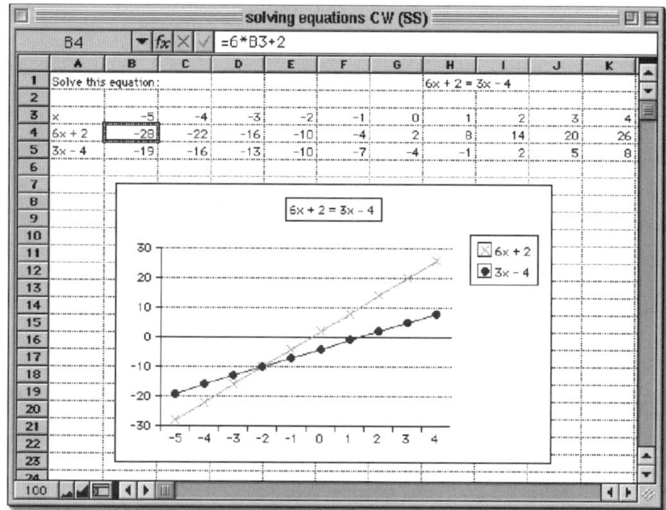

B4 | =6*B3+2

	A	B	C	D	E	F	G	H	I	J	K
1	Solve this equation:							6x + 2 = 3x - 4			
2											
3	x	-5	-4	-3	-2	-1	0	1	2	3	4
4	6x + 2	-28	-22	-16	-10	-4	2	8	14	20	26
5	3x - 4	-19	-16	-13	-10	-7	-4	-1	2	5	8

Now both the graph and spreadsheet values show that when x equals –2, the expressions are equal.

Solving for x in 3x – 4 = 14 – 2x. As in the previous problem, set up the spreadsheet and place the following formula in cell C22:

=B22+1

Copy this formula to cells D22 through K22. Instruct the spreadsheet to calculate the values for each of the following expressions:

=3*B22–4

in B23 and

=14–2*B22

in B24.

Copy these formulas from cells B23 and B24 to K23 and K24.

Create a chart of the information. The resulting graph shows:

- The expression $3x - 4$ is greater than the expression $14 - 2x$ for the values of x from a little more than 3 through 10. The data in the spreadsheet also shows this information.
- The expression $3x - 4$ is smaller than the expression $14 - 2x$ for values of 3 and smaller.
- The problem is to find the exact point at which these two expressions are equal.

The intersection point is somewhere between 3 and 4.

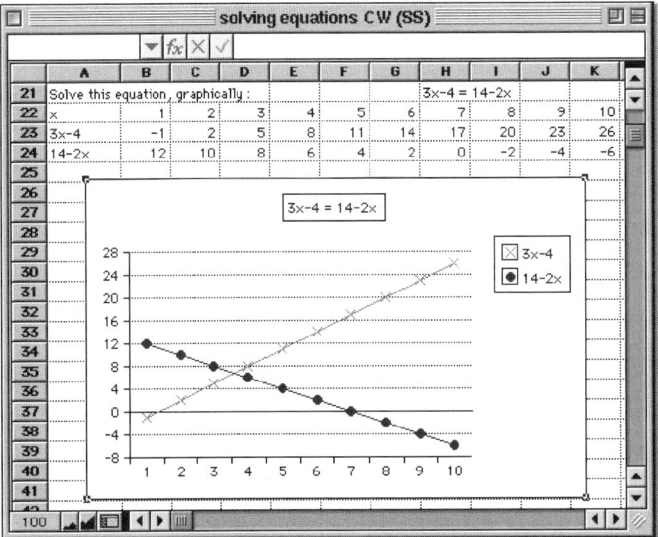

	A	B	C	D	E	F	G	H	I	J	K
21	Solve this equation, graphically:							3x-4 = 14-2x			
22	x	1	2	3	4	5	6	7	8	9	10
23	3x-4	-1	2	5	8	11	14	17	20	23	26
24	14-2x	12	10	8	6	4	2	0	-2	-4	-6

To make the increment smaller, change the formula in cell C22 to add .5 instead of 1. Enter

=B22+.5

into cell C22. The graph changes immediately, but the two lines are no longer straight. Why not?

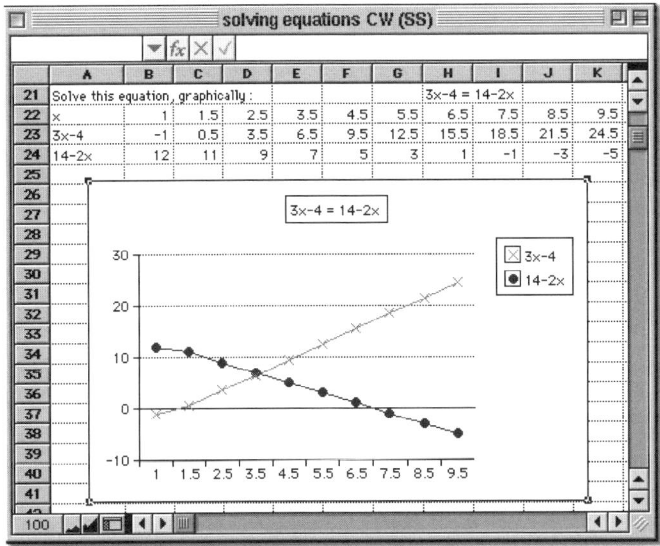

	A	B	C	D	E	F	G	H	I	J	K
21	Solve this equation, graphically:							3x-4 = 14-2x			
22	x	1	1.5	2.5	3.5	4.5	5.5	6.5	7.5	8.5	9.5
23	3x-4	-1	0.5	3.5	6.5	9.5	12.5	15.5	18.5	21.5	24.5
24	14-2x	12	11	9	7	5	3	1	-1	-3	-5

The change in the formula in C22 must be copied to the cells D22 through K22. Once that is done, the lines will straighten. But is the intersection point obvious?

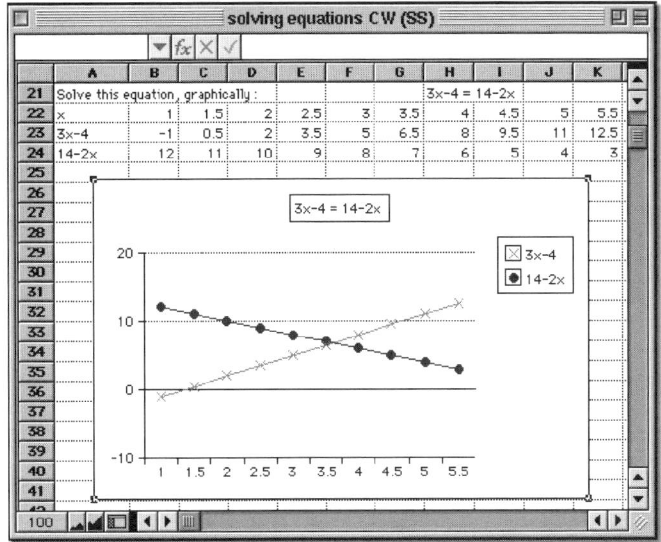

	A	B	C	D	E	F	G	H	I	J	K
21	Solve this equation, graphically:							3x-4 = 14-2x			
22	x	1	1.5	2	2.5	3	3.5	4	4.5	5	5.5
23	3x-4	-1	0.5	2	3.5	5	6.5	8	9.5	11	12.5
24	14-2x	12	11	10	9	8	7	6	5	4	3

No, it is somewhere between $x = 3.5$ and $x = 4.0$.

Change the information in row 22 to make x begin at 3 and the increment become .1. The new formula in cell C22 will be

=B22+.1

This formula must be copied to cells D22 through K22.

solving equations CW (SS)

C22 =B22+0.1

	A	B	C	D	E	F	G	H	I	J	K
21	Solve this equation, graphically:							3x-4 = 14-2x			
22	x	3	3.1	3.2	3.3	3.4	3.5	3.6	3.7	3.8	3.9
23	3x-4	5	5.3	5.6	5.9	6.2	6.5	6.8	7.1	7.4	7.7
24	14-2x	8	7.8	7.6	7.4	7.2	7	6.8	6.6	6.4	6.2

Now it is clear. The expression $3x - 4$ equals the expression $14 - 2x$ when $x = 3.6$. Both the graph and the spreadsheet data demonstrate this fact.

Using this process of surrounding the point of intersection with finer and finer ranges encourages students to estimate the intersection from the data and the graphical representations. More examples will enhance students' ability to work with formulas in spreadsheets. Consider these:

$$2(7 - 4x) = 6x - 7 \quad \text{and} \quad 4(x + 3) - 6 = 24$$

In this example, the formula for cell B47 requires parentheses:

=2*(7–4*B46)

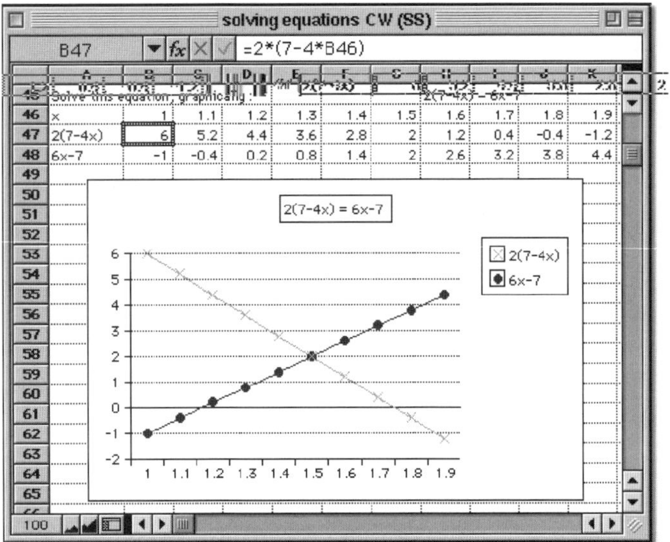

In the next example, the expression 4(x + 3) – 6 is equal to 24. This value must be entered in cells B66 through K66. The graph of this constant becomes a horizontal line.

The task now is finding where this horizontal line intersects with the line for $4(x + 3) - 6$.

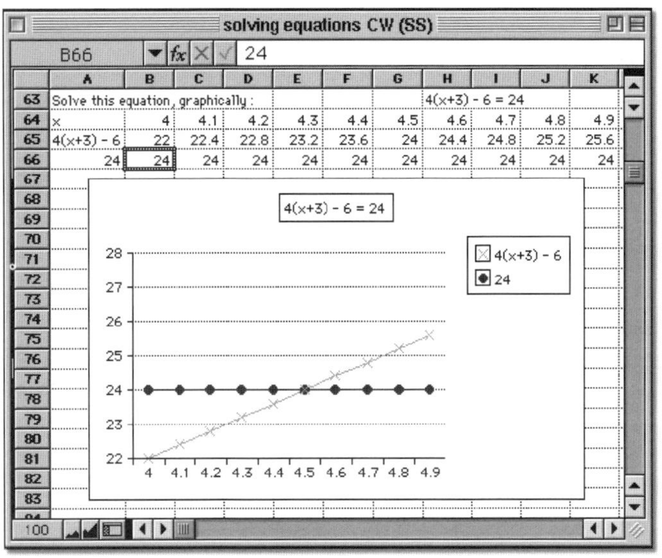

A Natural Extension: Inequalities

The algebraic solution of inequalities always presents special difficulties for students because of their reliance on rote methods. Their primary concern is, "When do I switch the sign?" A graphical approach gives clear visual clues to solve inequalities. For example, consider this question:

For what values of x is $9x - 2$ less than $x + 3(x + 5)$?

Using the same method as that used with equalities, students simply change the question from finding where the expressions are equal to finding where one is less than the other. The graphs are quite clear.

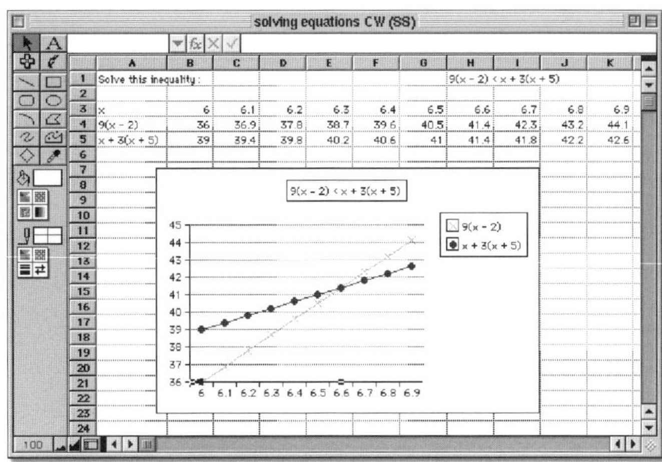

For values of $x < 6.6$, $9x - 2$ is less than $x + 3(x + 5)$. There the line for $9x - 2$ is *lower* than the line for $x + 3(x + 5)$. Using the tool bar, students can draw an arrow to show this solution.

As noted in the *Curriculum and Evaluation Standards for School Mathematics* (National Council of Teachers of Mathematics, 1989), more attention must be given to (1) developing an understanding of variables, expressions, and equations; (2) using a variety of methods to solve equations and informally investigate inequalities; and (3) using estimation to solve problems. Concurrently, less attention should be given to manipulating symbols, memorizing procedures, and drilling on equation solving.

Using spreadsheets provides a mechanism to make these changes in how students solve equalities and inequalities. As pencil and paper computation becomes less important, the skills and understandings supported by spreadsheet use to solve equalities and inequalities give students methods that support a conceptual understanding. Rote procedures do not. With this approach, students connect the idea of x varying throughout their work with the discovery of different values for the expressions and graphically observing numerical results.

Reference

National Council of Teachers of Mathematics, Commission on Standards for School Mathematics. (1989). *Curriculum and evaluation standards for school mathematics.* Reston, VA: Author

JULIO'S RUN

Studying Graphs and Functions

This spreadsheet and graphing activity uses a simple physical exercise routine to help algebra students learn to create functions and graphs in line with the NCTM standards.

By Azita Manouchehri and Lyle Pagnucco

Subject: Algebra

Grade Level: 7–11 (Ages 12–17)

Technology: Graphing calculators, spreadsheets

Standards: *NETS* 3 & 5. (Read more about NETS at www.iste.org—select Standards Projects.) *NCTM* 2, 5, 6, & 10. (Read the new mathematics standards at www.nctm.org.)

The National Council of Teachers of Mathematics (NCTM, 1989) has recommended that secondary students devote much attention to the study of graphs and functions. The NCTM standards also call for all students to:

1. model real-world phenomena with a variety of functions,
2. analyze and represent relationships using equations and graphs,
3. translate symbolic and graphical representations of functions,
4. recognize the effects of parameter changes on the graphs of functions,
5. represent situations that involve variable quantities with expressions, and
6. use graphs as tools to interpret expressions and equations.

One of the most essential requirements for fulfilling these recommendations is improving students' ability to build geometric models for analytical situations or vice versa. This activity allows students to experience visual and analytical domains related to the study of functions with the aid of graphing calculators. It was conducted with two groups of Algebra I and II students. We used TI-83s and spreadsheets, but any type of graphing calculator will work.

Consider Julio's Run

> Every morning Julio runs for approximately 45 minutes. For the first five minutes, he runs slowly and then increases his speed until he reaches full speed. Julio runs at that rate for about 10 minutes and then slows down gradually over five minutes until he is running at half speed. He continues the half-speed run for 20 minutes and then gradually slows to a stop.

We asked students to read the scenario and graph the pattern of Julio's daily run. They had to determine what form the graph took and what variables described the scenario. Furthermore, each student was asked to present his or her graph to others and to identify critical points while building the representation. Following these presentations, students examined the class graphs and decided in small groups which adequately represented the conditions of Julio's run and discarded the rest. We encouraged them to analyze and compare in great detail these graphical representations. This activity leads to two major investigations:

1. Find the best fit function for the data.
2. Investigate families of curves and the effects of parameters on the behavior of the defined function.

Later, we asked students to consider the graphs they initially discarded. Students modified the conditions of the original scenario to match them. This task helped students understand domain and range as well as the graphs themselves. Moreover, it helped them learn to pose situations that matched the restrictions demonstrated by graphs.

Investigate Julio's Run

Students decided that a plot of time against distance was an appropriate and convenient model for Julio's run. Figure 1 shows three representations.

One student focused on two features. The first was the initial five minutes. She described the gradual increase in speed by saying, "The curve needs to rise slowly at first because not much distance is covered. Then, it begins to rise quickly to correspond to the rising pace of the runner." Then, she focused on the end of the run, remarking, "When you stop, distance does not increase, so the curve should flatten to a horizontal line." Her explanation did not address the changing slope along her curve, but gave others the opportunity to ask, "Why does the slope change?"

The second student also provided an explanation for his curve by identifying that the data focused on the change between full and half speed. He claimed the curve showed slowing and thus needed to bend down for a short period and then back to half speed. This student also made the claim that the curve needed to bend back to a horizontal line when the runner stopped because "The runner doesn't cover any more distance." Although there was no indication of a conclusive statement defining half speed, this student's solution provided an opportunity to probe for a more precise definition. At that point, the question under investigation became whether the slope had something to do with defining half speed.

The third student generalized the lower the point on the curve, the slower Julio was running. This argument came under fire when other students asked, "How do you explain that the runner continues to increase speed over the entire distance?" The question was authenticated when a student inquired if anyone had seen a marathon where a runner continues to increase his or her speed over the entire race. Although there were truths and falsities in each claim, the substance of the conversation was focused on finding an acceptable explanation for a real-life phenomenon.

Find the Best Fit Function

After the discussion and students' analyses of each graph, we asked students to find the best fit function for the data. We allowed them to use graphing calculators, and we asked guiding questions that motivated their search of a variety of parent functions. Can a linear function describe the data adequately and accurately? Why or why not? How

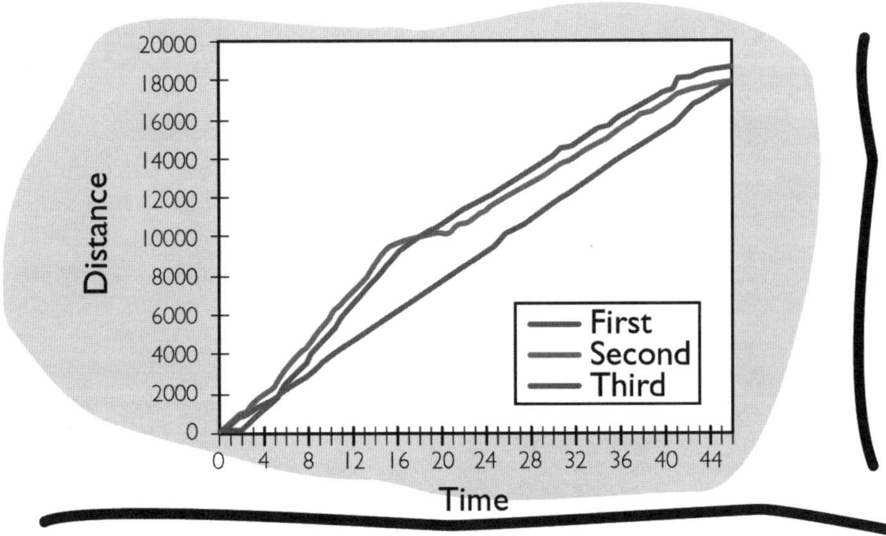

Figure 1. Three student representations of Julio's run.

about a quadratic function? A third-degree polynomial? An equation in exponential form?

Our goals were to increase students' intuitive knowledge of the relationships between the graphic and symbolic representations of the data and to improve their reasoning skills.

Using their graphing calculators, students examined different functions and developed a sense for the functional approximation of the curve representing the data. Making a sound and educated decision about the best fit function, however, required a more systematic search. We encouraged students to consider, for example, the first student's graphical representation of Julio's run $\{0 < x < 5\}$, to identify the best fit curve within this domain, and to pick critical points from which the best fit function could be referenced. The notion was not to use methods such as least squares to define the function but to use a more general definition that hinged on curve characteristics and critical points. For example, we suggested they use $x = 0, 5, 10, 15, 20, 25, 30, 35, 40,$ and 45, and then define a curve (perhaps polynomial in nature) that reflected the change in curvature and closely approximated these points (Figure 2). Our only concern at this point was the nature of the curve within the interval.

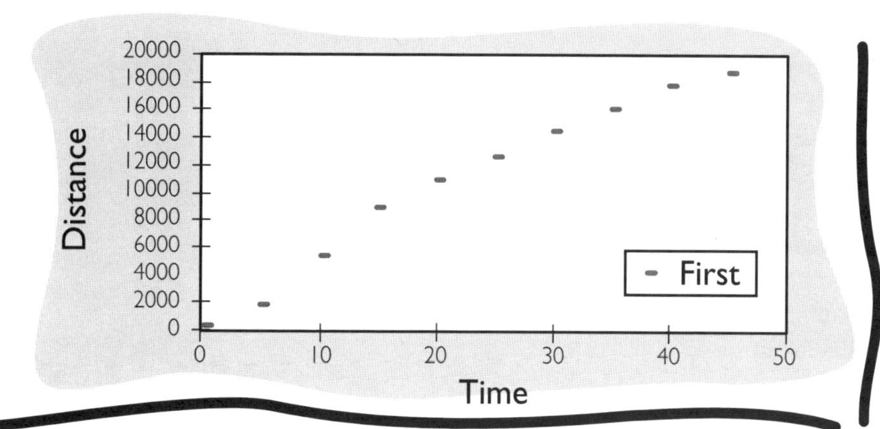

Figure 2. Reference points.

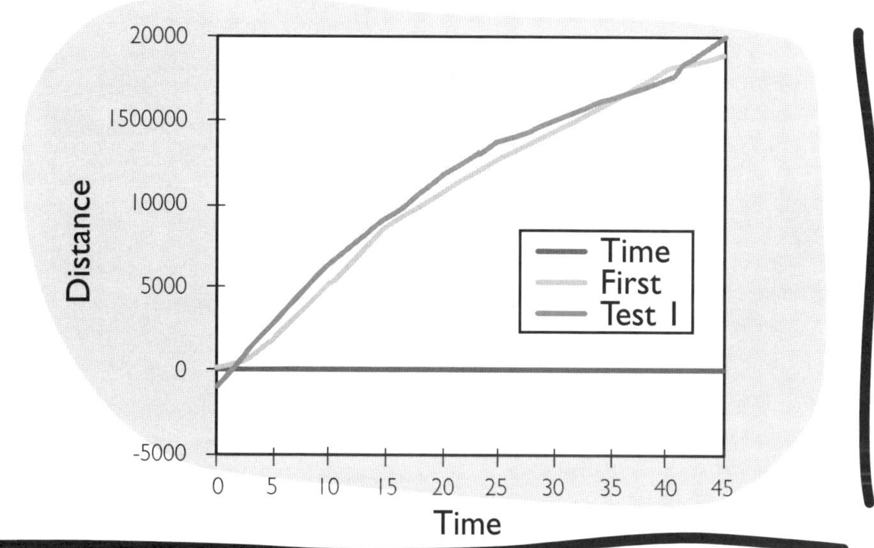

Figure 3. Graph of student's fourth-degree polynomial approximation of Julio's run.

The students claimed that the general distribution of points indicated at least a fourth-degree polynomial definition, based on the changing nature of the curve. This was a critical point in the discussion, as the fourth-degree polynomial could be a minimal choice in selecting a best fit function. Some students suggested that the function could be of higher degree because it could give a more precise match for the curve. After some argument, they decided to begin with the fourth-degree polynomial and investigate the appropriateness of a higher degree polynomial later. We agreed with the students' decision because they could take a standard fourth-degree polynomial function with variable coefficients and work to intercept as many of our reference points as possible. Although mathematically this was not the most efficient method, it emphasized the effect of various coefficients in generating our curve. That is, by considering various situations, students examined families of curves and the role of parameters on the corresponding graphs. Later in the discussion there was an opportunity to focus on making this task more efficient, creating the need for an algorithmic definition.

This point was when a spreadsheet and its graphing capabilities and a graphing calculator became important tools in the process. As illustrated in Figure 3, a function based on the standard form of a fourth-degree polynomial was used in a guess-and-check style to match a curve to the given data. The students made conjectures and tested those conjectures immediately. Figures 3 and 4 illustrate one such approximation based on a standard polynomial function:

$$f(x) = ax^4 + bx^3 + cx^2 + dx + e$$

After using a guess-and-check method in this manner, students should understand how algebraic and graphic representations relate to each other.

Extend the Activity

Julio persuades Mary and Nick, his two neighbors, to accompany him on his daily run. On the first day Nick, because of his bad back, is unable to run as fast as Julio and maintains a speed that is exactly one-third of Julio's throughout the 45 minutes. Mary, however, decides to outrun Julio and runs twice as fast as he does. On the second day, Nick increases his speed to exactly half of Julio's, and Mary runs three times as fast as Nick.

1. Graph Mary's and Nick's runs on each day. Compare and contrast these with Julio's graph.
2. What would be the best fit function describing each? How do these functions relate to one another?
3. What if on the third day Julio, Mary, and Nick decided to have a 15-minute race toward the end of their run. What would the graph look like if Nick were the winner?
4. What if the race was during the first 15 minutes of the run. How would the new set of graphs compare with the last ones?

With such extensions, we attempted to motivate the need for finding algorithms for transformation of functions as well as studying the effects of parameters on the behavior of different functions.

a	0.0104	
b	–0.74	
c	6.9	
d	705	
e	–1000	
Time	**First**	**Test 1**
0	0	–1000
5	1760	2611
10	5260	6102
15	8760	9149
20	10835	11587
25	12585	13405
30	14335	14747
35	16085	15916
40	17835	17370
45	18710	19720

Figure 4. Table of student's fourth-degree polynomial approximation of Julio's run.

Explore & Evaluate the Activity

The heart of the activities presented in this article is the generation of "what if" questions that motivate (1) the investigation of more sophisticated ideas, (2) the search for counter-examples, and (3) the modification of students' misconceptions about the nature of their solutions as well as evaluation of their peers.

The teacher's role is fundamental. Whole-class discussions focus on clarifying student responses by self-evaluation and on developing strategies for productive problem solving. Such discussions help students make their existing intuitive and informal knowledge more explicit.

In our classrooms, we have accompanied exploration of such activities with situations that involve data collection through direct experimentation by students. In such settings, students' enthusiasm for problem solving, forming and clarifying arguments, and persuading one another of the validity of their conjectures about symbolic and graphical representations of the data increases. Such behaviors are in line with the underlying themes of the NCTM standards (1989) related to mathematical reasoning, communication, and problem solving.

Reference

National Council of Teachers of Mathematics. (1989). *Curriculum and evaluation standards for school mathematics.* Reston, VA: Author.

Chaos, Fractals, and Polynomials

By J. Louis Tylee and Thomas B. Tylee

In Steven Spielberg's dinosaur movie *Jurassic Park,* Dr. Ian Malcolm bases his claim that the park was doomed from the beginning on his knowledge of chaos theory. He conjectured that small variations in certain inputs (butterfly wings flapping) would result in chaotic behavior in the park. Well, the chaotic behavior occurred, but was it due to chaos or just angry dinosaurs? Chaos has long been associated with population systems, so maybe Dr. Malcolm was right. If so, it's a great example of chaos in a real system.

In this article, we won't study dinosaurs, but we will study chaos and the way it results from something as simple as solving for the roots of a polynomial. We know that an Nth-order polynomial has, at most, N distinct roots, or N distinct solutions. We will use a numerical technique to find these solutions. Part of this numerical technique will involve making an initial guess at the solution. We will see that in some cases, very slight changes in this initial guess will yield quite different results; hence, the solution process is chaotic. However, we will also see that there is some order in this chaos, if that seems possible. By mapping out the solution space in a certain manner, spectacular, colorful patterns known as fractals appear.

Chaos—Numerical Solution of Polynomials

Solving a linear algebraic equation is a simple process; and to find the roots of a second-order polynomial, we use the quadratic equation. There are also specific procedures for finding the roots of a third-order polynomial. However, for fourth-order and higher polynomials, we need other ways to find roots. Usually a numerical approach is used. In general, for a polynomial of the form

$$f(z) = 0$$

we guess at the solution z and evaluate f(z). We then see how close we are to zero. Based on the distance from zero, we make a refined guess at the solution and continue until we have "converged" to a root, or solution.

A common algorithm for doing this solution adjustment is called the Newton-Raphson technique. It is illustrated for a simple curve in Figure 1. If z_i is our current guess at a solution, we compute $f(z_i)$ and $f'(z_i)$ (the slope or derivative of the function). Our refined (or next) guess at the solution is where the line tangent to the curve at z_i crosses the horizontal axis (a linear approximation to the solution). From Figure 1, we can show this value to be

$$z_{i+1} = z_i - f(z_i)/f'(z_i)$$

To solve for the roots of a polynomial using this technique, we:

1. Initialize i = 0. Make a first guess at the root z_0.
2. Compute z_{i+1} using the preceding formula.
3. Check if $|z_{i+1} - z_i|$ is a small number. (You must decide how small the number should be.) If it is a small number, the solution has converged. The root is z_{i+1}. If $|z_{i+1} - z_i|$ is not small, increment i and return to step 2.

A major difficulty in using the Newton-Raphson technique is making sure we find all the roots. An Nth-order polynomial has up to N distinct roots. The Newton-Raphson technique will converge to just one root, depending only on the initial guess at the root. We must find N different initial guesses that converge to the N different roots to completely solve an Nth-order polynomial. It seems logical that if our initial guess at the root is near the true value, that guess will converge to that root. Yes, it seems logical, but it's not always true! Using the

Newton-Raphson technique to find polynomial roots is a chaotic process. We will see (by knowing the correct answer ahead of time) that even if we make a first guess which is near a root, it is possible the converged solution is at another root, far away from the initial guess. And, even more bizarrely, we will see that the slightest change in an initial guess can result in a completely different converged solution! Let's look at one particular polynomial, where we already know the location of its roots.

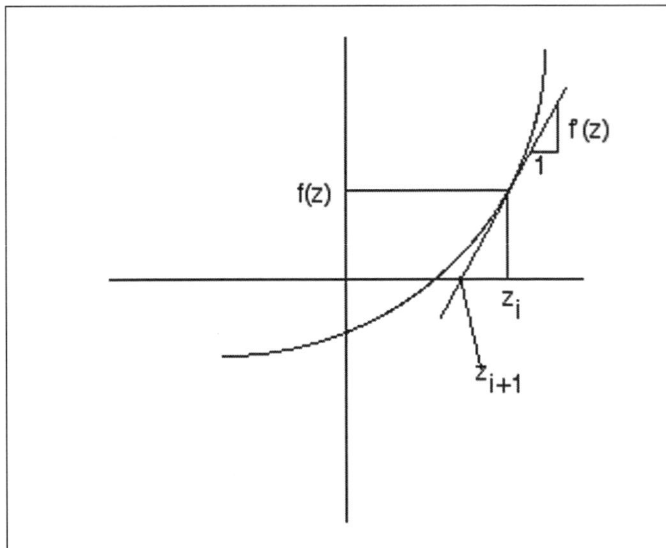

Figure 1. Newton-Raphson solution technique.

Fractals—Mapping Polynomial Solutions

Let's use $f(z) = z^N - 1 = 0$, where N is some positive integer. Complex number arithmetic is needed to find the roots of f(z). A major reason for choosing this polynomial is that its roots are easily found analytically. Using the complex plane as a reference, the roots will lie along a circle of radius 1 with its center at the origin (0, 0). The first root will always be at z = 1 + i0, i.e., on the real axis, with the remaining N - 1 roots on the "unit" circle, separated by 360/N degrees. For example, if N = 2, we have two real roots at z = 1 and z = -1. In Figure 2, we plot the roots for N = 3. (Note the 120-degree separation around the circle.)

First we need to calculate $f'(z_i)$, or the slope of our function.

$$f'(z_i) = Nz_i^{N-1}$$

Substituting the function and derivative into the Newton-Raphson equation yields

$$z_{i+1} = z_i - (z_i^N - 1)/Nz_i^{N-1}$$

which reduces to

$$z_{i+1} = (N - 1)z_i/N + 1/Nz_i^{N-1}$$

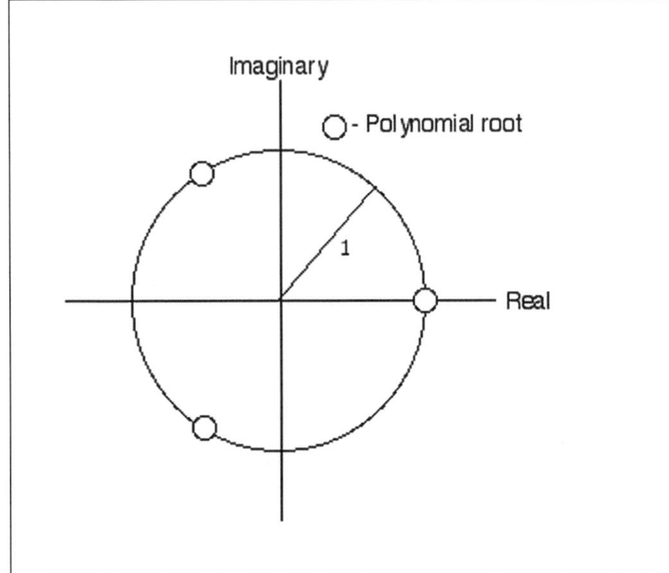

Figure 2. Roots of $z^3 - 1 = 0$.

This is the expression we will use to find the roots of our polynomial and, in the process, draw some fractals. Let's outline that procedure.

To use the Newton-Raphson method to find polynomial roots, we need an initial guess at a root, z_0. Every point in the complex plane is a potential solution, hence, a potential guess. If we use each point in the complex plane as an initial guess, compute the resulting converged root, and track which guess converged to which of the N possible roots, we obtain a mapping of initial guesses to final roots. This mapping, when drawn in color on a computer screen, can provide beautiful and surprising results. These mappings are fractals. We will develop a procedure for generating fractals on a computer for the selected polynomial. For each step in the procedure, the pertinent equations are provided to help you in programming the technique yourself on a computer. You can also use the provided equations to extend the technique to other polynomials.

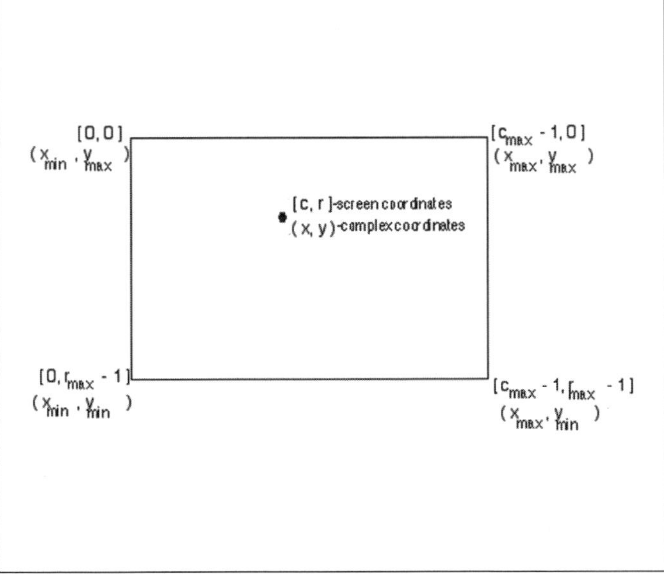

Figure 3. Screen and complex coordinates.

Search Region and Coordinate Systems

We can't search the entire complex plane for roots, so a first step is to determine the search region. We need to establish limits on the range of the initial guesses, z_0. Each z_0 has a real and imaginary part,

$$z_0 = x_0 + iy_0$$

so, equivalently, we need limits on x_0 and y_0. We assume that x_0 can range from x_{min} to x_{max} and y_0 can range from y_{min} to y_{max}, or

$$x_0: \{x_{min}, x_{max}\}$$
$$y_0: \{y_{min}, y_{max}\}$$

These limits cover an infinite number of potential initial guesses. How do we reduce this to a finite number? Fortunately, our desired display device, a computer monitor, has a finite number of points (pixels) for which solutions can be computed. Therefore, we need to be able to convert a point on the screen to the corresponding (x, y) pair in the complex plane. This is a straightforward process. Figure 3 relates screen coordinates to complex coordinates.

Screen coordinates are enclosed in square brackets, [c, r], where c is the screen column and r is the screen row. The upper left corner of the screen is [0, 0] while the lower right corner is [cmax - 1, r_{max} - 1], where c_{max} is the number of columns on the screen and r_{max} is the number of screen rows (meaning that the screen is c_{max} pixels wide and r_{max} pixels high). Complex coordinates are enclosed in parentheses, (x, y). The upper left corner of the complex plane is (x_{min} y_{max}) and the lower right corner is (x_{max}, y_{min}). Using Figure 3, we see that the point (x, y) is related to the screen coordinate [c, r] by

$$x = x_{min} + c(x_{max} - x_{min})/(c_{max} - 1)$$
$$y = y_{max} + r(y_{min} - y_{max})/(r_{max} - 1)$$

These two equations are different in form because of the way screen coordinates are defined. Note that x increases as the corresponding screen coordinate c increases, yet y decreases as its corresponding screen coordinate r increases.

Establishing an Initial Guess

To draw the fractal representing the solution of the chosen polynomial, we will assume that every point on the computer screen is an initial guess at a root. Hence, we will systematically cycle through every point on the screen, using the ranges

$$c: \{0, c_{max} - 1\}$$
$$r: \{0, r_{max} - 1\}$$

This means we will look at (c_{max}) x (r_{max}) potential guesses.

After we have selected the current screen coordinate [c, r], we use this to initialize our guess at a root. First, c and r are converted to the corresponding initial complex coordinates using the previously developed relations

$$x_0 = x_{min} + c(x_{max} - x_{min})/(c_{max} - 1)$$
$$y_0 = y_{max} + r(y_{min} - y_{max})/(r_{max} - 1)$$

These values then establish the initial root guess, z_0:

$$z_0 = x_0 + iy_0$$

and we also initialize the iteration counter i at zero (i = 0).

Iterating and Checking for Convergence

At this point, we use the Newton-Raphson equation to compute updated estimates of the polynomial root. Recall that this expression is

$$z_{i+1} = (N - 1)z_i/N + 1/Nz_i^{N-1}$$

Thus, we use z_0 to compute z_1, z_1 to compute z_2, and so on. A few words on evaluating this expression. To compute z_i^{N-1}, we multiply z_i by itself N - 2 times. Then, prior to addition to the $(N - 1)z_i/N$ term, we must divide Nz_i^{N-1} into 1. Also, for small z, this denominator term may become very small and cause numerical problems. To remedy this, restrict the denominator to a minimum value. It is suggested that in all operations, the real and imaginary parts of the computation be kept separate.

After computing each new value for z, we must check for convergence; that is, we must see if we have reached a root of our polynomial. A suggested convergence check is

$$|z_{i+1} - z_i| < e$$

where e is some small number. Therefore, if our newly computed value has not changed much from the previous value, we assume that we have reached a root, the Newton-Raphson iterations are stopped, and we color-code the solution as described in the next section. If convergence is not attained, we increment i and compute the next z value. In a computer implementation, it is suggested that convergence checks be made on both the real and imaginary parts of z. Some limit on the number of allowed iterations should also be in place.

Color-Coding and Generating Fractals

Once we have converged to a root, we need to color-code the corresponding screen point (pixel) to indicate which root was reached. This is a simple process. We assign each of the N possible roots a distinct color. Then, after determining which root the solution procedure has converged to, we color the corresponding pixel with the color assigned to that root.

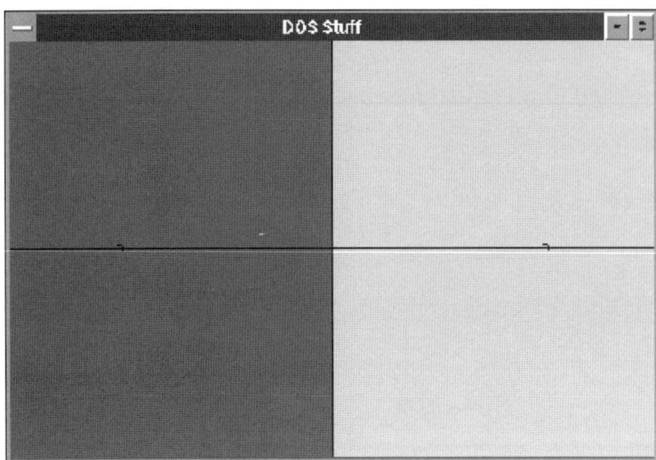

Figure 4. Fractal for N = 2.

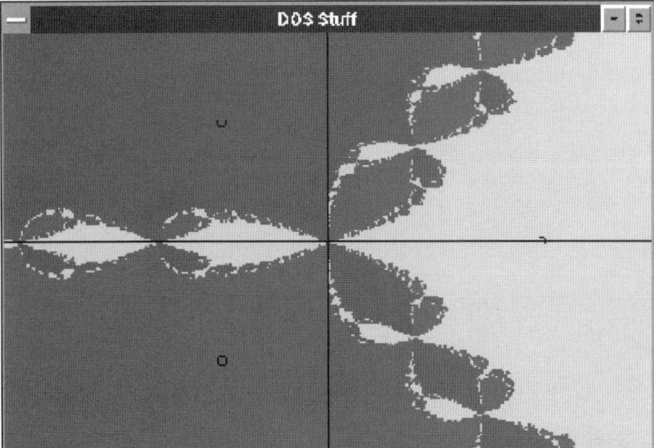

Figure 5. Fractal for N = 3.

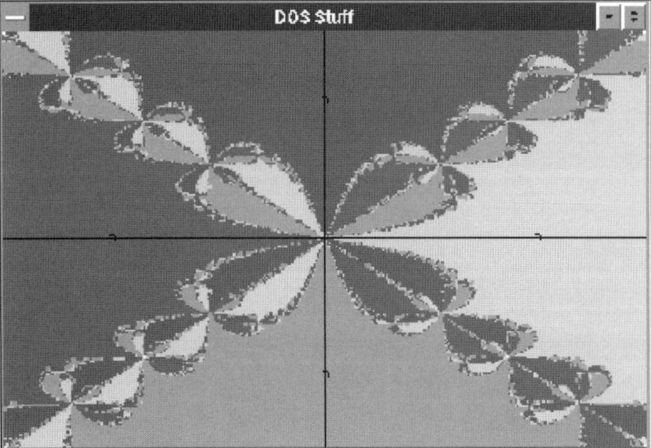

Figure 6. Fractal for N = 4.

After we have assigned a color to each pixel on the computer screen, the solution process is complete. At this point, a beautiful fractal will (or should) appear. Let's look at some fractals resulting from applying the solution procedure to our selected polynomial for values of N = 2, N = 3, and N = 4. The fractals shown in Figures 4, 5, and 6 were computed using a computer running MS-DOS and a code written in Microsoft Quick C. Screen dimensions used were 320 horizontal pixels by 200 vertical pixels, with up to 16 colors available. Convergence checks were made with a value of e = 0.0001. Iterations were limited to a maximum of 1,000. (This maximum was never reached, but it is needed to avoid infinite loops.) Also, in each fractal, the range of the initial root guesses was limited by

$$x_{max} = -x_{min} = y_{max} = -y_{min} = 1.5$$

Figure 4 shows computed results for N = 2. With N = 2, there are two roots, one at z = 1 and one at z = -1. (The roots are marked by circles in the three figures.) Points in the complex plane that converged to the negative root are colored magenta (purple), while points that converged to the positive root are colored yellow. The fractal in Figure 4 is quite boring. It simply shows that each initial guess converged to the root closer to that guess. Any guess with a negative real part converged to the negative root. Any guess with a positive real part converged to the positive root.

With N = 3, we get more interesting results. In Figure 5, initial points that converge to z = 1 + i0 are colored yellow, those converging to z = -0.5 + i0.866 are colored magenta, and those converging to z = -0.5 - i0.866 are colored red. We see that points around the roots tend to converge to the nearest root. However, some interesting behavior occurs along the lines equidistant from each root. Colorful, pear-shaped patterns appear, and guesses far from a particular root can still converge to that root. These colorful patterns are recognized as fractal. Let's look at an example of the chaotic behavior of the Newton-Raphson solution technique. If our first guess at a root of $z^3 - 1$ is $z_0 = 1.07 + i1.5$, the converged root is z = -0.5 + i0.866. If we change the real part of the initial guess slightly to 1.08, the converged root is completely different, that is, z = -0.5 - i0.866. And if the real part is 1.09, the converged root is z = 1! Hence, a simple 1 percent change in the initial real part can yield completely different polynomial solutions.

Figure 6 shows similar behavior along boundary lines separating the four roots when N = 4. In this fractal, complex plane points that converge to the root at z = 1 + i0 are colored yellow, those converging to the root at z = 0 + i1 are colored magenta, those converging to the root at z = -1 + i0 are colored red, and those converging to the root at z = 0 - i1 are colored cyan (bluish).

Conclusion

Now we can ask, "What good is all this?" Well, first it's fun to draw fractals and discover that math can generate art. Try different ranges on potential solutions and see how the fractals change as you expand/narrow the search range. Try different coloring schemes and see how the fractals change. For example, use different shadings depending on how fast (number of iterations) a first guess converges to a certain root.

More seriously, the chaos inherent in the Newton-Raphson technique should instill a bit of caution in anyone who uses numerical solutions. Most real-world problems are too difficult to attack directly—we often must resort to some numerical technique for a solution. And many real-world problems have multiple solutions. This brings up many interesting questions. If a problem has multiple solutions, how do we know when we've found them all? Nth-order polynomials have at most N distinct solutions. In complicated nonlinear problems, we may not know how many solutions there are. Once we're satisfied that we've found all the solutions, how do we choose the "best" solution? Some other criteria must be applied. And once we've found the solution we like, what if one of our initial parameters changes a bit? We've seen that slight changes in parameters can yield completely different results in our solution process. We may no longer get the desired answer. Such questions make real-world problem solving interesting when chaos becomes involved.

Try programming the fractal-drawing process on your own computer using whatever language you know. (See the note at the end of this article for information on obtaining a disk copy of the programs I've written in Microsoft QBasic and Quick C. They can be used and/or modified to make fractals work with your system.) Use the example polynomial or try one of your own. A forewarning: The program that generated the fractals for this article computes polynomial solutions at 64,000 (320 x 200) different points and hence, depending on the computer and language used, can take a long time to execute. Try modifying the fractal code. Can you make it more general, allowing the solution of a general polynomial of the form

$$f(z) = z^N + a_{N-1}z^{N-1} + a_{N-2}z^{N-2} + \ldots + a_1 z + a_0 = 0$$

or make it work with just any function f(z) (not necessarily a polynomial)? In such cases, you probably won't know the roots ahead of time, so how do you draw a mapping? I suggest that you keep track of each root found and, after finding a new root, assign a distinct color to that root. Look at various solution spaces for distinct behaviors. If you have a fancy graphics monitor (for example, an SVGA for an MS-DOS machine), change the display mode and screen dimensions and use more colors in drawing the fractals. I'm sure you can come up with other modifications. But the main point is to have fun!

Program Availability: If you'd like to receive a disk copy of the Microsoft QBasic and Quick C programs I've written to generate the fractals described in this article, send a blank, MS-DOS formatted 3.5" disk with a self-addressed, stamped disk mailer to J. Louis Tylee, 15600 NE 8th, Suite B1-314, Bellevue, WA 98008.

CHAPTER 9

Number and Operations

The Conversation Continues

Anne: Hey, Ivan, have you ever had this experience at a restaurant? You're dining with a group of friends, the waiter brings one bill, and everyone tries to figure out how to split the bill in a fair way. Often you see people within the group trying to figure out what they had to eat and how much it cost. Plus, they try to factor in their percentage of the total tax in addition to their portion of the tip that was automatically included in the table's total bill. It's often a confusing time with so many people trying to estimate or use mental mathematics to determine what they owe. In some cases, the group simply agrees to divide the bill equally among the number of people there, regardless of what each person ordered.

Ivan: Well, just figuring out the tip can be a challenging experience for most people. At least with the automatic tip being 15%, many people figure out what 10% is in their heads and then add half as much again to the 10% to get 15%. There's some talk about raising the automatic gratuity to 17%, which will pose new mental math challenges to restaurantgoers.

Anne: This simple example of going to a restaurant points out our need to develop strong number-sense skills in students so that they'll feel comfortable when addressing number issues that arise in their daily lives.

Ivan: And with the increase in international travel, people have to deal with the more complicated number issues of exchange rates, which change on a regular basis. At least European countries have gone to a single monetary system with the Eurodollar. This change was partially due to wanting to eliminate having to deal with many different monetary systems and exchange issues.

Anne: Speaking of traveling, I remember your telling me a story about a friend who was shopping while on vacation and was admiring a dress in a store window.

Ivan: Yes, and when she inquired about the price of the dress, the salesperson said it was "thirty-nine ninety-five." My friend said, "That's not bad at all." And the saleslady said, "That depends on where you put the decimal point." Apparently, the dress cost $3,995 and not $39.95.

Anne: Well, the restaurant scenario and the shopping stories are just two situations in which dealing with numbers can be baffling. The articles in this chapter address developing number sense through technology so that numbers will be less confusing when we encounter them.

INSIGHT

Numeracy and Estimation

By Anne Raymond

What is "numeracy"? Numeracy is mathematical literacy. It refers to the ability to understand basic mathematical ideas. It also deals with being able to grasp how mathematical concepts such as chance, logic, and graphs permeate our daily lives and the decisions we make (National Research Council, 1989).

In toda y's society, it is socially unacceptable to be illiterate. In fact, people who cannot read usually try to hide that fact. On the other hand, it seems perfectly acceptable for people to be innumerate. People actually find they are in good company and feel as if they are part of the "normal" group when they claim, "I've never been any good at math."

In an insightful and humorous book titled *Innumeracy,* Paulos (1988) provides many examples of how people misinterpret data and are often fooled by others (such as politicians and advertisers) who are trying to manipulate them with numbers. He expresses outrage that today's society is largely innumerate while at the same time it depends so much on mathematics and science it does not understand.

Bennett, Briggs, and Morrow (1996) suggest that part of the reason innumeracy persists is because society still holds many misconceptions about mathematics. The authors provide arguments as to why these commonly held beliefs are not true. Here are a few misconceptions they claim exist:

- Math requires a special brain
- Math in modern issues is too complex
- Math makes you less sensitive
- Math makes no allowance for creativity
- Math provides exact answers
- Math is irrelevant to my life

In response to an increasingly technological world that will require its citizens to become more and more savvy about numbers, the National Council of Teachers of Mathematics set standards for number and operations that address many issues related to numeracy.

NCTM Number and Operations Standard

Number and Operations: Instructional programs from prekindergarten through Grade 12 should enable all students to

- Understand numbers, ways of representing numbers, relationships among numbers, and number systems

- **Understand meanings of operations and how they relate to one another**
- **Compute fluently and make reasonable estimates**

According to the NCTM standard on number and operations, students need to be able to understand relationships among numbers and understand the operations. An activity involving "magic squares" can help students develop number sense in both of these areas.

MAGIC SQUARES AND SPREADSHEETS

A magic square is a square with cells such that each row, each column, and each diagonal add to the same sum. The cells are filled with a series of consecutive numbers. For example, consider the 3 x 3 square below.

Try to insert the numbers 1 through 9, with each number being used exactly once, so that each row, column, and diagonal sums to 15. Through trial and error, students may find this possible solution:

2	7	6
9	5	1
4	3	8

Magic squares can come in any dimension and can be developed to have magic squares within magic squares. They provide a means of having students build number sense through problem solving with the guess and check method. You can facilitate the guess and check by setting up a spreadsheet to find sums at all angles of the magic square and have students experiment with which arrangements would work. (Download the Magic Squares spreadsheet activity from the companion Web site for this book, http://education.bellarmine.edu/baugh/.)

Many moments throughout our day call for us to make connections to numeracy. In most real-life situations, we do not use exact numerical computations, we use estimates. We often have to estimate time, size, money, and distance. In forming estimates, we have to approximate whole numbers, fractions, decimals, and percents.

There are a variety of ways to develop students' ability to form reasonable estimates for the purposes of checking their arithmetic work or making decisions. Here are a few ideas.

IDEA 1—MAGAZINE AND NEWSPAPER HEADLINES

Look at each of the following headlines and determine from the context whether the number in the headline is an estimate or an exact value. Explain your decision.

- ❑ A record attendance of 50,000 fans at opening day!
- ❑ Eighteen seniors graduate with honors
- ❑ The school fundraiser netted $400
- ❑ The Backstreet Boys have their 19th top ten hit song
- ❑ There are 112 stores and restaurants at the new mall
- ❑ SAT scores increased 10% this year at local high school
- ❑ 100 cases of flu have been reported this week
- ❑ 650 students participate in state science fair
- ❑ New passenger plane seats 327 people

IDEA 2—OVERESTIMATION VERSUS UNDERESTIMATION

Sometimes it makes sense to overestimate an amount, and sometimes it's better to underestimate. For each of the following estimates, decide whether it is more likely to be an overestimation or an underestimation and explain why.

- ❑ Lunch should cost about five dollars.
- ❑ The movie starts in about 15 minutes.
- ❑ The restaurant is about four miles down the road.
- ❑ I have about one eighth of a tank of gas left.
- ❑ The cost of repairing the car is about 225 dollars.
- ❑ I'll be ready to go in about 1 hour.

IDEA 3—REASONABLE ANSWERS

For each of the following arithmetic problems, circle the estimate you think is closest to the correct answer.

Arithmetic Problem:	*Circle one of the estimates below:*		
566 + 444 + 112=	800	1000	2000
33 x 419 =	120	1200	12000
9241 - 3357 =	5700	6000	6300

Idea 3 is included because too often, computational estimation activities that students are asked to complete direct students to make estimations and then check their estimations by finding the exact answer. In the third idea the estimation, not the exact answer, is the important element.

The International Society for Technology in Education adopted technology literacy standards for students. Standard 6 provides a link between numeracy/innumeracy and technology.

NETS•S Problem-Solving Standard

6. Technology problem-solving and decision-making tools

- Students use technology resources for solving problems and making informed decisions.
- Students employ technology in the development of strategies for solving problems in the real world.

As students learn to use technology appropriately, they will employ numeracy skills as they use the "reasonableness of results" test to evaluate the answers they get when they use formulas and complete data analysis. It becomes imperative that we impress upon our learners that just because the computer gave an answer does not make the answer valid. When we enter a faulty formula or make errors when entering data, applying the "reasonableness of results" test will help us make decisions about the results presented for our use.

Resources

Bennett, J. O., Briggs, W. L., & Morrow, C. A. (1996). *Quantitative reasoning: Mathematics for citizens in the 21st century.* Reading, MA: Addison-Wesley.

National Research Council. (1989). *Everybody counts: A report to the nation on the future of mathematics education.* Washington, DC: National Academy Press.

Paulos, J. A. (1988). *Innumeracy.* New York: Vintage Books.

THEORY INTO PRACTICE

Calculators in Mathematics

Myths Versus Benefits

By Anne Raymond

Calculators and graphing calculators are useful tools for learning mathematics. Although a number of benefits can be named and illustrated, many people still worry that calculators pose more problems than solutions. The great "calculator debate" accelerated when the original National Council of Teachers of Mathematics Curriculum and Evaluation Standards promoted increased use of calculators. The debate continues today and has expanded to include graphing calculators in the discussion.

Within the ongoing discussion about the use of calculators, many myths have emerged and persist today. Making arguments in favor of appropriate calculator use (Fey, 1992; Reys & Arbaugh, 2001; Van de Walle, 2001), mathematics educators attempt to dispel the myths. Some myths and possible responses are:

Myth: *Calculators eliminate the need to learn the essentials of arithmetic.*

Response: *This has often been a concern of parents who believe that calculators will replace the learning of arithmetic. On the contrary, calculators are not meant to replace arithmetic, but to help build understanding behind basic computations. For example, when learning the basic multiplication facts, students can engage in calculator activities that expose them to patterns that ultimately lead to conceptual understanding and fact mastery.*

Myth: *Calculators decrease students' ability to assess the reasonableness of computational answers.*

Response: *Sometimes students use calculators to compute final answers to complicated mathematical problems. Calculator users often find themselves hitting the wrong button when keying in calculator problems. Although some are concerned that students will merely copy down the answer provided by the calculators, others contend that calculator errors provide a learning opportunity. In short, as we teach children to perform basic arithmetic, we ought to simultaneously instill in them the estimation or mental math skills to ascertain whether or not a given answer is reasonable.*

Myth: *Students will become too dependent on caculators, thus weakening their basic skills over time.*

Response: *Even as students move through middle school and high school and have already mastered basic arithmetic, there are times when such skills should be practiced. That is, calculators should not necessarily be available every minute of every class. Consequently, there should be times when calculator use is allowed* *and times when it is not. In fact, to make sure basic skills are in place, many middle and high school teachers develop tests or quizzes in two parts—one part allows calculator use and the other does not. The bottom line is teachers must implement calculator use in purposeful ways, not merely making them available at all times, but making them available when their use makes sense.*

Myth: *Calculators are not important tools anymore because of the increase in computer software capabilities.*

Response: *It is difficult to argue against the fact that computers and computer software are becoming more classroom accessible and that their capabilities are quite advanced. Even though new portable technologies are on the horizon, the calculator is still the most affordable and easily portable computational technology available today.*

Myth: *Graphing calculators allow students to perform complicated mathematical procedures without students developing the intuition behind the mathematics.*

Response: *Particularly in upper level classes where statistics and calculus concepts are taught, students love to learn about graphing calculator features that do the work for them. Again, the teacher has to be pedagogically aware in order to make the use of calculators an enhancement and not a replacement. Many teachers claim that once basic statistics and calculus techniques are mastered in a computational way, the graphing calculator allows for further exploration and graphical connections which enhance the understanding behind the computations. In this sense, the graphing calculator can help build intuition rather than hinder it.*

Some of the responses to the myths above imply benefits of calculator and graphing calculator use such as

- Calculators can be used to develop concepts
- Calculators can be used to practice skills
- Calculators enhance problem solving
- Calculators improve student attitudes
- Calculators are commonly used in society so ought to be understood

While talking about calculators and graphing calculators, we should mention the opportunities available with calculator-based instruction (CBI) technology. These systems allow students to generate and record motion and sound data that can be interfaced with graphing calculators. This data can also be interfaced with spreadsheets. Thus, with the CBI technology, students can generate their own data and then compute statistics, develop graphs, and learn to interpret the data.

References

Fey, J. T. (Ed.). (1992). *Calculators in mathematics education.* Reston, VA: National Council of Teachers of Mathematics.

Reys, B. J., & Arbaugh, F. (2001). Clearing up the confusion over calculator use in grades K–5, *Teaching Children Mathematics, 8*(5) 90–94.

Van de Walle, J. A. (2001). *Teaching mathematics developmentally* (4th ed.). New York: Addison Wesley Longman.

INSIGHT

Developing Fractions Concepts Using a Spreadsheet

By Ivan W. Baugh

In my experience, students grasp basic concepts of fractions (halves, quarters, thirds, etc.) with a minimum amount of difficulty. They can add fractions with common denominators with relative ease. But give them fractions without common denominators and problems surface.

You can use a spreadsheet to create your own manipulatives that students can use to help them understand the process they need to employ when adding fractions without common denominators. For example, how do you add 1/4 + 1/3?

To help students comprehend the difference in size between 1/4 and 1/3, create two pie charts like those in Figures 9.1 and 9.2.

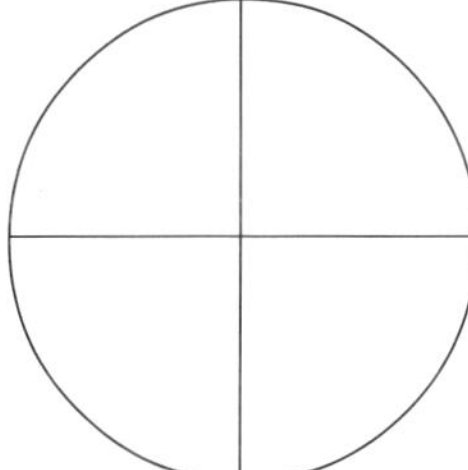

Figure 9.1.

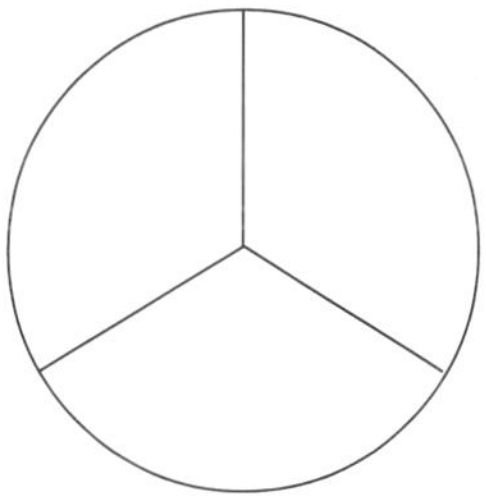

Figure 9.2.

Enlarge each pie chart to fit one sheet of 8 1/2 x 11-inch card stock. Make sure each of the pie charts has the same diameter. Print each pie chart on different color card stock and cut the pieces apart. When it becomes evident the pieces aren't the same size, we teach the principle of the least common denominator. We multiply the denominators to find a common denominator—12. Now create a pie chart with that number of pieces (Figure 9.3).

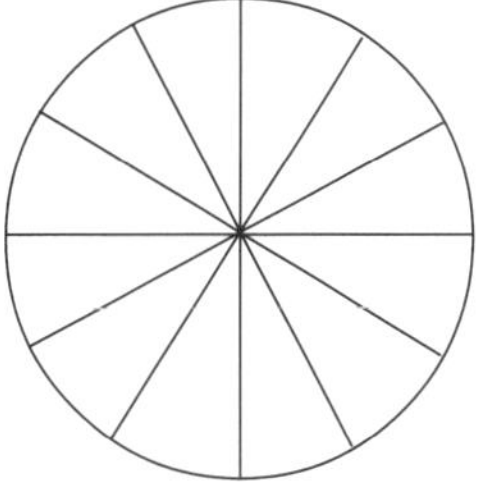

Figure 9.3.

Do not cut this chart apart. The students take a piece representing one fourth of a pie chart and lay it atop this pie chart. How many pieces does it cover? Take a piece representing one third of a pie chart and lay it adjacent to the piece representing one fourth? How many pieces does it cover? Together the two pieces cover a total of 7/12 of the whole.

Ask the students to create their own problem without common denominators. Use the pieces to solve the problem. Then have them share their problem with a classmate and discuss the solution.

How did I make the pie charts? In a spreadsheet, I typed 1 in 4 cells. I selected the cells, created the chart, and selected the option to label the parts. I repeated the process placing 1 in 3 cells. Then I put 1 in 12 cells. When I made this chart I selected the label option that shows value to put the 1 beside each piece of the pie. For an example of how this is done, I have included a file titled Fraction Concepts on the companion Web site, http://education.bellarmine.edu/baugh/.

You can create other fractional pieces following this same principle until your students develop a mastery of the concept. I would suggest that you keep the number of pieces to as small a number as possible: 1/3, 1/4, 1/5, 1/6. By not cutting apart the chart with the largest number of pieces, you make it easier for the students to manipulate the pieces.

Do you have access to an ink jet color printer? If so, you can print your various charts on a transparency, cut them out, and model using them on the overhead, just as the students will use them as manipulatives.

When you are ready to teach converting fractions to decimals, a spreadsheet makes it easy. After teaching the process the students will use to convert fractions to decimals and they demonstrate mastery of the principle, allow them to use a spreadsheet. Enter a fraction as a formula: =1/2 or =1/4. Computers understand the / to represent division. The spreadsheet immediately divides 2 into 1 or 4 into 1 and puts the results in the cell as a decimal. Then sum the numbers in the cells to get the total of the decimals.

INSIGHT

Fraction Sense and Manipulatives

By Anne Raymond

Fractions are often difficult for students to understand. Concrete models provide a key means of helping students make connections between a physical or pictorial image of fractions and the numerical representation of them. Concrete models also help students build an understanding of operations with fractions because they allow students to "see" how fractions relate to one another.

Students need to be exposed to and experience three categories of models of fractions. "Models of fractions" refers to the images or contexts that depict the "whole" in the "part of the whole" definition of the word fraction. The three models of fractions are: the area model, the linear model, and the set model. These three models of fractions are described in detail below:

MODELS OF FRACTIONS

Area model: A model (sometimes called a region model) in which the "whole" being described is an enclosed, two-dimensional area.

Example of the fraction 1/4 using an area model.

Linear Model: a model (sometimes called a number line model) in which the "whole" being described is a one-dimensional length or measurement.

Example of the fraction 1/4 using a linear model.

Set Model: a model in which the "whole" being described is a group of distinct objects.

Example of the fraction 1/4 using a set model.

Many materials around us can be used to model fractions. For example, when introducing students to the "part of a whole" definition of *fraction* in the context of an area model, teachers can have students share a candy bar or graham cracker that may be broken into equal-sized pieces to illustrate part of the whole. Teachers can also make sense of what "part of the whole" means in the context of a set model by having students with brown eyes stand up while the rest of the class sits down. In this case, *the whole* is the entire set of students, and *the part* is the set of students who have brown eyes. Thus, if there are 20 students, and 7 of them stood up, then we can write the fraction 7/20 to represent the number of the students who have brown eyes.

Manipulatives of all varieties have been developed to help students engage in concrete mathematical experiences that have many connections. Most mathematics manipulatives have a range of applications. However, some make better models for developing fraction sense than others. Here we provide some examples of mathematics manipulatives that are available for purchase in physical form and also available for computer manipulation as students interact with different models of fractions.

Two well-known manipulatives for illustrating the area model are color tiles and pattern blocks. Color tiles are 1-inch square tiles that come in four colors: red, yellow, blue, and green. Although color tiles can be used in many ways, building two-dimensional shapes showing particular fractional values is one popular use.

For example, the rectangles in Figure 9.4 are 1/2 red, 1/3 green, and 1/6 blue. Not only do we see multiple ways of representing these fractional values, but we also see opportunities to illustrate equivalent fractions and proportional reasoning.

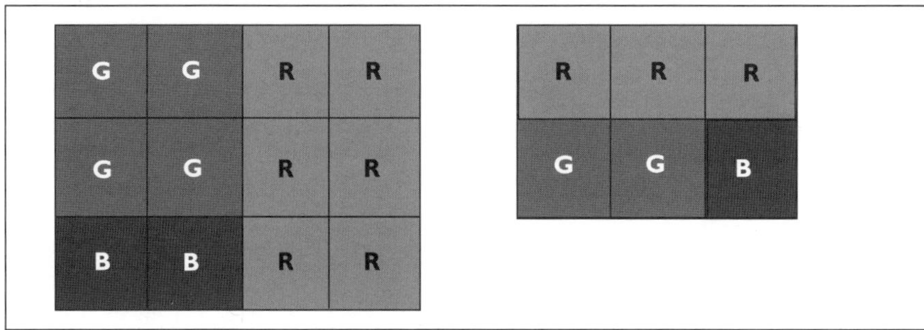

Figure 9.4. Color tiles representing fractional values (G = green, R = red, B = blue).

Pattern blocks (Figure 9.5) are composed of six geometric shapes whose side lengths are related to allow for many opportunities for manipulations, substitutions, and exploration of relationships between the shapes. The shapes of the pattern blocks include a square, a regular hexagon, an equilateral triangle, two different parallelograms, and a trapezoid.

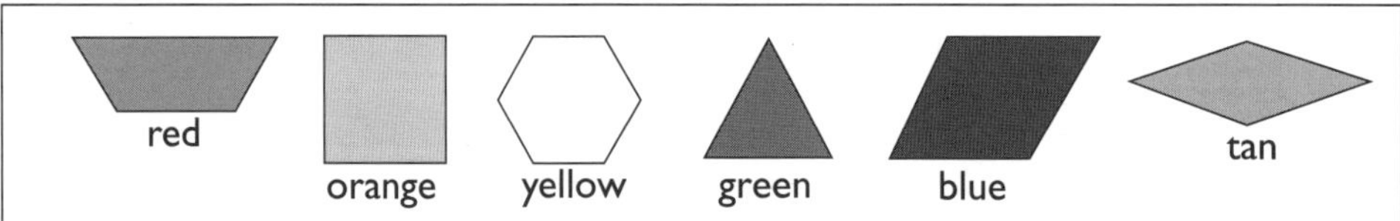

Figure 9.5. Pattern blocks.

Pattern blocks are wonderful manipulatives for exploring a range of geometric relationships. The example in Figure 9.6 demonstrates how, using the yellow hexagon as the whole, we can illustrate the fractional values of the green triangle and the blue parallelogram.

The linear model is a vital one to develop in students because it is so directly linked to students' work with number lines, which will ultimately lead to their understanding of coordinate graphing in four quadrants of the plane. A mathematics manipulative that was initially developed specifically for the exploration of fraction sense from a linear perspective is the set of cuisenaire rods (Figure 9.7). Cuisenaire rods are sets of rods of 10 lengths ranging incrementally from 1 cm to 10 cm.

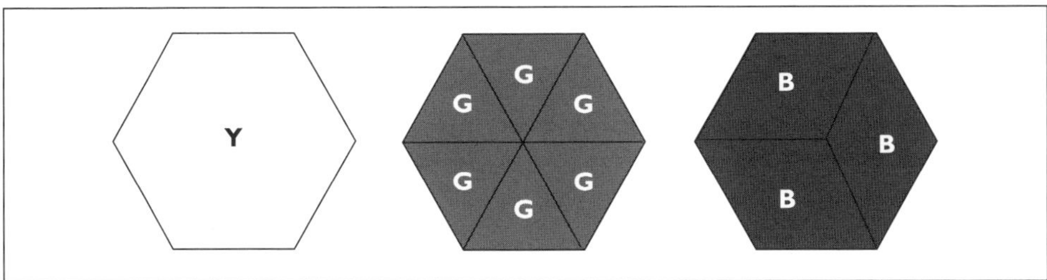

Figure 9.6. The triangle is 1/6 of the hexagon. The parallelogram is 1/3 of the hexagon (G = green, Y = yellow, B = blue).

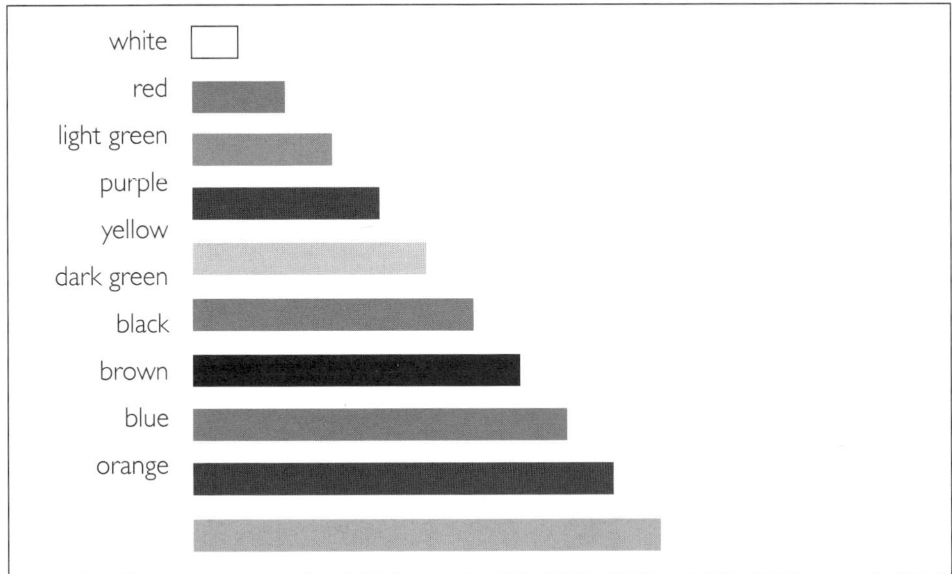

Figure 9.7. Cuisenaire rods.

Figure 9.8. Given the light green rod is the whole, the white rod is seen to be 1/3 as long.

If the purple rod is the whole, the dark green rod is 1 1/2 times longer.

With these rod lengths, we can demonstrate many fractional relationships, depending on which rod we designate as the whole. Cuisenaire rods can help develop the "part of the whole" image of a fraction, and also provide a concrete means of examining the ratio meaning of fraction. Figure 9.8 shows two examples.

The set model can be demonstrated with a variety of manipulatives. Any manipulative that can represent different collections of same-sized objects can provide good images of the set model. Color tiles can be a solid manipulative for representing sets of objects of different fractional values.

For example, in Figure 9.9, 3/8 of the color tiles are red, 1/4 of the tiles are blue, 1/4 of the tiles are green, and 1/8 of the tiles are yellow.

Another mathematics manipulative that provides a meaningful image for the set model are two-color counters. Two-color counters are circular chips that are red on one side and yellow on the other (some sets of two-color counters come in red and white instead of red and yellow). Besides being manipulatives to illustrate fractions, two-color counters are clear models for students to use when working with positive and negative numbers.

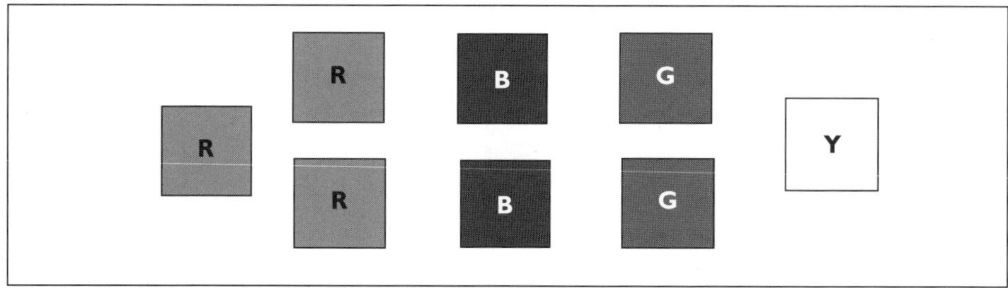

Figure 9.9. Color tiles (R = red, B = blue, G = green, Y = yellow).

Figure 9.10. An example using two-color counters to model fractions within sets (Y = yellow, R = red).

Figure 9.10 shows a simple example of using two-color counters to model fractions within sets, where 3/5 of the set is yellow and 2/5 of the set is red.

Creative teachers who do not have access to two-color counters or color tiles may use items such as M&Ms, marbles, or buttons to form sets of fractional proportions.

Another source for working with mathematics manipulatives is through software programs that provide graphic images of given manipulatives. All of the manipulatives mentioned in the previous examples can be found on software programs (see Resources below.)

Resources

Software and Web Sites

MindWare: www.MINDWAREonline.com

Sunburst, Networks, School Licenses, Lab Pack, and More: sunburst.com

Tom Snyder Productions: Software for Teachers Who Love to Teach: www.tomsnyder.com

Resources for Manipulatives and Activity Books Related to Manipulatives

Dale Seymour Publications, Pearson Learning: www.pearsonlearning.com/dalesey/dalesey_default.cfm

Delta Education Hands-On Math: www.delta-ed.com

Educators Outlet: www.educatorsoutlet.com

ETA Cuisenaire Publications: www.etacuisenaire.com

Manipulatives Webquest: http://matti.usu.edu/nlvm/nav/sitegd.html

Nasco Math: Nasco.com

NCTM Catalog Resources for the Mathematics Educator: www.nctm.org

Summit Learning Math Manipulatives: www.summitlearning.com

Teaching Resource Center: www.trcabc.com

Exploring Number Structures With Spreadsheets

262626 ÷ 37 = ?

Spreadsheets and the power of the computer to do mathematical computations combine to help your students get beyond the obvious in this math activity. The focus is on the development of higher level skills; by getting students interested in number structures they will start asking the interesting questions rather than the obvious ones.

By Azita Manouchehri

If you take any number and repeat it three times, the new number is evenly divisible by 37. Do you agree or disagree with this statement?

How do you think your students would test the statement? Typical middle school and high school students begin by testing it for a few single-digit or double-digit numbers. These instances do support the assertion—therefore, most students readily accept the statement as true. But this is clearly an overgeneralization because the statement has been tested for only two classes of numbers.

Other classes of numbers, namely three-digit and four-digit numbers, need to be considered. However, if students are encouraged to test the statement for larger numbers, they often either lose interest in the problem or arrive at ill-structured conclusions as a result of careless computational errors. In either case, students think they have reached closure on the problem even though they haven't paid adequate attention to the underlying number structures involved in the situation or asked the important "why" questions.

One way to keep students engaged in significant mathematical discourse and help them discover important numerical and functional relationships is to have them use spreadsheets to explore these kinds of problems. In my own experience in the classroom, I have found that the use of spreadsheets encourages students to explore number structures and generate other related conjectures and mathematical problems.

A Classroom Scenario

Here is a typical example of how I have students investigate the "evenly divisible by 37" statement. After I pose the problem in class, each student selects a single-digit number to test as the starting point for the activity. Because these calculations are relatively simple, most students are able to test the statement for all single-digit numbers and then conclude that the statement is true for those numbers.

I then ask each student to choose a two-digit number to test. With a class of 25 students, this provides a considerable pool of cases to consider. With the aid of calculators, the students examine these cases and hypothesize that the statement is true for two-digit numbers.

Testing Larger Numbers

At this point, students are usually ready to accept the statement as valid for any class of numbers. When this happens, I suggest that they need to test the assertion for larger numbers. This creates a dilemma for students. The screen on their calculators only shows up to seven digits and thus is inadequate for investigating the problem.

Some students begin using paper and pencil to start the long-division process for testing the statement for three-digit numbers. I then suggest that they use the calculator on the computer at the workstation in our classroom. This computer is connected to an LCD overhead panel, so all of us can see the screen. One student is appointed by the rest of the class to do the keyboarding while others view the calculations on the overhead projector screen.

300300 ÷ 37 = ?

A description of one of my classroom sessions serves as a good indicator of how the process works. In one of my experiences in teaching this activity, the first three-digit number selected was 100, which offered the students an immediate exception to their previous findings. Everyone was puzzled because they all had been convinced that the statement was valid prior to this discovery. Interestingly enough, one student suggested verifying the statement for several other three-digit numbers. Five other three-digit numbers (123, 859, 789, 413, and 456) were tested, each of which resulted in yet another exception to the students' previous assumption that the pattern of even divisibility would hold.

At this point, the students could clearly see that the initial statement had been found to be false, but nevertheless, no one quit working. One student tested the statement for the three-digit number 111 and noticed that the resulting number was evenly divisible by 37. This caused a lively discussion among the group. I asked students to take notes on all the questions and observations being made and to try to pose as many questions as they could about the problem under investigation. Here are some of the questions they generated:

- Is 111 the only three-digit number that when repeated three times is evenly divisible by 37? If not, what other three-digit numbers work?
- We know that the statement holds for all single-digit numbers and double-digit numbers. Why doesn't it hold for all three-digit numbers?
- Does the statement hold true for four-digit numbers? (The students often assume that it won't.)

The students now decided to pursue the truth of the original statement in a more organized fashion. They first attempted to test the statement for four-digit numbers using the calculator on the computer. Because the first exception they had found for three-digit numbers was 100, they decided to test the statement for 1,000, expecting this to also provide an exception. They were pleasantly surprised when they noticed that 100,010,001,000 was, in fact, evenly divisible by 37. Moreover, all five other examples they randomly selected (1,265, 2,954, 8,371, 4,572, and 3,911) were evenly divisible by 37. At this point, the question was: We know that the statement holds for a few four-digit numbers, but will it hold for all four-digit numbers? This time, my students were not so quick to make overgeneralizations.

One student predicted that because the statement was true for 1, 11, and 111, it would also be true for 1,111. This prediction was immediately tested and verified. When some students began to randomly select and test a variety of other four-digit numbers, I suggested that they try to devise a more systematic approach by recording what they knew thus far.

The students' first observation involved the behavior of numbers that were of powers of 10. One student observed that while the statement held for 1, 10, and 1,000, it failed to hold for 100. After some thought, the students formed the following hypothesis: If the number is an odd power of 10 and is repeated three times, the resulting number will be evenly divisible by 37. If it is an even power of 10 and is repeated three times, it will not be evenly divisible by 37.

The Power of Spreadsheets

Once again, the absence of a good tool to study this problem got in the students' way. They couldn't test their conjecture with the calculator on the computer because it only displayed a maximum of 11 digits. A more powerful tool was needed to make the test more efficient. I suggested using a spreadsheet to further explore the problem. The students appointed me to do the keyboarding while they suggested strategies and methods for analyzing the problem.

The first question was how to design a spreadsheet that would allow the students to test their conjectures for a large number of cases without having to do too much keyboarding. A portion of the spreadsheet that was designed for this purpose is shown in Figure 1. The students decided to use column A for the set of whole-number factors that would be multiplied by 37. Column B was used as the placeholder for the product of 37 and the corresponding number in column A. Column C lists the numbers students create when the entries from column B are repeated three times, and column D was reserved for finding the quotient of entries in column C when divided by 37.

A=Whole numbers	B=A*37	C=BBB	D=C/37
1	37	373737	10101
2	742	747474	20202
3	111	111111111	3003003
4	148	148148148	4004004
5	185	185185185	5005005
6	222	222222222	6006006
7	259	259259259	7007007
8	296	296296296	8008008
9	333	333333333	9009009
10	370	370370370	10010010
11	407	407407407	11011011

Figure 1. Spreadsheet for analyzing multiples of 37 for one- and two-digit numbers.

Once we filled in the appropriate functions for columns A and B, we needed to define the functions for cells in column C. One student suggested that we manually enter the numbers in each cell and then calculate each quotient individually for each number in column C. Even though this is an inefficient way to use a spreadsheet and does not take full advantage of its power, I encouraged my students to perform this task as suggested, hoping that their findings would enable them to build their intuition about the structure of the numbers under investigation. As they completed the task and obtained the sequence of numbers in column D, they began noticing the pattern among those numbers.

Furthermore, several students noticed that when entries in column D were divided by the whole numbers in the corresponding cells in column A, the quotient for all three-digit numbers in column B was 1,001,001. Figure 2 shows column E, which was added to the original spreadsheet and reflects the results of the calculations that led to this finding.

E=D/A
10101
10101
1001001
1001001
1001001
1001001
1001001
1001001
1001001
1001001
1001001

Figure 2. Column E.

I asked the students to make a few more conjectures about the behavior of the three-digit numbers and also about which three-digit numbers they thought would be evenly divisible by 37. One student commented that when a three-digit number is repeated three times, it is the equivalent of multiplying the number by 1,001,001. We used this observation to generate the spreadsheet shown in Figure 3. In this spreadsheet, column A contained three-digit numbers, column B was used to multiply the number in each cell in column A by 1,001,001, and column C was used to divide each cell in column B by 37. The students designed the spreadsheet so that a set of 50 numbers at a time could automatically be entered into column A. Then by simply changing the initial value in cell A1, they could efficiently analyze the behavior of all three-digit numbers.

They observed from the data in this spreadsheet that, when repeated three times, any three-digit number that itself had repeating digits (e.g., 111, 222) was evenly divisible by 37. Moreover, they discovered that beyond the set of multiples of 37 generated previously (a few of them are shown in column B of the spreadsheet in Figure 1), there would not be any other three-digit numbers evenly divisible by 37.

Analyzing Even Larger Numbers

I now asked my students to make conjectures about the behavior of four-digit numbers. They suggested that all four-digit numbers with repeating digits would be evenly divisible by 37. Moreover, they claimed that, when repeated three times, any number with repeating digits, regardless of its number of digits, would be evenly divisible by 37. One student reasoned that "there would probably be more four-digit numbers evenly divisible by 37 because there are more multiples of 37 among whole numbers between 1,000 to 9,999." This in itself was powerful reasoning that enriched our investigation of the pattern of behaviors of four-digit numbers.

A=Numbers	B=A*1001001	C=B/37
100	100100100	2705408.108
101	101101101	2732462.189
102	102102102	2759516.27
103	103103103	2786570.351
104	104104104	2813624.432
105	105105105	2840678.514
106	106106106	2867732.595
107	107107107	2894786.676
108	108108108	2921840.757
109	109109109	2948894.838
110	110110110	2975948.919
111	111111111	3003003
112	112112112	3030057.081
113	113113113	3057111.162
114	114114114	3084165.243
115	115115115	3111219.324

Figure 3. Spreadsheet for analyzing three-digit numbers.

They anlayzed their findings with a spreadsheet adapted from the one used for three-digit numbers. In designing the spreadsheet, they concluded that since four-digit numbers occupy three additional place values compared to three-digit numbers, then the multiplication factor for four-digit numbers to be used in column B had to be 100,010,001. Moreover, they predicted that in the case of five-digit numbers, the process would be similar, and the multiplication factor would be 100,001,000,001. Ultimately, they made the following generalization: As we increase the number of digits and repeat the number three times, the number of zeros in the multiplication factor has to increase accordingly.

The students then tested their conjectures for four-digit and five-digit numbers. One finding truly surprised them: They were startled to find that all four-digit numbers repeated three times were evenly divisible by 37. At this point, some students decided to statistically analyze the number of five-digit numbers that, when repeated three times, were evenly divisible by 37. They reported that 7% of numbers from 10,000 to 10,099 were evenly divisible by 37. For the numbers from 10,100 to 99,900, the percentage grew to 25%, and for the numbers between 99,901 and 99,999 it reached 57%. They felt that this was a rather remarkable and significant finding.

Generalizing the Results

Their discoveries about number behavior caused the students to pose the magic question: Why does it work like this? After reviewing the information that had been collected while they were examining the behavior of numbers repeated three times, they formed the following conclusions:

- Single-digit numbers. When a single-digit number is repeated three times, it is the same as multiplying the number by 111. Because 111 is evenly divisible by 37, the statement holds true for all single-digit numbers.
- Two-digit numbers. When a two-digit number is repeated three times, it is the same as multiplying it by 10,101. Because 10,101 is evenly divisible by 37, all two-digit numbers are evenly divisible by 37.
- Three-digit numbers. When a three-digit number is repeated three times, it is the same as multiplying it by 1,001,001. Because 1,001,001 itself is not evenly divisible by 37, the original three-digit number has to be a multiple of 37 if the number repeated three times is to be evenly divisible by 37. Because 1,001,001 divided by 37 is not a whole number, the original number ABC has to be evenly divisible by 37 in order for the number repeated three times to be evenly divisible by 37.
- Four-digit numbers. When a four-digit number is repeated three times, it is the same as multiplying it by 100,010,001. Because 100,010,001 is evenly divisible by 37, all four-digit numbers repeated three times are evenly divisible by 37.
- Five-digit numbers. In the context of the statistics reported for numbers with one to four digits, we have no explanation for why five-digit numbers behaved as they did beyond the trivial observation that in order for a five-digit number repeated three times to be evenly divisible by 37, the original five-digit number itself must be evenly divisible by 37. We can also conclude that because of the distribution of such numbers among all five-digit numbers, there will be more five-digit numbers evenly divisible by 37 when compared to three-digit numbers.
- Six-digit numbers. Similar to four-digit numbers, all six-digit numbers repeated three times are divisible by 37 because the multiplication factor for a six-digit number repeated three times—1,000,001,000,001—is evenly divisible by 37.

More Questions, More Learning

At this point the class discussion took a different direction when one student asked, "What if instead of repeating the numbers three times we had repeated them four times? Would the results be the same?" Another student asked, "What if we had repeated the numbers five times? Is the pattern related to whether we are repeating the number an even number of times or odd number of times?" These questions generated the opportunity for further extensions of the problem, including the following:

- What if instead of dividing by 37 we had divided by 27 or 17? For which set of numbers does the statement hold true? How many times do we have to repeat the number to make the statement true for the new divisors?
- What if instead of 37 we had used a number that was not a prime number. Does the pattern exist only for prime numbers?
- What numbers repeated four times yield only prime numbers? Only odd prime numbers?
- What if we had repeated the number only twice? Would the outcomes be similar?
- Is there a divisor for which the statement always holds regardless of the dividend selected?

A Discussion

As the questions the students posed clearly illustrate, the combined power of computers and spreadsheets coupled with challenging mathematical content can develop students' problem-solving skills while encouraging them to explore mathematical structures in depth. Students can develop a conceptual understanding of mathematical ideas first and then are better able to construct powerful algorithms for describing them.

A natural consequence of using technology-supported problem solving in the classroom is the emergence of "what if" and "what if not" questions about the situation under investigation. Consequently, solving the initial problem becomes secondary compared to the much wider range of problems that are generated. This facilitates students' skills in making connections among various math problems as well as math topics.

As demonstrated by the classroom scenario, the use of spreadsheets and powerful computing capabilities fosters an experimental attitude toward learning mathematics. As a result, students develop the disposition to go beyond the obvious and deepen their understanding of the similarities and differences inherent in various mathematical problems and structures. These outcomes are consistent with the goals of mathematics instruction at all levels of education as identified by recent reform movements (NCTM, 1989).

Reference

National Council of Teachers of Mathematics. (1989). Curriculum and evaluation standards for school mathematics. Reston, VA: Author.

Number Patterns in Repeating Decimals

1/7 = .14285714285714285714...
2/7 = .285714285714285714 28...
3/7 = .428571428571428571 4...

Graphing calculators overcome the display limitations of scientific calculators and the expense of computers, helping middle school students to explore number patterns in repeating decimals. By inputting a program into a graphing calculator, you can increase the power of this tool for teaching concepts related to fractions and decimals.

By Leon Roland

There are several ways that middle school students can be introduced to the idea of repeating decimals. The most obvious way is to use long division to change fractions to decimals, but for the average student, using long division to produce, for example, the decimal representation for 7/29, which repeats in 28 places, is time consuming, subject to error, and boring. Both computers and scientific calculators can help solve these problems, but each tool still has its limitations. The display face on a scientific calculator is usually limited to showing only eight digits, which diminishes its value for studying lengthy repetitions. Computers, although they can display an almost endless series of digits, are expensive to acquire and maintain.

Graphing calculators provide an affordable alternative to these other methods. Although graphing calculators have their own set of drawbacks—occasionally they operate slowly and their display face is still somewhat limited—they can nevertheless be used effectively to help students make important discoveries about fractions and repeating decimals.

This article provides a program that can be used with the Texas Instruments TI-82 graphing calculator to customize student explorations of this topic, along with some ideas for teaching concepts related to fractions and decimals.

The Program

As noted, the program provided at the end of this article is for the TI-82, which should probably be called a "hand-held, dedicated mathematics computer" and not a calculator. The program will run more slowly on the TI-82 than a similar program would run on a computer, but it nevertheless provides a good tool for beginning students to use. (You can adapt it for use on other graphing calculators.) The program contains some extra lines, which indicate the individual sections for the following five steps, which are covered in the code:

1. Inputting values
2. Reducing fractions
3. Getting prime factors of the denominator
4. Determining the number of

repeating and nonrepeating digits
5. Displaying decimals

Getting Started

There are a number of ideas you can encourage students to think about as they begin studying patterns in repeating decimals. The list of ideas will grow as students look at various fractions, but the following five topics offer a good place to begin:

1. The denominator is the number that determines the number of places that repeat.
2. It is important to used reduced fractions (3/6 will behave like 1/2 and not like other nonreducible fractions having 6 as their denominator).
3. An understanding of prime factorization is important to the study of repeating decimals.
4. Prime factorization is especially important in relation to the denominator.
5. Terminating fractions have some power of 2 and 5 in their prime factorizations.

If your students have a basic understanding of fractions and long division, you can introduce them to the use of the graphing calculator while exploring the first three points. Then you can have them work on the more demanding fourth point, which will help them recognize certain repeating patterns, and the fifth point, which will further increase their understanding of repeating fractions and fractions that terminate.

The fifth point is effective for building on the first four points. Try having students look at all fractions having a numerator of 1 and denominators from 2 to 50. If they learn that the numerator does not affect the number of places that repeat, they should also realize that using 1 for the numerator eliminates any chance that a fraction can be reduced. Challenge them to look for other patterns also.

The amount of help students will need depends largely on their experience in looking for patterns. If they are strongly driven to find the "right" answer, this activity can be very effective in awakening them to the fact that not every question has only one "right" answer.

Experimenting and Exploring

Using the program with the TI-82 allows students to experiment and explore quickly and easily. For example, they can enter the fraction 3/56 (Figure 1). The calculator then computes and displays the prime factorization and the number of repeating and nonrepeating digits (Figure 2). In this case, the line {2 3 7 1} is the prime factorization of the denominator: $2^3 \times 7^1$. The NONREPEAT 3 indicates that there are three digits in the decimal representation before the number starts to repeat, while the REPEAT 6 indicates that there are six digits in the repeating sequence. Figure 3 shows the decimal expression of 3/56 with a space between each set of repeating digits.

The Value of Discovery

What do students discover as they use graphing calculators to analyze decimals? The answer is: a lot of things. Some things they discover will be the result of working with introductory concepts. Other discoveries will be surprises that result from more complex experiments and explorations. Many of their initial "insights" may later turn out to be false. But the most important thing is that students will be learning as they are "discovering" mathematics.

Here are a few of the discoveries they might make:

- The number of nonrepeating digits is related to the exponents 2 and 5.
- The repeating sequence will always begin repeating at a place that is at least one less than the denominator.
- All fractions with the denominator of 7 repeat using the same six digits in the same order. The only difference is that they start in a different place.
- The fraction with the denominator of 7^1 repeats in 6 places, 7^2 in 42 places, 7^3 in 294 places, each of which is 7 times the previous number of places. This pattern holds for other prime numbers.

Many discoveries will be simple and will hold only for a limited set of circumstances. Nevertheless, they are still excellent discoveries, and you should encourage your students to discover their value as they make them.

```
NUMERATOR 3
DENOMINATOR 56
REDUCES TO:
Ø  3/56
```

Figure 1. Entering the fraction 3/56.

```
            {2 3 7 1}
NONREPEAT      3
REPEAT         6
```

Figure 2. Prime factorization and the number of repeating and nonrepeating digits.

```
.Ø53 571428 5714
28 571428 571428
571428 571428 571
428 571428 571428
571
```

Figure 3. The repeating sequence.

Program for the TI-82 Graphing Calculator

```
(Input of Fraction)
:ClrHome
:0>S
:Input "NUMERATOR ",N
:Input " DENOMINATOR
",D
(Reduces Fraction)
:0>W
If N>D
Then
:iPart (N/D)>W
N-W*
End D>N
N>Z
:While (fPart
(N/Z)≠0 or (fPart
(D/Z)≠0)
:Z-1>Z
:End
:N/Z>N
:D/Z>D
:Disp ("REDUCES TO:")
:Output(5,1,W)
:Output(5,5,N)
:Output(5,6,"/")
:Output(5,7,D)
:Pause
(Prime factors of denominator)
:D>Z:1>S
:ClrList L5
:2>F
:0>E
:√Z>M
:While F≤M
:While fPart (Z/F)=0
:E+1>E
:Z/F>Z
:End
:If E>0:Then
:F>L5(S)
:E>L5(S+1)
:S+2>S
:0>E
:√Z>M
:End
:If F=2
:Then:3>F
:Else:F+2>F
:End
:End
:If Z≠1:Then
:Z>L5(S)
:1>L5(S+1)
:End
:ClrHome
:Disp L5
(Repeating & Nonrepeating)
:D>Z
:0>S
:Lbl 1
:If fPart (Z/2)=0
:Then
:1+S>S:Z/2>Z
:Goto 1
:End
:0>T
:Lbl 2
:If fPart (Z/5)=0
:Then
:1+T>T:Z/5>Z
:Goto 2
:End
:If T>S
:T>S
:1>A:9>Y
:If Z=1
:0>A
:If fPart (Y/Z)>0
:Then
:Repeat fPart (Y/Z)=0
:1+A>A
:(Y-(iPart (Y/Z)*Z))*10+9>Y
:End
:End
:Output(3,1,"NONREPEAT")
:Output(3,11,S)
:Output(5,1,"REPEAT")
:Output(5,11,A)
:Pause
(Display decimal fraction)
:ClrHome
:Disp "NUMBER OF PLACES"
:Input "(<125)",P
:N*10>N
:2>C:1>L
:ClrHome
:Output(1,1,".")
:For(J,1,P,1)
:iPart (N/D)>T
:(N-T*D)*10>N
:Output(L,C,T)
:C+1>C
:If C=17
:Then
:1+L>L
:1>C
:End
:If A>0
:Then
:If ((J-S)/A=iPart ((J-S)/A))
and (J≥S)
:Then
:Output(L,C," ")
:C+1>C
:End
:End
:If C=17
:Then
:1+L>L
:1>C
:End
:End
```

Logo & Negative Numbers

Logo explorations can help students apply mathematical concepts to the real world in a way that textbook assignments cannot. Candace Strawn helps students explore negative numbers using Logo.

By Candace A. Strawn

In *The Children's Machine: Rethinking School in the Age of the Computer,* Seymour Papert (1993) claims that the startling growth of science and technology in the 20th century has caused massive changes in areas such as telecommunication, entertainment, transportation, and medicine. Schools, however, have not changed much in the past 100 years. In particular, Papert thinks the traditional approaches to teaching and learning mathematics do not prepare students for the sophisticated quantitative skills that they will need in the Information Age.

Papert (1993) discusses how mathematics is still largely taught in a rote, meaningless manner without any association to the real world. He believes that the Logo computer language can bridge the gap that exists between textbook learning and the real world as well as facilitate student control over the cognitive process. See "What Is Logo?" at the end of this article for more information about Papert's programming language. In this article, I deal only with Logo's turtle graphics capabilities; for information about Logo's other capabilities, see the *L&L* online article supplements page at **www.iste.org/publish/learning/supplement.html**.

To examine how effectively Logo bridges the gap between mathematics and the real world, I visited a sixth-grade classroom to observe how students applied what they had learned about negative numbers to Logo programming. I then interviewed the teacher to determine the usefulness of Logo in her classroom.

Classroom Application:
The Negative Number Assignment

Students in the sixth-grade classroom had just spent two weeks studying positive and negative numbers. They were working on several projects, such as locating coordinates on a plane. They had also used Logo on earlier projects—such as drawing regular polygons and developing a general formula—but did not know all the capabilities of the programming language. They did not know, for example, that they could use negative numbers in Logo—that is, by stating "FORWARD –100" (FD –100), they could make the turtle move backwards. The teacher's goal was to see whether students could connect the information from their math assignments on negative numbers to the Logo project. Consequently, she gave them this problem: "I want you to draw a square using 'turtle graphics.' Have the turtle make the square using only FORWARD (FD) and RIGHT (RT) commands. However, the turtle's head can only face up (north) or to the right (east) during any of the procedures. The turtle's head cannot face down (south) or to the left (west) at any time, nor can the REPEAT command be used."

The students were eager to work on the computers and waved their arms frantically so that the teacher would choose them. Once they began their work in the lab, it was a different story.

Although they were still enthusiastic about working on the computers, none of them could make the turtle draw a square without turning the turtle's head to the left or downward. They began by moving their turtles forward and to the right but could not figure out what to do next. At that point, I heard one boy tell a girl that the project was impossible to complete in that manner. A few students came up with interesting alternative solutions such as having the turtle rotate on its axis 270 degrees, but the teacher informed them that this rotation would cause the turtle's head to go in the wrong direction.

When the students seemed completely perplexed, the teacher told them, "Imagine yourself sitting in a saddle on the turtle's back." The students then went through a series of fantastic body contortions as they imagined how the turtle would move. Fifteen minutes later, the teacher gave them another hint: "Would anything you worked on in the past two weeks help you with this assignment?" Two students immediately wondered aloud whether negative numbers would work in Logo; they began to experiment with the "FORWARD –100" command. They were delighted to see the turtle move backward. Not long after this, a girl completed the assignment with this long procedure:

```
TO SQUARE
FD 30
RT 90
FD 30
FD -30
RT -90
FD -30
RT 90
FD 30
RT -90
FD 30
END
```

The teacher then asked her whether fewer commands could be used and if the turtle could draw a square with one continuous movement. She also told the girl that she could try the REPEAT command. Within a few minutes, the student accomplished the procedure using only negative numbers:

```
TO SQUARE
REPEAT 4[RT -90 FD -30]
END
```

(It is interesting to note that the girl called her first attempt a "success" and her second attempt a "real success.") The teacher's main goal for this assignment was for students to draw a square using only negative numbers and with the fewest possible number of commands.

Next the teacher instructed her students to draw triangles and five- and six-sided polygons using the same commands. Because this assignment was easy to complete, the students began writing their initials with the same restrictions. Fifteen minutes later, the first half of the class was over, and the other students rushed into the computer lab to work on the same assignment.

As the students were working at their computers, the teacher asked them to periodically stop and write down which ideas did and did not work. She later asked them if they would have been able to figure out that Logo could use negative numbers without her hints. Most students answered, "No," but a few commented that they might have gotten it on their own given a little more time.

What was most interesting, however, was how students quickly grasped the concept of programming with negative numbers. One student said that negative numbers are just the opposite of positive numbers and, therefore, "RT –90" was equal to "LT 90." Another student stated that the degrees of angles were determined by dividing 360 degrees by the number of sides and then drew a 13-sided polygon within five minutes. The girls and boys were having so much fun with the turtle graphics that it was difficult for them to leave at the end of the period.

The Power of Logo

The teacher later talked about digital technology and how Logo provided a learning environment in which students and teachers learn to restructure their knowledge. "In addition to providing unique ways to study mathematics," she said, "Logo helps students organize information, draw conclusions, analyze problems, and delve deeper and more broadly into advanced mathematical concepts." With its concrete applications, Logo is especially good at helping students develop spatial awareness. She continued, "Teachers also learn more about their students because Logo shows what is going on in students' heads; teachers can understand what kind of learners their students are and more effectively teach each girl and boy." In this learning process, the teacher is more facilitator than instructor. He or she asks meaningful questions that guide students through a problem rather than tell them how to solve it. Students are then stimulated to respond creatively and develop a deeper understanding of mathematics. In the negative number assignment, for example, the teacher guided students' explorations of negative numbers, but the students assisted in the discovery process.

Logo also offers a friendly, pupil-oriented learning environment in which students are not afraid to make mistakes. In fact, they love to experiment with the "turtle graphics" and deliberately make changes just to see what would happen. An unexpected Logo drawing also provides natural feedback for users. When the turtle does not perform as they expect, pupils simply revise the procedure. This process is easy, fun, and non-judgmental. Students view mistakes as a positive part of the learning process, which fits Papert's definition of education. For example, Papert (1993) uses the concept of a microworld to describe a Logo-like environment. A microworld is an environment for learning in which an individual is free to experiment, test and revise theories, and invent her or his own activities. In addition, the microworld must produce a product, and this product must be created by the student. If more students could experience Logo's meaningful mathematical environment, Papert believes, fewer individuals would develop what he calls "mathophobia," or the hate and fear of mathematics.

As with most programming languages, Logo does have drawbacks. First, teachers need to understand the cognitive process to guide their students effectively. In the case of the negative number assignment, for example, the sixth-grade teacher thought that many students would be able to bridge the gap between the classroom activity and the computer room. When this transfer of knowledge did not take place, she was forced to provide more support to her students. Second, teachers should be comfortable with mathematical concepts that go beyond the typical elementary school curriculum. Third, Logo requires a significant amount of time to plan meaningful individual and group activities. Finally, because Logo presents knowledge in a different way, some teachers who prefer the traditional way of teaching mathematics are not comfortable with it.

Despite these drawbacks, the positive results from using Logo far exceeded the negative in this case.

Reference

Papert, S. (1993). *The children's machine: Rethinking school in the age of the computer.* New York: Basic Books.

The author wishes to thank Eleanor Thomas for allowing her to visit her classroom and for providing first hand information about Logo's usefulness in the classroom.

What Is Logo?

Let's take a brief look at Logo. Developed by Seymour Papert and his colleagues at MIT in the late 1960s, Logo is a procedural computer language that provides a unique learning environment for students of all ages. It can be used in many ways.

In turtle graphics mode, a student sees a small turtle on the computer screen. The student can direct the turtle around the screen by using a series of simple commands called primitives. The words in the command indicate in which direction the turtle should move, and the numbers tell the turtle how far to travel. For example, the command "FORWARD 100" (FD 100) causes the turtle to move 100 paces straight ahead; it leaves a line as it travels. "RIGHT 90" (RT 90) will cause the turtle to turn 90 degrees to its right.

After the student learns how to manipulate the turtle one move at a time, he or she can learn a higher order of command called a procedure. A procedure can be accomplished by using the "TO" command followed by the procedure's own name. Students must also learn the "REPEAT" command at this level; the command causes the turtle to repeat a certain sequence of moves. For example, the following short program instructs the turtle to make a square:

```
TO SQUARE
REPEAT 4[FD 100 RT 90]
END
```

This command tells the turtle to repeat the two moves, FD 100 and RT 90, as a sequence four times.

To teach Logo effectively, teachers should instruct students to put themselves in the turtle's place, so that they can see everything from the turtle's perspective. The students then program the turtle to draw a certain procedure such as a square. Having the students teach the turtle to perform such tasks is the concept that makes Logo effective as a classroom learning tool. Reading and studying about a square does little to help a student internalize and understand the concept. By actually creating a square, however, a student can begin to comprehend that polygon's properties. In addition, the student has learned to control the computer rather than be passively controlled by it.

CHAPTER 10

Data Analysis and Probability

The Conversation Concludes

Anne: How do you know when a claim is true?

Ivan: It's hard to tell, especially when you're listening to an advertisement, to facts about a political race, or to someone promoting a cause.

Anne: You know, advertisements are interesting in many ways. First, some of them try to sell a product using an actor or actress who is familiar to the public and well liked. The idea is, if we like the person we'll like the product.

Ivan: I love the "I'm not a doctor, but I play one on TV" ads that advertise medicines such as aspirin. Since viewers know the actor as "a doctor," his claim that a certain medicine is best seems more believable. Then there are the commercials that suggest more hospitals use a certain brand of pain reliever than any other. When they make this claim, they're telling the public the truth. However, what they don't say is that hospitals choose a particular brand of pain reliever because the makers of that brand provide their product free to hospitals simply so the advertisers can make the claim.

Anne: Another way advertisements try to influence us is by using statistics. One of my favorites is when a toothpaste advertisement claims "9 out of 10 dentists surveyed" recommend their brand of toothpaste. This leads one to think that 90% of all dentists prefer this brand. I question, "How many dentists did they survey?" It could have been only 10 for their claim to be true, and this would hardly be representative of the entire population of dentists. Plus, they might have only asked specific dentists, not a random sampling of dentists. I mean, if they survey dentists who live in the town where that brand of toothpaste is produced, the participants may not be an unbiased sampling of dentists. The bottom line is, advertisements make statistical claims all the time. We have to look closely at those claims to determine just how valid they are.

Ivan: In one of the Insights in this chapter, we further discuss misleading statistics. Also, in the articles that follow, we see examples of how we can manipulate numbers by using spreadsheets to create statistics.

THEORY INTO PRACTICE

How Statistics Can Be Misleading

By Anne Raymond

Often, statistics are used to convince us to buy a certain product or vote for a particular candidate. Those developing the statistics are sometimes unclear, whether purposefully or not, about how they determined their statistics or how they arrived at their particular interpretation of the numbers. The National Council of Teachers of Mathematics believes the mathematics curriculum ought to prepare students to interpret data and make appropriate predictions based on data. If students are prepared to do this, they are less likely to be misled or fooled by statistics.

NCTM Data Analysis and Probability Standard

Data Analysis and Probability: Instructional programs from prekindergarten through Grade 12 should enable all students to

- Formulate questions that can be addressed with data and collect, organize, and display relevant data to answer them
- Select and use appropriate statistical methods to analyze data
- Develop and evaluate inferences and predictions that are based on data
- Understand and apply basic concepts of probability

Many people have heard of the book *How to Lie with Statistics* (Huff, 1993). The crux of the book's message is that if you use the right statistics, you can convince people of almost anything. The following examples illustrate how statistics can be misleading in several different contexts.

A Principal and SAT Scores

Consider this scenario. The principal at a local high school addresses the school board. He says SAT scores were down 10% in 2000 but were up 15% the following year. He is pleased with the 5% net gain in SAT scores over the 2-year period. Is he correct in his claim?

The answer is, Not exactly. He should be pleased that scores went up over the 2-year period, but they did not go up as much as 5%. Rather, the scores were up a net amount of 3.5%. What the principal did not take into account was that the increase of 15% the second year was based on 15% above the lower score from the year before. For example, suppose for ease of computation that the average SAT scores were 1,000, but they went down 10% in 2000, to 900. Then, when you compute a rise of 15%, you are computing a rise of 15% from the 900 score, not the 1,000 score. Thus, the 15% increase of the 900 average score would be 1,035, or only a 3.5% increase from 2 years ago. So even though the principal may not have intended to be misleading, it is easy to convince people your statistics are true because we often aren't savvy enough to know the proper way to compute statistics.

Advertisements for Margarine

A leading brand of margarine, Brand A, came out with a "light" margarine that had 1/3 the fat of the regular margarine—quite a selling point in this age of health-conscious consumers. A competing brand of margarine, Brand B, came up with a light version as well, but it had 2/3 the fat of regular margarine. Most people can easily understand that the one with 1/3 the fat was a better health choice than the one with 2/3 the fat. So to compete but still provide accurate information, Brand B began advertising its light margarine as having "1/3 less fat than regular margarine." This is simply a different way of saying 2/3 the fat, but the advertisers likely assumed that most consumers wouldn't see the difference between "1/3 the fat" and "1/3 less fat." In this case, the way you state the statistic can influence decisions.

NETS•S Communication Standard

4. **Technology communication tools**
   - Students use telecommunications to collaborate, publish, and interact with peers, experts, and other audiences.
   - Students use a variety of media and formats to communicate information and ideas effectively to multiple audiences.

NETS•S Research Standard

5. **Technology research tools**
   - Students use technology to locate, evaluate, and collect information from a variety of sources.
   - Students use technology tools to process data and report results.
   - Students evaluate and select new information resources and technological innovations based on the appropriateness to specific tasks.

Technology provides powerful tools for communication and research. Students can use technology to gather data from many online resources. It can also help compile data from surveys. Spreadsheets offer students the tools with which to complete statistical analyses. Because of integrated applications software and office suites, students can incorporate appropriate sections from the data analyses into their word processing documents or Internet documents when they publish the results of the study. Each of these tools empowers the student to spend time analyzing the data rather than manually completing the complex calculations. When data are entered accurately and formulas reference appropriate cells, more time becomes available for reporting the results in a more efficient manner.

References

Huff, D. (1993). *How to lie with statistics* (Reissue ed.). New York: W. W. Norton.

Resources

Best, J. (2001). *Damned lies and statistics: Untangling numbers from the media, politicians and activists.* Berkeley, CA: University of California Press.

Jaisingh, L. R. (2001). *Statistics for the utterly confused.* New York: McGraw-Hill.

INSIGHT

Developing Appropriate Graphs to Display Statistics

By Anne Raymond

There are many types of graphs and we encounter them regularly. We see graphs in the newspaper, in magazines, and in television news reports, and they are used primarily as a mean of providing information. Specifically, graphs supply another way of reporting statistical information. Graphs, like all statistics, can be a vehicle for convincing someone to take a certain stance. Sometimes graphs are accurate, and other times they can be either misleading or flat out wrong interpretations of data (Jones, 2000).

The type of graph you choose to illustrate your data often depends upon the data you have. For example, when you have discrete data, such as data that can be put into specific categories, some sort of bar graph would be appropriate. However, when your data is continuous, such as changes over time, a line graph is appropriate.

Bar Graphs

Bar graphs come in several varieties. They can be vertical or horizontal. They can be single, double, or triple bar graphs. Generally, bar graphs illustrate the frequency with which something occurred. For example, suppose marketing executives conducted a survey of 10,000 people to determine which brand of popcorn they preferred among Brands A, B, and C. They could illustrate their findings as in Figure 10.1.

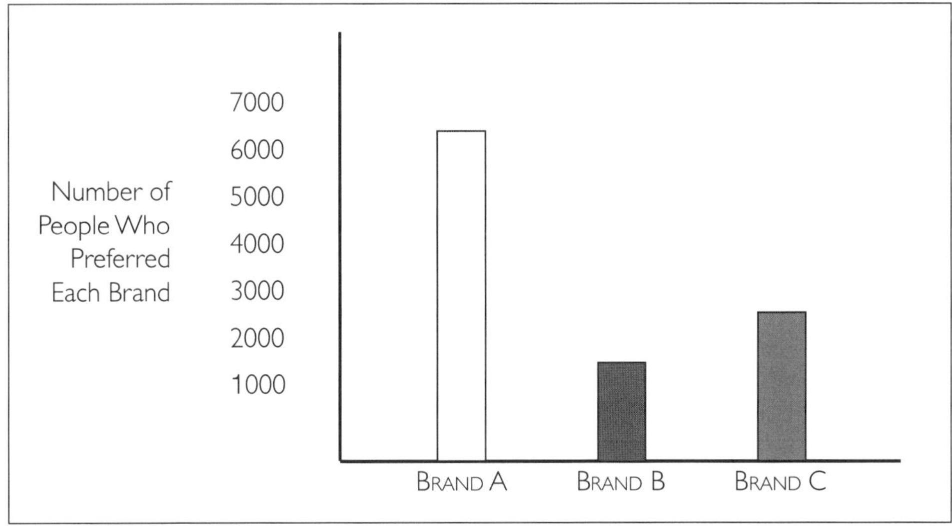

Figure 10.1.

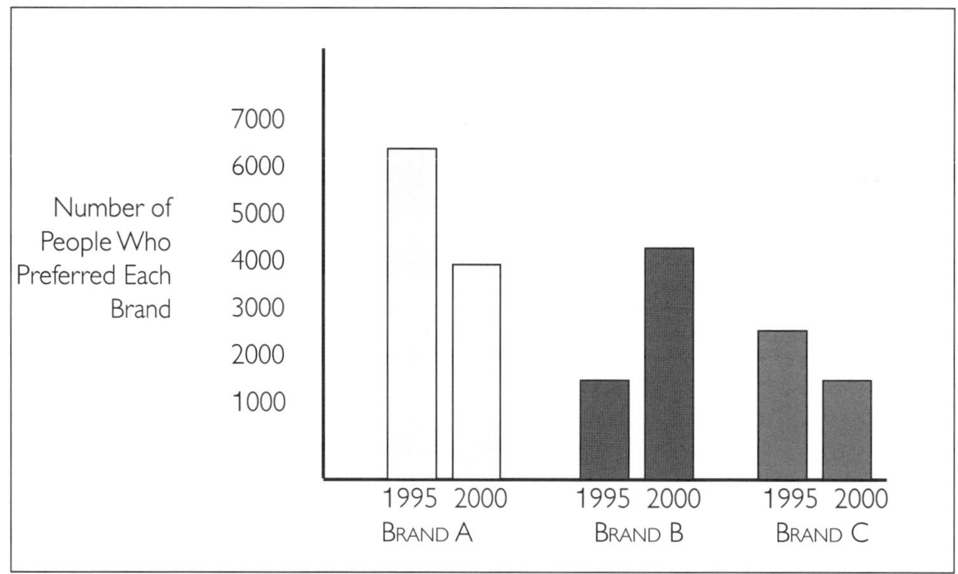

Figure 10.2.

Now suppose the marketing company for Brand B popcorn wants to illustrate how effective its campaign has been over the past 5 years. They can show comparative data in a multiple bar graph as seen in Figure 10.2.

Line Graphs

Suppose you want to lobby your congressional representatives for stricter drunken driving laws. Providing them with graphical information about the increase in DWI arrests over a 10-year period could be convincing data. Since the data is continuous over time, a line graph is appropriate. Consider the graph in Figure 10.3.

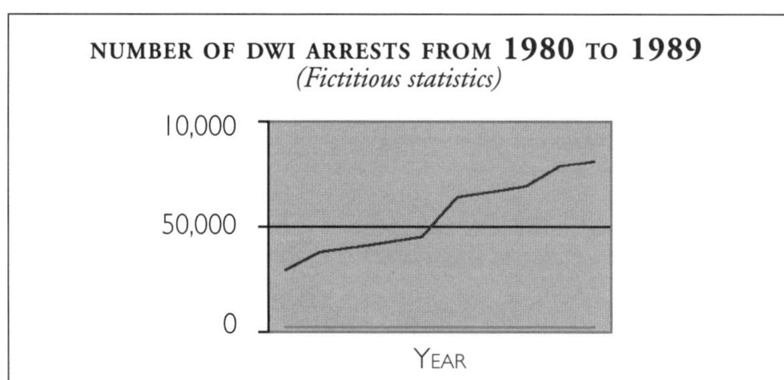

Figure 10.3.

Pie Charts or Circle Graphs

Pie charts are vivid, usually colorful representations of how data from a population break down across categories. We usually develop a pie chart by figuring out what percent of the whole population surveyed fell into each category. This percent is then translated to representative wedges within the pie chart.

Pie charts are probably one of the most misused forms of graphs. Consider the data in Figure 10.4 regarding the percent of men from New England states who played hockey as a child.

PERCENTAGE OF MEN WHO PLAYED HOCKEY AS A CHILD IN NEW ENGLAND		
Maine	25%	(Data is fictitious.)
New Hampshire	30%	
Massachusetts	51%	
Connecticut	22%	
Rhode Island	34%	
Vermont	40%	

Figure 10.4.

Because the data are given in percents, some people want to display the data in a pie chart, which is known for illustrating "percent" data. However, the problem is that pie charts illustrate "what percent of the whole" each wedge of the pie represents. The total of all the percents should equal the whole, or 100%. If you add up the percents in the data above, they equal more than 100% because they are not representing parts of a whole.

A more appropriate scenario for a pie chart would be in illustrating the following data. What are high school students' favorite types of music to listen to? Of the set of students surveyed, the following (Figure 10.5) represents the percentages of the whole group surveyed that preferred each type of music.

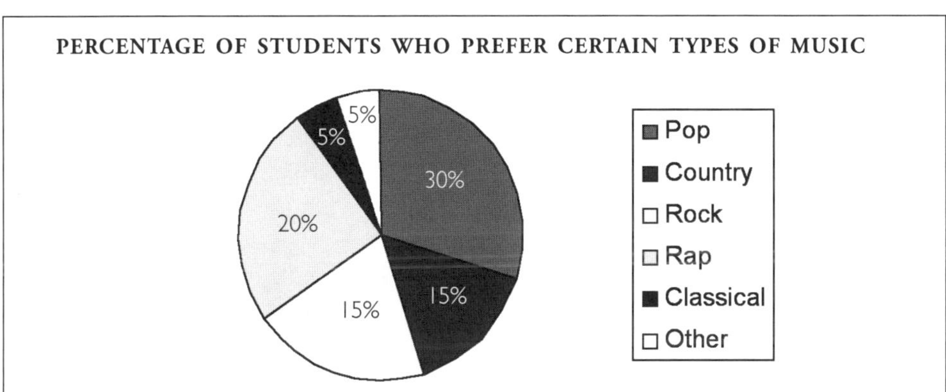

Figure 10.5.

Scatterplots

At some point, we want students to be able to predict and determine whether there is a relationship between two variables. For example, does smoking cause cancer? Does the amount of time listening to classical music increase your IQ? Does the amount of time you spend studying for a test affect how well you do on the test? To determine if there is a correlation between two items, we can gather pairs of data and form a scatterplot to see if the data relate. A scatterplot is made by plotting coordinates in a plane, similar to plotting points on a set of x and y axes.

Suppose we asked 20 students two questions each: "How long did you study for your math test?" and "What was your test score?" Figure 10.6 shows the data we received from each of the students. These data result in the scatterplot in Figure 10.7.

The scatterplot indicates that as the amount of study time increases, a student's test score improves. Of course, there are exceptions in the data, and these exceptions provide the opportunity to discuss other factors that influence test scores that may explain why these exceptions exist.

STUDY TIME (HRS)	1	1.5	2	2.3	4	3	2.5	3.5	1	0.5	2.7	5	3.3	4.5	7	6	1.2	2.2	2.5	3
TEST SCORE	65	68	70	70	85	83	86	87	50	48	75	90	85	82	96	99	45	70	87	84

Figure 10.6.

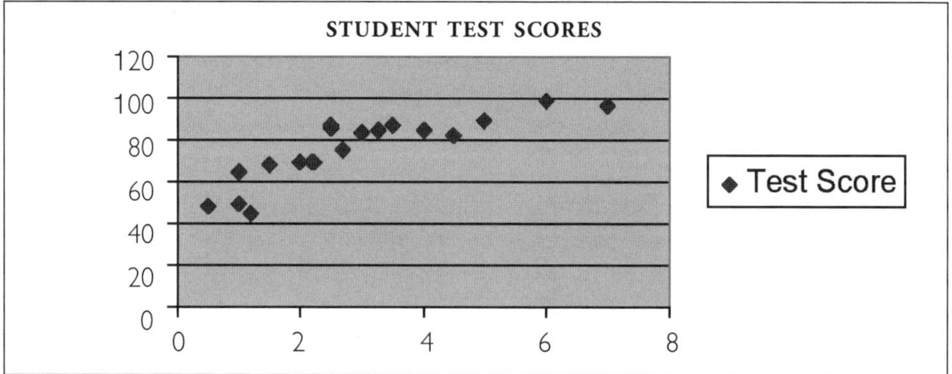

Figure 10.7.

Axes Labels Can Change How A Graph Is Interpreted

One final thing to consider: How you choose to label your axes can result in very different interpretations of the data. Suppose we reconsider the graph on popcorn brand comparisons. If we simply change the frequency axes to represent a different scale but still plot the same data, we get the graph in Figure 10.8. Because the scale is much more compact, the disparity in the preferred brands is not as evident as it was in the previous graph. Therefore, the people who work for Brand B popcorn would still see themselves in third place, but they may not see themselves as far behind the other brands as the first graph implies.

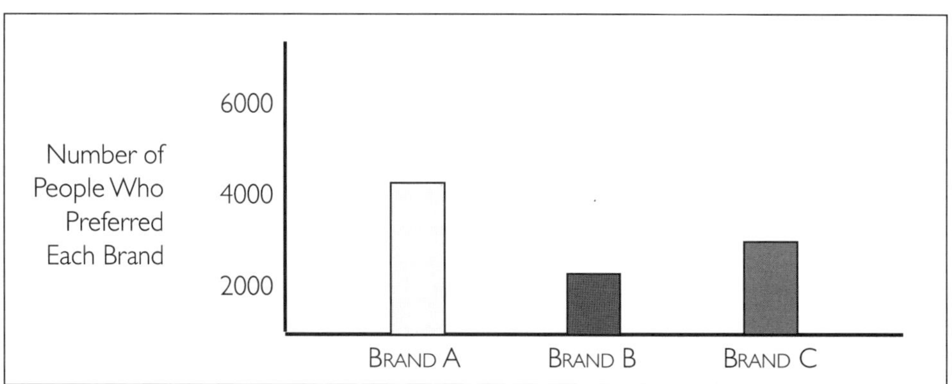

Figure 10.8.

Many other types of graphs can be used to illustrate data. Most software programs include graphing components. Students should have many opportunities to play with these graphing features to experiment with graphs and figure out what they think are the best representations of data in a given context.

Resources

Dixon, J. K., Falba, C. J. (1997). Graphing in the information age: Using data from the World Wide Web. *Mathematics Teaching in the Middle School, 2*(5), 298–304.

Jones, G.E. (1995). *How to lie with charts.* Berkeley, CA: Sybex International.

INSIGHT

Simulations

By Ivan W. Baugh

Spreadsheets can help students conduct investigations involving statistics. Two familiar spreadsheet activities include "rolling the dice" and "coin toss." Before using a spreadsheet, however, students will need hands-on experiences to understand what the spreadsheet can do for them.

Depending on students' level of knowledge regarding spreadsheets, you can provide them with a prepared spreadsheet such as the Coin Toss—Roll the Dice file available on this book's companion Web site at http://education.bellarmine.edu/baugh/. Or, have them enter the equations that will empower them to examine the probabilities investigated in these activities. The spreadsheet in Figure 10.9 shows a comment explaining the formula and another comment explaining the formula in cell A3. A comment could also pose questions for the students to consider as they conduct their investigations. Another way to view the formula is by clicking in a cell and looking at the formula in the Formula bar at the top of the screen, or you can have the spreadsheet show all formulas. This can facilitate class discussion as you guide your students to an understanding of how the spreadsheet simulates the action they have completed (tossing the coin or rolling the dice).

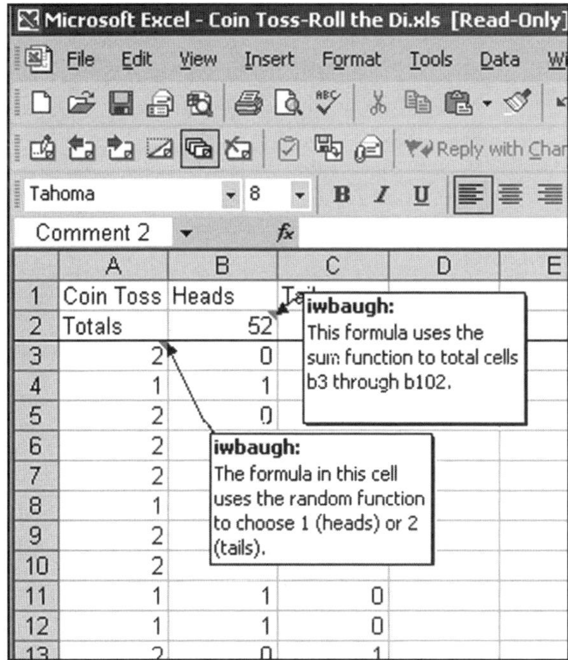

Figure 10.9.

Let's examine the coin toss first. When we toss a coin, only two outcomes exist. Since a spreadsheet works more easily with numbers than words, we assign 1 to represent heads and 2 to represent tails. The software you are using will determine the structure of the equations. In Microsoft Excel, enter this equation: =INT(RAND()*2)+1.

◇	A	B	C
1	Coin Toss	Heads	Tails
2	Totals	55	45
3	1	1	0
4	2	0	1
5	2	0	1
6	1	1	0
7	1	1	0
8	2	0	1
9	2	0	1
10	1	1	0
11	2	0	1
12	2	0	1
13	1	1	0
14	1	1	0
15	2	0	1
16	2	0	1
17	2	0	1
18	2	0	1
19	2	0	1
20	1	1	0

Figure 10.10.

As seen in Figure 10.10, row 1 identifies the information appearing in each column. Row 2 totals the number of times heads and tails occurred. Have the students recalculate the spreadsheet. Have them count the number of times they must recalculate (toss the coin) to get the 5-5 split. Ask them to predict whether they will ever have a back-to-back 5-5 split.

After completing the above investigations, invite students to increase the number of rows by groups of 10. Select rows 12–22, and from the Edit menu use Fill Down to increase the number of coin tosses by 10. Watch how the totals change. Remember to adjust the formula in row 2 to include the added cells. Have students compare results with investigators on either side of them. Ask the students to predict what they think the results will be each time they add a group of 10. Follow this with a discussion of the accuracy of their predictions and the factors that made it challenging to accurately anticipate the results.

You can also have students create a chart of the totals, placing it beside the columns where you simulate the coin tosses. They can visually monitor the results as well as see the numbers. Each time they recalculate the spreadsheet, the chart will redraw to reflect the new values.

Continue adding groups of 10 (rows 22–32, 32–42, etc.) until a total of 100 coin tosses is reached. Ask your students to hypothesize whether they will get a 50-50 split when they toss the coin 100 times. Ask them to recalculate the spreadsheet 10 times and tabulate the number of 50-50 splits they get (Figure 10.11).

	A	B	C	D	E	F	G	H	I
1	Di Roll	One	Two	Three	Four	Five	Six		Number of Rolls
2	Totals	20	17	25	18	16	24		120
3	2	0	1	0	0	0	0		
4	3	0	0	1	0	0	0		
5	3	0	0	1	0	0	0		
6	1	1	0	0	0	0	0		
7	3	0	0	1	0	0	0		
8	6	0	0	0	0	0	1		
9	1	1	0	0	0	0	0		
10	5	0	0	0	0	1	0		
11	6	0	0	0	0	0	1		
12	1	1	0	0	0	0	0		
13	6	0	0	0	0	0	1		
14	3	0	0	1	0	0	0		
15	2	0	1	0	0	0	0		
16	5	0	0	0	0	1	0		
17	4	0	0	0	1	0	0		
18	2	0	1	0	0	0	0		
19	1	1	0	0	0	0	0		
20	2	0	1	0	0	0	0		
21	4	0	0	0	1	0	0		
22	6	0	0	0	0	0	1		
23	2	0	1	0	0	0	0		
24	1	1	0	0	0	0	0		
25	5	0	0	0	0	1	0		
26	5	0	0	0	0	1	0		
27	2	0	1	0	0	0	0		
28	3	0	0	1	0	0	0		
29	1	1	0	0	0	0	0		
30	6	0	0	0	0	0	1		
31	2	0	1	0	0	0	0		

Coin Toss | Roll the Di | Sheet3

Figure 10.11.

Would you get a 50-50 split if you tossed the coin 1,000 times? You could continue adding rows until you get to row 1002. Another way to achieve the same results has the students copy the totals in row 2, go to an empty sheet in the workbook, click in a cell, then from the Edit menu choose Paste Special. Choose the option to paste values only. Recalculate the spreadsheet and repeat this process 10 times. Total the results and test the accuracy of the hypothesis.

When the students have a thorough understanding of how this spreadsheet works and the formulas they used in the spreadsheet, challenge them to adapt it to Roll the Dice. Lead the class in a discussion of how they will need to alter the formula in column A. What changes will they need to make to the formula in column B? How many columns will they need to use to tabulate the results? Ask the question, "Will 100 rolls give an opportunity for a 10-10-10-10-10-10 split? If not, how many rolls (rows in the spreadsheet) will they need to use to have an opportunity for an even split of numbers?

After setting up 120 rows, have the students recalculate the spreadsheet and observe the results they get. Ask them to determine the greatest number variance in the results. How many times did they have an even split among the numbers? What change will they need to make in the formulas in row 2? They may make the change in one cell, then select the cells to which they want the formula to apply and from the Edit menu use Fill Right to replicate the formula. This provides an opportunity to discuss relative cell references by examining how the formula changed as it filled across.

Spreadsheet simulations allow students to focus attention on results rather than the mechanics of the learning activity. Your students can write up their results using a word processor and insert a copy of the saved spreadsheet file in their document to facilitate a reader's understanding of the analysis. As part of this analysis, ask students to think of other simulations where they could use the skill they acquired while completing this learning activity. This author has created a spreadsheet simulation of the work of renowned geneticist Gregor Mendel. I often say, "It took Mendel 9 years to complete his work; your students can simulate his work in 9 minutes using a spreadsheet." I included a copy of the file titled Coin Toss on our companion Web site at http://education.bellarmine.edu/baugh/.

INSIGHT

Fair Games and the Role of Probability

By Anne Raymond

Playing games can be a lot of fun, but only if there is a sense the games are fair. That is, most people probably wouldn't play a game they didn't think they had a fair chance to win. Whether you are playing a game at the carnival, a board game, or the lottery, you play because you believe you can win. Let's look at some of these games and how probability plays a role in determining how likely you are to win.

Yahtzee

The game Yahtzee is all about probability. You roll five dice and try to come up with different number combinations on three rolls of the dice. You try for such results as three of a kind, four of a kind, full house, and a "Yahtzee." Each combination is worth a different number of points. A Yahtzee is the most valuable roll because it is the most difficult to achieve.

A Yahtzee is a combination of rolls of the dice that results in all five dice showing the same number value. Probabilistically, this is the most unlikely outcome to occur. Therefore, getting a Yahtzee (which most people do not get in a given game of Yahtzee) is worth a lot—50 points! Just to give you some sense of how unlikely a Yahtzee is, of the 7,776 possible outcomes when rolling five dice, you can get a Yahtzee only six ways. Of course, in Yahtzee, you get multiple rolls and can choose only a subset of the five dice to roll on subsequent rolls. However, even with additional rolls, the odds of winning do not increase so dramatically as to make getting a Yahtzee very likely.

Spin the Wheel

At most carnivals, you can go up to any number of booths and try to win a prize by selecting a number and hoping the booth worker spins your number on the wheel. There can be any total of numbers on the wheel. Regardless, the game is fair because there is an equal chance that any number on the wheel can be spun on a given spin. Thus, if a wheel has 40 numbers and you put your money on 1 number on the board, you have a 1/40, or 3%, chance of winning. People often improve their odds of winning by spending more money on a single spin. That is, they put their money on 10 numbers at a time, increasing their chances of winning to 10/40, or 25%.

The Lottery

Thousands of people play the lottery every week. And yet, what are the chances of winning? Even though the chances are slim, the game is a fair game, and the payoff is so large most people are willing to play regularly even though the likelihood of being a big winner is low.

Just how unlikely is it that you will win the lottery if you buy one ticket? Most lotteries include the option of a player selecting the six numbers that will be drawn in the game. The order in which the six numbers are drawn doesn't matter. You merely have to pick the correct combination of six numbers. So if the numbers you can choose range from 1 to 60, there are (60 x 59 x 58 x 57 x 56 x 55) * (6 x 5 x 4 x 3 x 2 x 1) = 40,495,840 possible combinations of six numbers. If you buy one ticket with one set of six numbers, you have a 1 in 40,495,840 chance of winning the lottery. Not very promising! However, since the numbers are chosen at random, you at least know you have as fair a chance of your numbers being selected as does another person choosing a set of six numbers.

It's interesting to note that when the lottery payoffs are higher, more people play, even though the odds of winning are not any higher than with a small payoff. It is also interesting that some people religiously play "their numbers" for the lottery rather than purchasing a "quick-pick" set of lottery numbers that are computer generated. They believe their numbers are luckier than a quick-pick, even though, again, the odds are the same regardless of how the numbers are selected.

There is an advertisement for a lottery in one state that claims, "Somebody's going to win, it might as well be you." This is a bothersome advertisement because there is no guarantee that for any given drawing someone will definitely win. However, it sounds to the general public as if someone will win that lottery drawing, when in reality all the claim can really mean is that eventually someone will win the lottery.

Rock, Paper, Scissors

You may remember from your childhood the game Rock, Paper, Scissors. On the count of three, you and your friends would each thrust out a hand showing either the sign for paper, the sign for scissors, or the sign for rock. The winner was determined in this way: paper beats rock, rock beats scissors, and scissors beats paper. Sometimes decisions about who was to go first or who had to do an unpleasant chore were made by playing this game. The game is fair because you have an equal number of ways to win and an equal number of ways to lose. The game is also random, so the randomness combined with equal opportunity to win makes this a fair game.

An Unfair Version of Paper, Scissors, Rock

Suppose the rules to the game were a little different. Suppose three people played the game and each person was designated as being either Player A, Player B, or Player C (roles are assigned before the rules of the game are revealed). The players are to make the paper, scissors, rock signs of their choice each time they play a game. However, this time the rules say Player A wins if all three people show the same sign, Player B wins if exactly two people show the same sign, and Player C wins if all three players show a different sign.

If you play this game repeatedly, it will soon become clear this is not a fair game. Player B wins the majority of the time, and Player A wins the fewest times. Probabilities help explain why this occurs and why this game is not fair.

First, we list all the possible ways for Player A to win. Recall that this happens only when all three players show the same sign. Therefore, Player A wins when the following combinations of signs occur:

Paper	Paper	Paper
Rock	Rock	Rock
Scissors	Scissors	Scissors

There are only three ways Player A can win.

Now let's consider all the ways Player C can win, that is, all the ways each of the three players chooses a different sign:

Rock	Paper	Scissors
Rock	Scissors	Paper
Paper	Scissors	Rock
Paper	Rock	Scissors
Scissors	Paper	Rock
Scissors	Rock	Paper

There are six ways Player C can win.

Finally, lets figure out all the ways Player B can win. Let's first figure out all the ways Player B can win if the first person always chooses rock.

Rock	Scissors	Scissors
Rock	Paper	Paper
Rock	Rock	Scissors
Rock	Rock	Paper
Rock	Scissors	Rock
Rock	Paper	Rock

The above combinations represent only one third of the ways Player B can win.

Notice there would be 6 ways for Player B to win if the first person always chooses rock. Similarly, there would be 6 ways for Player B to win if the first person always chooses scissors, and there would be 6 ways for Player B to win if the first person always chooses paper. So Player B has a total of 18 ways to win the game.

If we add it all up, the game has 27 possible outcomes. Since Player A has only 3 ways to win, we can say the probability that Player A will win is 3/27, or 1/9. Further, the probability that Player C will win is 6/27, or 2/9, and the probability that Player B will win is 18/27, or 2/3. Thus, the game is incredibly unfair, since Player B has 6 times as many chances of winning as Player A has and 3 times as many chances of winning as Player C has. Also, Player C has twice as many chances of winning as Player A has. In all, the game is unfair, and as such, it is not as much fun to play (unless you're Player B!).

Resource

Masse, L. N. (2001). The possibility of perfection. *Mathematics Teaching in the Middle School, 6*(9), 500-505.

MATHEMATICS AND M&MS

An old "staple" in mathematics classes introduces spreadsheets with graphing capabilities as tools for mathematics investigations.

By Margaret L. Niess

Mm mm good, an edible manipulative for the mathematics classroom that melts in your mouth and not in your hands!

That's right, M&Ms are useful in many ways in the mathematics classroom. How many M&Ms are in a typical bag? How many different colors of M&Ms are in a typical bag? What is the fraction of red M&Ms? What is the percent of green M&Ms? What decimal part of the whole bag do the yellow M&Ms comprise? These questions are but a few of the many possible questions for students to consider.

The beginning of a new school year is a time to refresh the students' memories and to instruct them in the use of some of the tools they will be using throughout the year. The spreadsheet with a graphing option is a valuable tool for the mathematics classroom. Here is an activity that I have used many times to introduce students to the idea of a spreadsheet. The activity does require that each student has a bag of M&Ms.

Warm-up

Find out what the students know about a bag of M&Ms. Accept their responses without saying "Right" or "Wrong." When students open their own bag, they can adjust the responses.

What is in a typical bag? (M&M candies.) Can you be more specific? (They are round candies with an M on them.) Do they all look exactly alike? (No, there are different colors of candy.) How many different colors? (Answers typically vary from five to seven.) What are the colors? (Red, yellow, green, orange, tan, and brown are the colors.) Are there the same number of each color in a bag? (No.) Which color has the most? (Let them guess.) How many total M&Ms are in a typical bag? Record all responses.

To clarify the responses, open a sample bag of M&Ms and demonstrate what the students will be doing with their bag. Place a transparency on the overhead, pour the candies on the sheet, and form a histogram with the candies as in Figure 1.

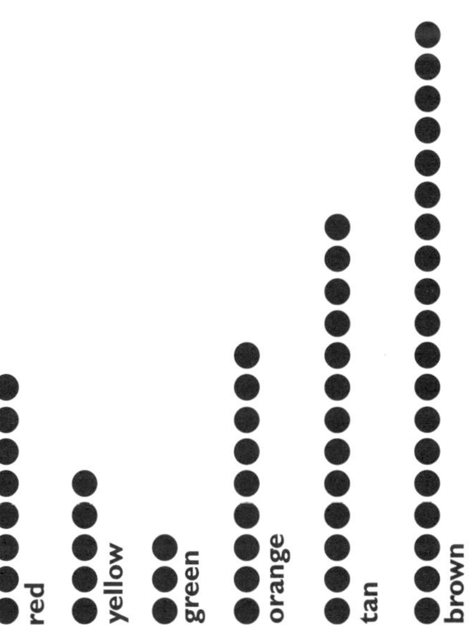

Figure 1. Sample candy histogram.

(M&Ms do melt so don't leave the overhead light on for the whole class period!) Ask students to count the numbers of each color and find the total number of M&Ms in the bag. Compare these numbers with the student guesses.

Data Collection

Students are now prepared to estimate the numbers of each color and the total number for their own bags of M&Ms. Have students record their estimates (guesses based on information from the warm-up) on a record sheet similar to Figure 2. Be sure to give clear instructions before handing out the bags of M&Ms. Students need to complete the following activities *before* they eat the data.

1. Open the bag of M&Ms on a sheet of paper.
2. Move the M&Ms to form a candy histogram as in Figure 1.
3. Count and record the actual number of each color of M&Ms.
4. Record the total number of M&Ms.
5. Using six 3X5 cards, record the number of each color on a separate card. Make the number large and dark. On the reverse side of the card record the corresponding color for that number.
6. Using one more 3X5 card, write the student name in large, dark letters.
7. Complete the record sheet by finding the difference between your guess and the actual number found in the bag. This difference might be negative.

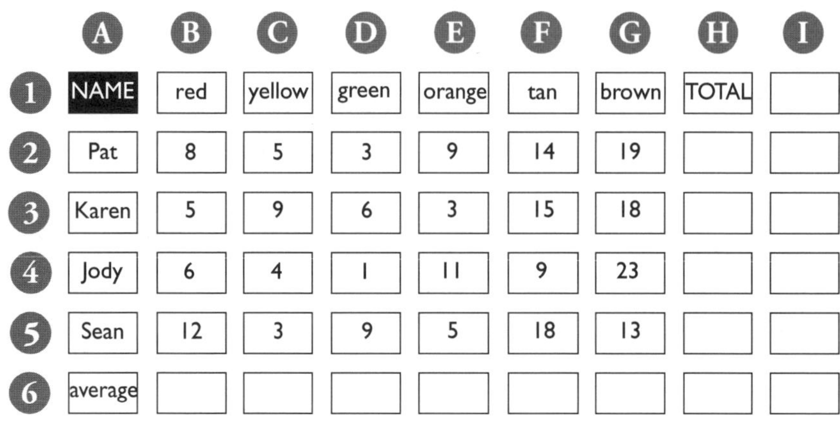

	A	B	C	D	E	F	G	H	I
1	NAME	red	yellow	green	orange	tan	brown	TOTAL	
2	Pat	8	5	3	9	14	19		
3	Karen	5	9	6	3	15	18		
4	Jody	6	4	1	11	9	23		
5	Sean	12	3	9	5	18	13		
6	average								

Figure 3. Sample floor spreadsheet.

A "Floor" Spreadsheet

When all students have finished, a "floor" spreadsheet is used to display the class data. Place column (marked A, B, C, D, E, F, G,H,I) and row (1, 2, 3, 4, 5,...) identifiers (two more than the number of students in your class) made out of 3x5 cards on the floor, marking out the area for the spreadsheet. Next, place labels for each of the columns that you will use, naming the cells as you store the information (see Figure 3). Instruct the students to place their records of information in particular rows (assign the rows to ease congestion around the spreadsheet).

After all data records have been entered into the spreadsheet, have the students find the cell that contains:

- the least number of green M&Ms (D4).
- the most number of red M&Ms (B5).
- the least number of orange M&Ms (E3).

Next, consider the differences between *labels* and *values*. Cells D4, B5, and E3 have values in them whereas cells A2, A3, and A4 have labels. The spreadsheet cannot add, subtract, multiply or divide labels, but the spreadsheet can operate on values.

COLOR	GUESS	ACTUAL	ACTUAL-GUESS
Red			
Yellow			
Green			
Orange			
Tan			
Brown			
Total			

Figure 2. Sample data collection sheet.

	A	B	C	D	E	F	G	H
1	Name	Red	Yellow	Green	Orange	Tan	Brown	Total
2	Pat	8	5	3	9	14	19	58
3	Karen	5	9	6	3	15	18	56
4	Judy	6	4	1	11	9	23	54
5	Sean	12	3	5	9	13	13	55
6	Average	8	5	4	8	13	18	56
7								

Figure 4. Completed spreadsheet.

One value of the spreadsheet is that formulas may be entered into cells to have the computer compute other values. Cell H1 has the label Total. Ask the students how they would find the total, but instruct them to tell you by referring to the cell names. In the example, Pat's total is determined by "adding the values in cells B2, C2, D2, E2, F2, and G2." As the formula is identified, write the equivalent formula, (B2+C2+D2+E2+F2+G2), for the spreadsheet on a 3x5 card and place the card in cell H2. Provide the students with the spreadsheet shortcut for this formula (for Microsoft *Works* on Macintosh, =SUM(B2:G2) and for *AppleWorks*, =SUM(B2...G2)). Have each student prepare a formula on a 3x5 card and place it in the appropriate cell in the spreadsheet. They should also act as the computer and prepare a card with the actual value, placing that card over the formula in the same cell.

As suggested by the label in cell A6, the next problem is to find the average number of M&Ms for each color. Ask the students how they would compute the average for the number (B2:B5). Assign students to place the appropriate formulas in the cells, and then have other students compute the value using the formula. When this task is completed, review the concepts of labels, values, and formulas referring to the specific information in the floor spreadsheets.

A Computer Spreadsheet

The next task is to enter the data into a computer spreadsheet similar to the floor spreadsheet (Figure 4). To assist in this task, I copy the floor spreadsheet for the students to use at their computers at the next class period; copy only the student data, not the values determined by formulas. Instruct students to enter the formulas, as they did for the floor spreadsheet. Experience with the floor spreadsheet simplifies this activity.

Demonstrate the use of the graphing tools to have the computer create a histogram of the average bag of M&Ms using the data in the spreadsheet (Figure 5). If you are using *AppleWorks*, you'll need an add-on graphing package such as *TimeOut Graph* from Beagle Bros. Challenge the students to create a histogram that compares the average bag of M&Ms with their personal data.

An excellent completion to the M&Ms activity is to have the students prepare a report using a word processor, copying the data and the graph into the report. For the report, students describe the experiment that was conducted and show the results, including the data and histograms. In the summary, they must respond to the following questions. From this sample of M&Ms:

- Which color(s) appear least?
- Which color(s) appear most?
- What is the ratio of red:yellow:green: orange:tan:brown M&Ms in an average bag?
- Based on this sample, if a bag of M&Ms contains 100 M&Ms, how many reds are expected? How many yellows? How many greens? How many oranges? How many tans? How many browns?

Summary

According to the National Council of Teachers of Mathematics (NCTM) *Curriculum and Evaluation Standards for School Mathematics* (1989), students should learn to use the computer as a tool for processing information and performing calculations to investigate and solve problems. And, as indicated in the NCTM (1991) *Professional Standards for Teaching Mathematics*, teachers not only need to use technology to perform mathematical investigations, but they need to help students do the same. The beginning of the school year is the best time to begin experiences in learning to use tools such as the spreadsheet. M&Ms

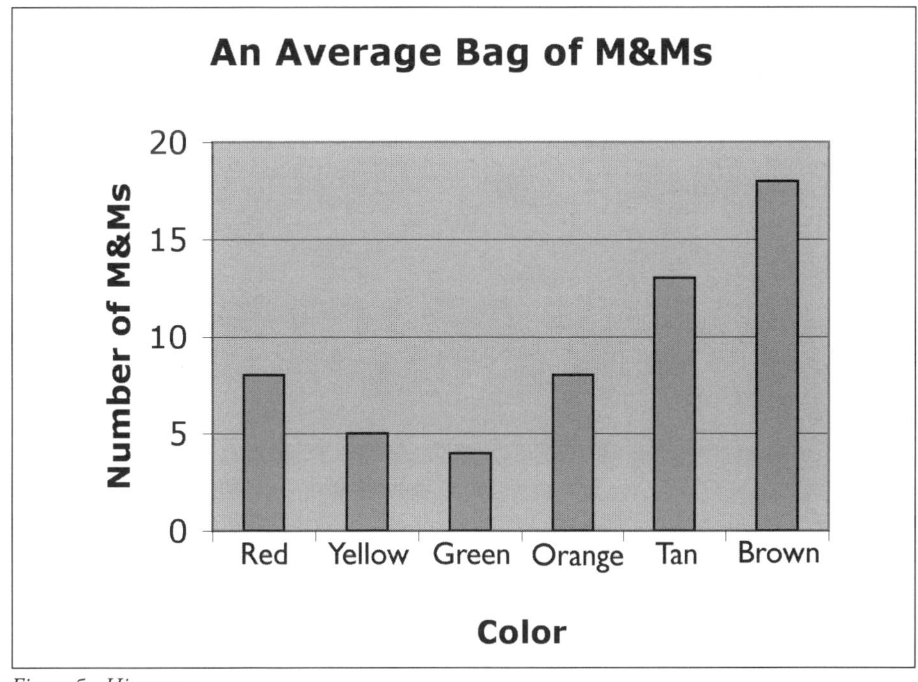

Figure 5. Histogram.

provide a "rich" environment for students to systematically collect, organize, and describe data using the appropriate tool, the computer. The floor spreadsheet provides a natural transition for students to learn about the computer spreadsheet. Learning to use the tool and using the tool to pursue mathematical investigations can and should be continued throughout the year.

References

National Council of Teachers of Mathematics, Commission on Standards for School Mathematics (1989). *Curriculum and evaluation standards for school mathematics*. Reston, VA: Author.

National Council of Teachers of Mathematics, Commission on Teaching Standards for School Mathematics (1991). *Professional standards for teaching mathematics*. Reston, VA: Author.

EXPLORING DATA WAREHOUSES

Finding good data can sometimes be a problem for even the most experienced researcher. In this month's Mining the Internet, though, the authors show how to find and use good, educationally sound data sets that are made available by corporations and educational institutions.

By Glen Bull, Gina Bull, and Hollylynne Stohl

Subject: Mathematics, Science, Data Processing

Grade Level: All

Technology: Internet/Web, spreadsheet software (e.g., Microsoft Excel or Works, AppleWorks)

One of the most promising uses of technology in the classroom involves the analysis of real data. Businesses often place data related to their corporate activities into online data warehouses. The World Wide Web also has many sources of educational data that can be readily accessed. This information can be useful in social sciences and humanities as well as in mathematics and science. This type of activity can facilitate interdisciplinary connections and engage students in interpretation of real data.

For example, folklore has numerous tales of natural world phenomena—if caterpillars are particularly fuzzy, expect a cold winter, it is possible to tell the current temperature by listening to the rate at which crickets chirp, and so on. Often natural world phenomena from folklore can be examined from a scientific or mathematical perspective, and the data needed to investigate such phenomena may already be available on the Web. The "Exploring Data" Web site developed by the Education Services Directorate in Queensland, Australia, is one example of this type of data resource. The data sets range from Oscar winners to world population data to the price of diamond rings.

Education Services Directorate, Queensland, Australia (http://curriculum.qed.qld.gov.au/kla/eda)

1970 US Draft Lottery	Fleas
1971 US Draft Lottery	Galileo's Experiments
AIDS / HIV	Global Temperature 1
Air Pollution	Global Temperature 2
Alligator!	Metric Estimates
Anscombe's Dataset	Oil Production
Bradman - an Outlier?	Old Faithful
Bradmanesque	Olympic Gold
Carbon Dioxide	Oscar Winners
Carbon Emissions	Pecking Order
Challenger	Pottery
Cloud Seeding	Smoking and Cancer
Codeine Concentration	Speed of Light
Cricket (the Insect)	Stride Rate
Density of the Earth	Wild Horse
Density of Nitrogen	World Population
Diamond Rings	Year 10 Certificates

Transferring the Data

The data sets available at this site include information relating temperature to the rate of cricket chirps. To use the data, you will need to transfer it from the Web site to a local computer. The specific transfer method will depend on which Web browser and type of spreadsheet you use.

If you have Internet Explorer and the Excel spreadsheet program, then the data sets can be directly opened in the native Excel format from within Internet Explorer. If you are using Netscape, the browser will offer to save the data file directly to disk. (If you are using a PC, you will need to use the ".xls" file extension in the name of the file to be able to access it from Excel.)

If you are using a different spreadsheet, such as ClarisWorks, then the data can be saved to a local disk in "tab-delimited format"—that is, with each number separated by a tab character—and then imported into your spreadsheet.

Numeric Display of Data

Once the file has been transferred from the "Exploring Data" Web site to a spreadsheet, the table of data will resemble the following example. In this instance, the data could also have been retyped. However, the skills acquired in downloading data from a Web site can be used to acquire and analyze much larger data sets than one would want to retype, which also introduces the likelihood of typographical errors. For these reasons, it is worth taking the time to perfect the data-transfer process with a small data set such as the cricket chirp set.

Once the data is in the spreadsheet, you may want to ask the class whether the rate of cricket chirps seems to be related to temperature. Sorting the data from highest to lowest temperature may make it easier to look for a pattern. This can be accomplished with the "Sort" option in the spreadsheet.

Graphical Display of Data

Plotting the data also will make it easier to examine the relationship. The "Insert Chart" option in later versions of Excel will produce a chart wizard that will guide you through the process. (Other spreadsheets have similar charting capabilities.) Select "X-Y Scatterplot" to produce a plot of temperature versus "chirps per second."

The axes must be scaled appropriately when a graph is analyzed. Setting the minimum temperature to 65 degrees Fahrenheit (°F) and the maximum temperature to 95 °F on the y-axis will display the range of temperatures for which data were collected. Similarly, setting the minimum rate to 14 chirps per second and the maximum rate to 21 chirps per second will display the relevant range of temperatures. Double-click a number on the x or y axis to produce an Excel dialog box that will allow you to adjust the scale. When you have adjusted the scale to display the most relevant section of the graph, the result may resemble the graph shown in Figure 1.

A rough relationship can be seen between temperature and the rate of chirping. After the graph has been printed, the class can place a string in a straight line over the data points to develop an estimate of the line that best fits the data. The line that best fits the relationship between temperature and chirps per second is known as a *trend line.*

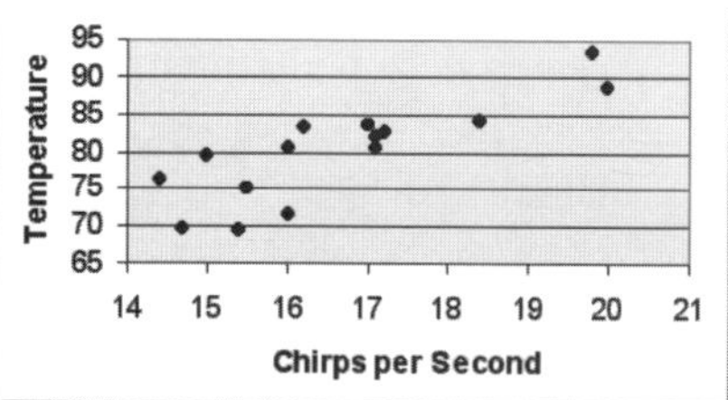

Figure 1. Crickets and temperature.

Cricket Chirps and Temperature

Chirps per Second	Temperature (degrees F)
20	88.6
16	71.6
19.8	93.3
18.4	84.3
17.1	80.6
15.5	75.2
14.7	69.7
17.1	82
15.4	69.4
16.2	83.3
15	79.6
17.2	82.6
16	80.6
17	83.5
14.4	76.3

Computer Analysis of the Data

Once the class has looked for a pattern in the data, the computer can be used to analyze the data and add a trend line. The students' estimates can then be compared with the mathematical analysis conducted by the spreadsheet program. To perform this analysis in Excel, choose "Add Trendline" under the "Chart" option on the menubar when the chart is selected (Figure 2).

Figure 2.

In this case, we chose to add a linear trend line, but other options such as a logarithmic or curvilinear trend line are also available. The "Linear Trendline" option finds the straight line that most closely fits the data (Figure 3).

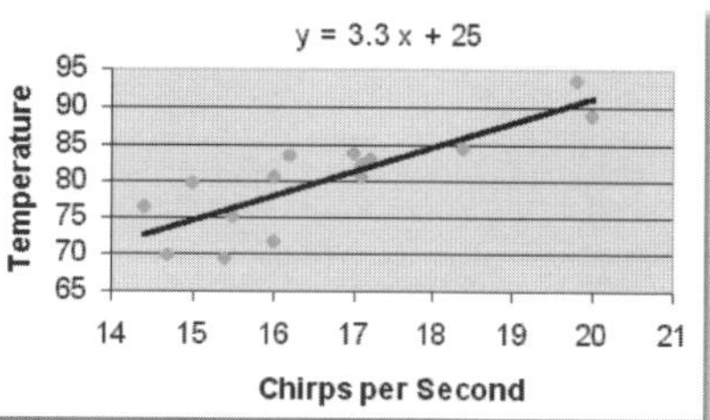

Figure 3. Crickets and temperature.

Algebraic Descriptions of the Data

Upper elementary or middle school classes should be able to explore relationships in the data by examining graphical displays. Students in middle school or high school who have had algebra also may explore algebraic descriptions of the data.

The "Options" tab of the "Add Trendline" dialog box in Excel includes a "Display Equation on Chart" option that shows the equation of the trend line on the graph. In this case, the equation suggests that the temperature can be estimated by multiplying the number of cricket chirps per second by three and then adding 25:

$y = 3x + 25$

In other words, if the cricket is chirping 20 times per second, then the temperature will be approximately 85°:

$3 \text{ (20 chirps)} + 25 = 85°$

The Education Services Directorate suggests that students not only work with data from the directorate's Web site but also collect their own data. A sound recorder utility (usually found under the Accessory menu) can be used to display the number of chirps in a second. Students could tape record crickets chirping and then digitize the recording with sound recorder (if the computer has a sound card). The data-collection process could be the subject of another article on microcomputer-based laboratories. We hope to listen for crickets later this summer, and we'll report on the results if we succeed.

Tab-Delimited Data

The Data and Story Library at Carnegie Mellon University is another example of an online educational data warehouse. The information in that library also addresses a wide range of subjects from the social sciences to music to biology. The data in this warehouse is available in the form of a tab-delimited text file rather than in a native spreadsheet format.

Because the data is in tab-delimited format, one extra step is required to download and import the data into a spreadsheet. Modern spreadsheets, however, automate the import process, making it straightforward.

Data and Story Library, Carnegie Mellon University (http://lib.stat.cmu.edu/DASL)

Archaeology	Health
Automotive	Legal
Biology	Medical
Census	Miscellaneous
Consumer	Music
Demography	Nature
Economics	Nutrition
Education	Physics
Energy	Psychology
Engineering	Science
Environment	Social science
Europe	Sociology
Finance	Space
Food	Sports
Government	Weather

We examined national birthrate data found in the Demography category of the data warehouse. To access that data once you have reached the Data and Story Library, click on the "Birthrates Datafile" link in the Web site's Demography section. The data will be displayed in your Web browser.

Because the data is stored as a tab-delimited text file, use the "Save As" option in your Web browser to save the displayed data file to disk; be sure to select "text file" rather than "HTML" format (Figure 4).

To display a text file when opening it within a spreadsheet, select "All File Types." In Excel, a wizard will appear that will ask whether you wish to import a delimited file. Select the default options presented to import the birthrate data (Figure 5).

Once the birthrate data has been imported, it can be graphed in much the same way as the cricket data. In the sample graph in Figure 6, the data points have been connected with a line to make it easier to follow trends over time.

Many educational objectives are related to national birthrate and population growth. For example, students studying U.S. history could ask questions such as:

- What event corresponds with the drop in birthrate in the mid-1940s?
- What period on the graph corresponds to the so-called Baby Boom?
- What factors might be responsible for

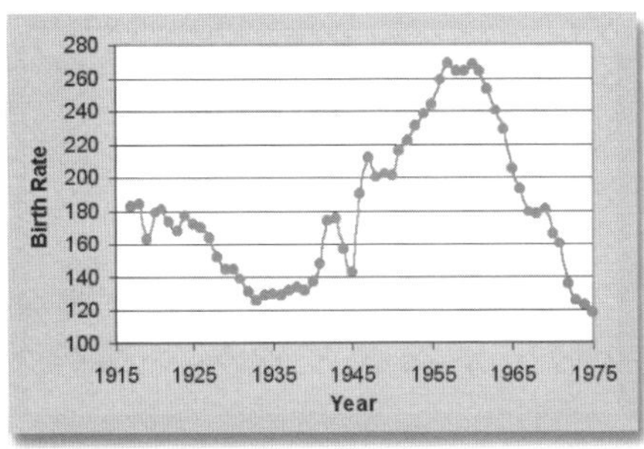

Figure 6. Birthrate per 10,000 23-year-old women.

Figure 4.

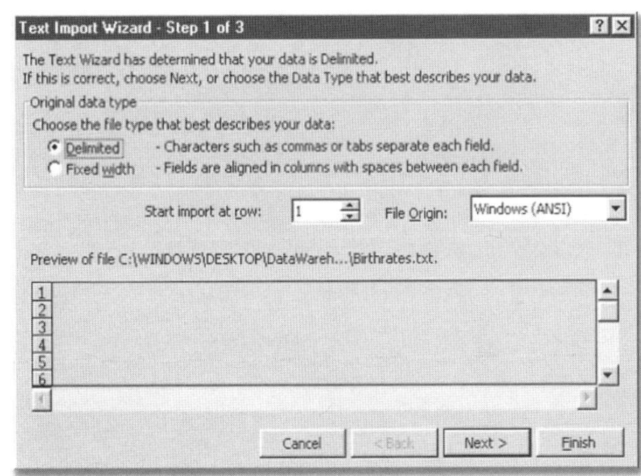

Figure 5.

the decline in birthrate since 1960?

Other Data Sources

Many other educational data sources can be found on the World Wide Web, often in formats that can be directly imported into spreadsheets for display and analysis. The table at right shows several other sources as starting points for exploration.

When data is acquired from the Web, students must be taught how to evaluate the source's credibility. Government and educational data sources from sites with ".gov" or ".edu" suffixes are less likely to have deliberate biases. Unintentional data errors are always possible, though, even with data compiled by neutral sources. Projects that allow students to collect their own data can help start discussions about how errors can occur.

Summary

Textbooks frequently contain charts and graphs, but these representations of data are static. Live data permits students to reanalyze the information in many different ways. Students can change the scale of the x and y axes to experiment with the different effects such changes might have on their interpretations of the data. Multiple representations of the data can also be used.

The example in this article examines four different forms of data representation: verbal, numeric, graphical, and algebraic. The *verbal* representation of the data can be expressed as follows:

> Crickets chirp faster as the temperature increases.

The *numeric* representation of the data consisted of a table of numbers with temperature in one column and chirps per second in a second column. The *graphical* representation of the data consisted of a scatterplot with temperature on the y axis and chirps per second on the x axis. The *algebraic* representation of the data consisted of the equation $y = 3x + 25$.

Students who begin working with small, easily comprehended sets of data will develop the skills needed to explore more complex data sets that involve more data points. In the future, citizens will have ready access to the tools needed to explore data from multiple perspectives. If we prepare students to routinely make such tools part of their problem-solving repertoire, then they will be more likely to use them after they graduate.

Data Warehouse Web Addresses

U.S. Census Bureau	www.census.gov
National Climatic Data Center	www.ncdc.noaa.gov
National Center for Health Statistics	www.cdc.gov/nchswww
Bureau of Labor Statistics Data	stats.bls.gov/datahome.htm
Math Forum Data Collections	forum.swarthmore.edu/workshops/sum96/data.collections/datalibrary/
Journal of Statistics Education Data Archive	www.stat.ncsu.edu/info/jse/datasets.index.html
Dartmouth Chance Database	www.dartmouth.edu/~chance/teaching_aids/data_sets.html

Rolling the Dice

Developing an Understanding of Experimental and Theoretical Probability

Merely memorizing formulas and theory isn't enough to get most students to understand the underlying concepts of statistics and probability. By taking hands-on data manipulation and combining it with real-world examples, though, many studentss may better grasp the basic ideas. In this article, author David Pugalee suggests ways to provide that "A-ha!" experience through the better use of educational technology.

By David K. Pugalee

Subject: Math, Probability

Grade Level: 7–9 (Ages 12–14)

Technology: spreadsheet software (e.g., Microsoft Works or Excel, AppleWorks)

Often we ask students to make decisions under uncertain conditions that call for an understanding of probability. To resolve these situations, students develop probabilistic strategies that may contain both appropriate and erroneous conceptions (Hartman, 1997). Part of this problem likely results from the lack of study of probability and statistics in mathematics classes, where the curriculum should contain opportunities for students to examine topics related to data and chance. Such experiences with probability should contribute to students' conceptual knowledge of working with data and chance, and technology offers powerful opportunities for students to develop their understanding of probability. Recognizing the need for developing these skills in students, the National Council of Teachers of Mathematics (NCTM, 1989) calls for more theoretical and experimental models of situations involving probabilities.

As you develop students' understanding of probability, don't emphasize memorizing formulas or computing the likelihood of events. Instead, engage students to simulate and experiment with situations that will develop ways of dealing with problems and experiences related to chance. Students should be able to construct sample spaces, understand models of experimental and theoretical results, make predictions based on such models, and extend this understanding to appreciate applications of probability in the real world (NCTM, 1989).

A Model for Conceptual Development

Here's one way to develop these objectives: Have your students work in groups of four to roll two dice and com-

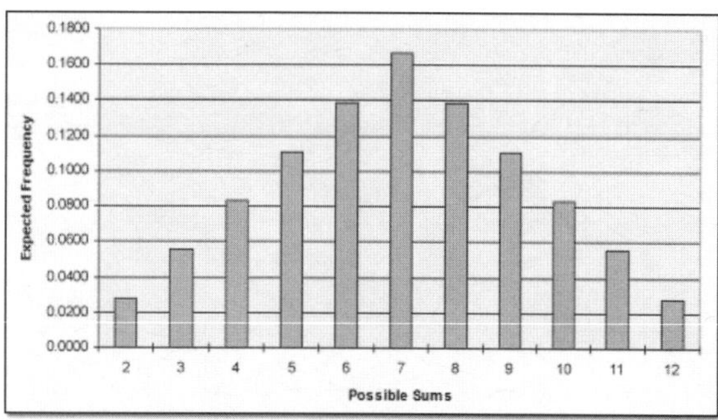

Figure 1. Graph of theoretical probabilities for sums in rolling two dice.

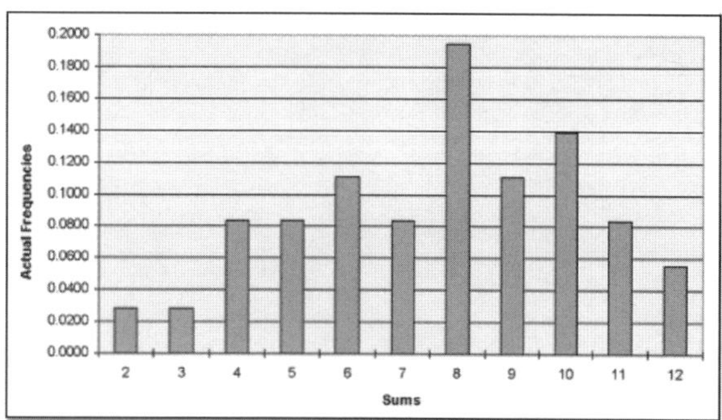

Figure 2. Graph of one group's experimental probabilities of sums in rolling two dice.

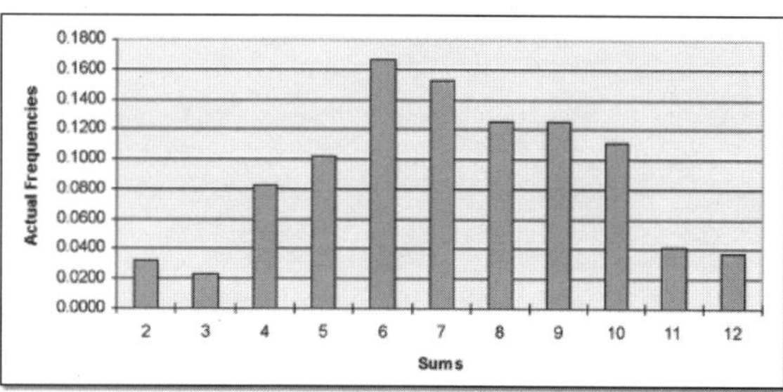

Figure 3. Graph of class experimental probabilities of sums in rolling two dice.

pute the resulting sums. By conducting trials and representing data, the students will learn about the underlying concepts and ideas of probability. First, however, you should develop and make sure your students understand the difference between theoretical and experimental probability. *Theoretical probability* is based on all possible outcomes of a simple experiment. Think of it as the ratio of the number of outcomes of an event to the number of possible outcomes. *Experimental probability*, on the other hand, is based on the actual results of an experiment or trial. Table 1 shows all possible sums from rolling two dice.

The list of all possible outcomes of an event is called the *sample space*. Ask your students to list all possible sums from rolling two dice. After they have time to conjecture, introduce a table for students to complete to help them visualize the possibilities. The table also will show students the importance of such organizational devices.

Next, quiz your students about the probability of certain sums. For example, the probability of rolling a sum of 2 is 1/36, while the probability of rolling a sum of 7 is 6/36 or 1/6. This allows students to expand their proportional reasoning in comparing probabilities and to understand that probabilities can be written as decimals and percentages. Understanding equivalent representations of values is important at this point, because comparing probabilities is a vital feature of this exercise. This reasoning will involve comparing fractions or decimals for both certain events (that is, experimental probability) and uncertain events (for theoretical probability). The probabilities presented in Table 1 are theoretical because they involve a logical analysis of the likelihood of certain events given all possible outcomes. The probabilities are not based on actual trial runs. This occurs as the next part of the exploration.

We also discussed the total number of possible outcomes. The table helps the students see that if a 1 is rolled on one die, the other die could show a 1, 2, 3, 4, 5, or 6. From this comes the intuitive sense that the total number of outcomes equals 36. Although investigations with probability at the middle school level should not focus on the use of formulas, an opportunity to help students connect formulas with meaningful events should not be lost. At this point, I usually point out that we have six possible outcomes on each die, and that 6 × 6 yields 36 possible outcomes. The arrays presented in tabular form are effective in helping students see why multiplying the two numbers to find the total number of possible outcomes makes sense and why the formula is meaningful.

Using Technology to Organize Probability Data

Next, I have students quickly graph the information from the chart using a spreadsheet (see Figure 1). Because they should be developing and understanding the ideas and mathematics related to probability instead of belaboring the best way to graph the data, they should have access to technology that speeds up the task. The graphs are primarily key reference points for students to refer to later in conjecturing and drawing conclusions about experimental and theoretical probability.

In groups, students should now roll the two dice and record their outcomes. Placing the information as ratios into a spreadsheet and setting the formatting so that the data will be transposed to decimal form will help students compare the data from the various parts of the exercise. The information presented in Table 2 shows the exercise results from one class. After completing their rolls, the groups each should construct a graph of outcomes. Once again, using a spreadsheet facilitates such graphing production. Figure 2 displays a graph of the outcomes from one group in the class. The teacher should again ask students questions about their data, such as how they would identify the experimental probability of rolling sums of 2, 7, and so on. Ask them to make comparisons between the numbers they derived in their experiment with the numbers from the samples presented in Table 1. How do they compare? The value of using decimal notation for the probabilities becomes evident when students begin to compare the various pieces of information. This could be extended effectively into a writing assignment in which students compare and contrast the two sets of data. Writing can be an effective tool for students to discover meaning from mathematics activities (Pugalee, 1997).

The next leg of the exploration collates the data from the different groups. Students should already have tabulated their outcomes. On the board or some other central place, gather all of the group information that identifies the number of times each sum was rolled. Once the tabulations are done, students should enter the information into a spreadsheet and construct a graph that depicts the class data (see Figure 3). Ask them how the graph of their group's data differs from the data graph for the entire class. Then compare and contrast these graphs with the class graph and tables from the theoretical probability models.

Given access to a color or a black-and-white printer and transparencies, these graphs can be displayed to the entire class to help the discussions about the representations. Encourage the students to conjecture about the differences in the graphs. By having the information in both graphical and tabular form, the students gain insights into the various ways to represent data. With directive questioning, they will begin to see that the data graph from the entire class approximates the theoretical data more closely than do the graphs of individual groups (see Table 2). Ask them to consider what the data from 100 or even 1,000 groups might look like; they will soon begin to understand the relationship and predict that the graphs would become more similar with increased experiments or trials.

Table 1. Possible Sums from Rolling Two Dice

+	1	2	3	4	5	6
1	2	3	4	5	6	7
2	3	4	5	6	7	8
3	4	5	6	7	8	9
4	5	6	7	8	9	10
5	6	7	8	9	10	11
6	7	8	9	10	11	12

Table 2. Summary of Probabilities Expressed as Decimals

Sum	*Theoretical*	*Group Data*	*Class Data*
2	0.0278	0.0278	0.0324
3	0.0556	0.0278	0.0231
4	0.0833	0.0833	0.0833
5	0.1111	0.0833	0.1019
6	0.1389	0.1111	0.1667
7	0.1667	0.0833	0.1528
8	0.1389	0.1944	0.1250
9	0.1111	0.1111	0.1250
10	0.0833	0.1389	0.1111
11	0.0556	0.0833	0.0417
12	0.0278	0.0556	0.0370

Conclusions

This activity is especially important because it includes many of the goals for probability instruction with middle school and high school students. It is effective because it involves the students in an experience that they find interesting and exciting. Using technology provides students with several types of data representation, so they can be asked to make important associations between various representations of data. More important, the activity helps them distinguish between theoretical and experimental probabilities. As a result of such exploration, students develop important conceptual and intuitive ideas about probability and build a foundation for additional inquiry into probability and data.

Using technology also provides students rich opportunities to develop computer skills as they work together to define problems, gather information, analyze and represent data, and interpret and generalize from various forms of data representation. In such learning environments and through such explorations, students may well understand the importance of technology as a tool for inquiry—a necessary goal for education in this technological age.

References

Hartman, A. (1997). On the knowledge of high school mathematics teachers for teaching probability. *Proceedings of the 21st Conference of the International Group for the Psychology of Mathematics Education*, Vol. 1(p. 238). Lahti, Finland: Psychology of Mathematics Education.

National Council of Teachers of Mathematics. (1989). *Curriculum and evaluation standards for school mathematics.* Reston, VA: Author.

Pugalee, D. K. (1997). Connecting writing to the mathematics curriculum. *Mathematics Teacher, 90,* 308–310.

The Data They Are A-Changin'

Changes are integral processes of science and can be measured to produce information for a wide variety of investigations. In this article, the author shows how students can use the World Wide Web to gather real-time earth and space science data and analyze it to understand and describe changes of interest to earth and space scientists.

The Internet provides a vast network of re sources for schools, and for most users it seems like a limitless and living encyclopedia. But it wasn't designed that way; it was intended to help scientists collaborate remotely. Fortunately for us, the World Wide Web gives schools access to the same high-quality data and computer-analysis tools used by research scientists. Much of this data is brilliantly colored, immediately relevant to students, and automatically updated every few minutes. It is this real-time, up-to-the-minute data that seems most effective at captivating students.

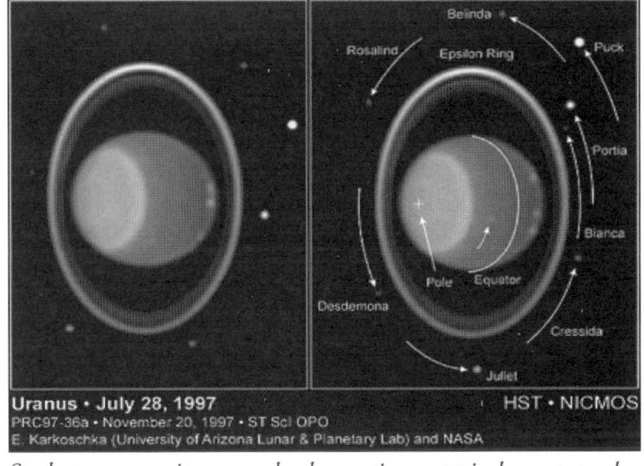

Students can monitor ozone levels over time, a typical way to track change (ftp://jwocky.gsfc.nasa.gov/pub/eptoms/images/npole/y98/POL_LATEST.GIF). Graphic courtesy of the Ozone Processing Team at NASA/Goddard Space Flight Center.

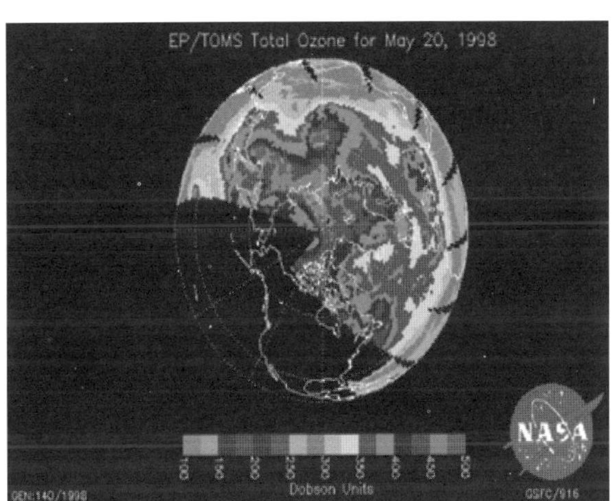

Photos from the Hubble Telescope, such as this one of Uranus, are available on the Web (http://oposite.stsci.edu/pubinfo/PR/97/36/content/9736aw.gif). Photo courtesy of E. Karkoschka (University of Arizona Lunar and Planetary Lab) and NASA. © 1997, Association of Universities for Research in Asttronomy, Inc. All rights reserved.

By Tim Slater

Subject: Earth and space science, math

Grade levels: 5–12 (Ages 10–18)

Technology: Web, Netscape Navigator, Microsoft Internet Explorer, PaintShop Pro, GraphicConverter, NIHImage, Scion Image, Microsoft Excel, GifConstruction

This article describes good data resources now available, briefly describes how to use Internet data and analysis software, and suggests ways to integrate real-time scientific investigations into your classroom instruction.

Curriculum Resources

The *National Science Education Standards* (National Research Council, 1996) clearly emphasize that science instruction should use unifying themes and concepts. Change is one such theme. Most things in the universe are in the process of changing. Changes are common in the properties of materials, positions of objects, and functions of systems. Changes in nature can vary widely in rate, scale, and pattern.

Scientists consider most patterns of change to be either trends or cycles. Trends are unidirectional changes that can be *linear* (such as continental drift), *exponential* (such as population growth), or *sequential* (such as the metamorphosis of caterpillar to butterfly). Cyclical changes are repetitions such as the phases of the moon and the recurrence of Earth's seasons.

Mathematics as the Primary Tool

Mathematics is the language used to describe changes in nature, and its use is essential to the measurement process. Using change as a theme naturally integrates the skills and concepts of science and mathematics. Moreover, analyzing such data allows students to experience scientific discovery firsthand.

The Web provides hundreds of resources that are immediately useful in teaching students about patterns of change. These resources come from a wide range of enterprises: government agencies, commercial industries, and colleges and universities. See Table 1 for examples of locations and lesson ideas.

TABLE 1. REAL-TIME DATA RESOURCES.

Web Site	*URL*	*Suggested Uses*
GEOLOGY		
Earthquakes Recorded by CNSS Networks	www.geophys.washington.edu/us.epi.gif	Use outline patterns of earthquake locations to map major fault lines.
Snow Water Equivalent	ftp://ftp.nohrsc.nws.gov/pub/products/west/wuw _new.gif	Predict when and where flooding is likely to occur.
SPACE SCIENCE		
Today's Space Weather or Space Weather Today	www.sel.noaa.gov/today.html or www.windows.umich.edu/spaceweather/	Predict the best time to observe the aurora borealis.
Current Solar Images	http://umbra.nascom.nasa.gov/images/latest.html	Measure the rotation rate of our nearest star.
Space Weather Current Conditions	http://space.rice.edu/ISTP/rt.html	Create a multimedia presentation to accompany a report on auroras.
J-Track 2.0 Satellite Locater	http://liftoff.msfc.nasa.gov/RealTime/JTrack/Spacecraft.html or http://liftoff.msfc.nasa.gov/RealTime/JTrack/NOAA.html	Measure the orbital period and calculate orbiting altitudes for major satellites.
Earth Probe TOMS Data & Images	http://jwocky.gsfc.nasa.gov/eptoms/ep.html	Determine the extent of seasonal variation in the ozone layer.
Mars Today	www-mgcm.arc.nasa.gov/	Track weather systems and monitor temperature changes on Mars.
Solar System Live	www.fourmilab.ch/solar/solar.html	Determine the best time to view or travel to other planets.
Virtual Reality Moon Phase Pictures	http://tycho.usno.navy.mil/vphase.html	Predict the moon's appearance on successive nights even when cloudy.
Yohkoh Public Outreach Project	www.lmsal.com/YPOP/Classroom	See animations of the solar cycle.
WEATHER		
The Weather Channel	www.weather.com/weather/maps/	Create weather forecasts and describe why certain predictions are made.
Weather Net / Weather Cams	http://cirrus.sprl.umich.edu/wxnet/wxcam.html	Using weather reports, guess what the sky and ground look like in various U.S. cities and then check the accuracy of the weather predictions.
Space Science and Engineering Center	www.ssec.wisc.edu/data/	Determine patterns of changing ocean temperatures to predict the best fishing locations.
WORLD POPULATION		
U.S. Census Bureau: The Official Statistics	www.census.gov/	Use mathematics to model national and world population growth.

Note: The Physics and Astronomy Education Group at Montana State University maintains an updated real-time data resource Web site at www.math.montana.edu/~tslater/real-time/

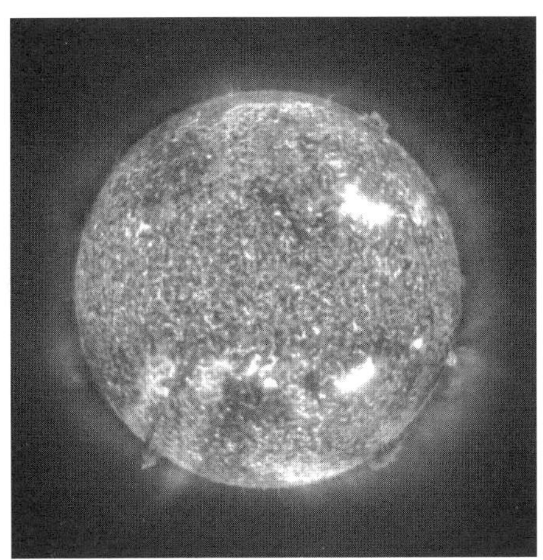

Current space weather images can be used to track sunspots. Photo courtesy of SOHO/EIT consortium (www.windows.umich.edu/spaceweather/latest_eit_304_gif_image.html). SOHO is an international collaboration between ESA and NASA.

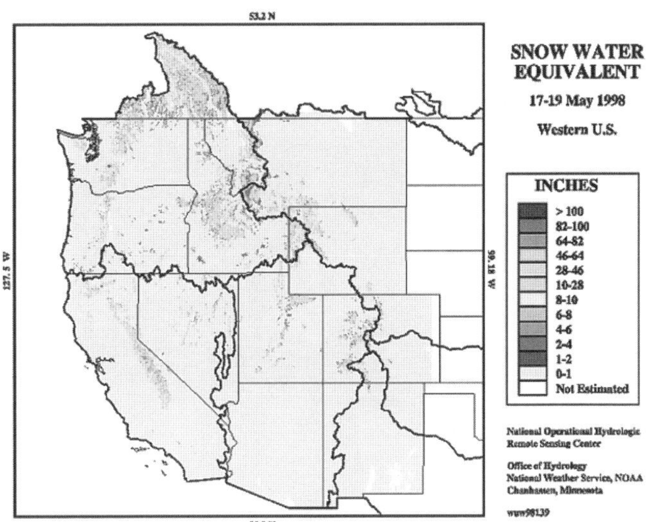

Students can use Web-based information to make predictions. This screen (courtesy of the National Operational Hydrologic Remote Sensing Center, Office of Hydrology, National Weather Service) shows water saturation in the western United States (ftp://ftp.nohrsc.nws.gov/pub/products/west/wuw_new.gif). Students can use this information to predict the likelihood of flooding.

Instructional Notes

All of the principles that identify high-quality, hands-on, and minds-on science education apply to Web-based science investigations. The best projects are relatively unstructured and will most certainly extend beyond the teacher's immediate scientific background. The opportunity to learn along with students should not be missed. Sometimes, working with real data is a leap of faith, but most teachers who try open-ended science investigations find them incredibly exciting.

As a first step, you'll find it helpful to provide students with a time line, milestones, and a reasonably well-defined task.

Highly structured projects are best for schools with extremely limited Internet access. They work well for two reasons. First, they describe the number of images to be analyzed and how and when the analysis is to be done. Second, they provide a strict template for completing a final report. When warranted, a teacher can access data and save it to disk or print images for students to analyze at their desks. This is the only option for teachers who use their home Web access or a school's single connection before the school day begins. Whenever possible, it's preferable to make students responsible for gathering their own data on a regular basis, storing it on a floppy disk, using Web resources to gather information, and immediately analyzing their data on a computer.

Real-Time Data Analysis. Such data analysis is appropriate for individual students who are working on complex science fair and multiple-classroom projects. Depending on the students' ages, they generally need to be introduced to data sources and their interesting attributes. As with science-fair projects, providing springboard project ideas for students is more useful than asking them right off the bat to come up with their own.

One advantage to assigning projects is that a teacher can select interrelated projects for groups of students. For example, one group might monitor changing sunspot activity. The teacher would need to ensure that students could identify a sunspot and access solar images and factual information (current and archived data are available at several Web locations). A second group might monitor the intensity of the northern lights (the aurora borealis). It doesn't take much data to demonstrate that solar activity often precedes the appearance of the northern lights by approximately three days. Students may do most of the legwork to find resources themselves. It's not surprising that they're often better at this than adults.

The Nuts and Bolts of Data Manipulation

Once a series of images has been obtained over several days or weeks, four powerful procedures can be done with even the smallest computers: (1) trend analysis, (2) animation, (3) enhancement, and (4) measurement. These processes often lead students to investigate more questions or conventional graphing or modeling approaches.

Trend Analysis. Trend analysis is the most basic form of analysis. Students look for changes in two or more images. Most atmospheric features over the continental United States, for example, move from west to east but at highly varying rates and evolutions. Thus, when looking at real-time U.S. weather maps, students can calculate the rates of change by measuring the number of miles a weather front moves and dividing the rates by the number of hours between images. Geography, map reading, and arithmetic skills are all enhanced by such activities.

Animation. Animation requires a computer. At the technique's most basic level, students can stitch together a series of images into an animated GIF using a program such as GifConstruction. Image-processing programs provide more user control. With animations, students can make movies just as good as—and sometimes better than—those on local TV weather reports. The Network Montana Project provides step-

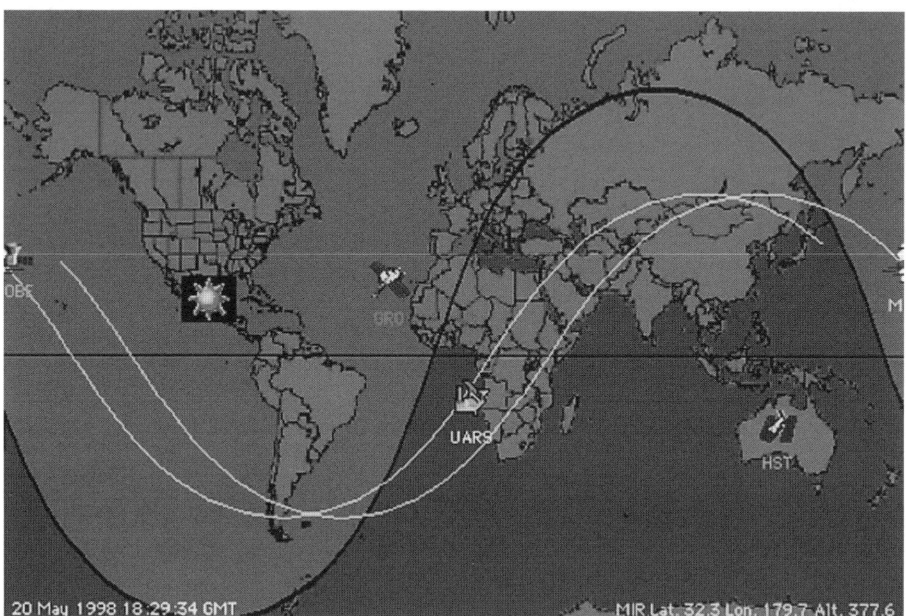

Some Web sites contain their own animations. This screen shows the J-Track satellite tracking system for monitoring the locations and movement of satellites over the United States (htpp://liftoff.msfc.nasa.gov/RealTime/JTrack/Spacecraft.html). Courtesy of Mission Operations Laboratory, Marshall Space Flight Centter, NASA.

by-step instructions on how to use image-processing
programs. Some Web sites, such as the Yohkoh Public Outreach Project and The Weather Channel, provide daily data already in movie format.

Enhancement. Computers also allow students to enhance their data images. By changing the colors they use to display their images, students can make different aspects and features more prominent. Most graphics and image-processing programs allow users to change image colors fairly quickly.

Measurement. Measurement is at the very heart of integrated mathematics and science. Sometimes scales are provided on maps, but students often must indirectly determine an image's scale. To do this, the student should measure a known distance in centimeters (cm). This distance might be anything: the diameter of a planet or the distance across a continent. The image scale is then calculated as the known distance divided by the ruler distance. To measure any feature in the image, multiply the measured distance in centimeters and the image scale to get the actual size. For example, if a picture of the sun is 15 cm across, then the image scale is the diameter of the sun divided by 15 cm (1,400,000 km ÷ 15 cm = 93,333 km per cm). Accordingly, if the size of a sunspot is 1.2 cm, then the actual size of the sunspot is 1.2 cm × 93,333 km per cm, or 112,000 km.

Similarly, on the computer, any digital image must be calibrated before its features can be measured. For NIHImage or Scion Image, the image must be converted to either the PICT or TIFF format using a graphics program. After the picture is opened, the user draws a line across a known distance (e.g., the map scale, the diameter of a planet, or the distance between two cities). Then the known units followed by the known value are entered so any feature can be measured simply by drawing a line across it. In image-processing software, the scaling holds true no matter how much the user magnifies or enhances the image.

Assessment

Conducting scientific investigations with real-time data allows many entry points for students with different backgrounds and aptitudes. Students can tackle projects at varying levels of complexity and duration. Accordingly, assessment strategies probably will not be conventional multiple-choice tests that focus on facts. The easiest way to assess student learning is to have students create reports that describe what they did and what they learned. These can be done as written research reports, colorful posters, electronic presentations, or oral reports. Using a checklist, the teacher can evaluate the degree to which students have used appropriate research strategies, the progress students made from their starting points, and the extent to which conclusions are based on the data presented. This is a great opportunity to use a robust combination of performance- and portfolio-assessment techniques.

Discussion

When we think about using Web resources to teach science and patterns of change, we should consider that not all of our classroom students will learn exactly the same list of facts. Scientific data from the Web allows students to explore real scientific data and test real hypotheses in meaningful ways. It encourages students to "do science" rather than simply memorize scientific terms. Knowing scientific vocabulary may be important, but students naturally learn vocabulary in the process of conducting a guided scientific investigation. The goal of increasing the cognitive level of learning science is met by students who actively build their knowledge through motivated investigation.

References

National Research Council. (1996). *National science education standards.* Washington, DC: National Academy Press.

APPENDIX A

National Educational Technology Standards for Students (NETS•S)

The National Educational Technology Standards for students are divided into six broad categories. Standards within each category are to be introduced, reinforced, and mastered by students. These categories provide a framework for linking performance indicators, listed by grade level, to the standards. Teachers can use these standards and profiles as guidelines for planning technology-based activities in which students achieve success in learning, communication, and life skills.

1. *Basic operations and concepts*

- Students demonstrate a sound understanding of the nature and operation of technology systems.
- Students are proficient in the use of technology.

2. *Social, ethical, and human issues*

- Students understand the ethical, cultural, and societal issues related to technology.
- Students practice responsible use of technology systems, information, and software.
- Students develop positive attitudes toward technology uses that support lifelong learning, collaboration, personal pursuits, and productivity.

3. *Technology productivity tools*

- Students use technology tools to enhance learning, increase productivity, and promote creativity.
- Students use productivity tools to collaborate in constructing technology-enhanced models, preparing publications, and producing other creative works.

4. *Technology communications tools*

- Students use telecommunications to collaborate, publish, and interact with peers, experts, and other audiences.

- Students use a variety of media and formats to communicate information and ideas effectively to multiple audiences.

5. *Technology research tools*

- Students use technology to locate, evaluate, and collect information from a variety of sources.
- Students use technology tools to process data and report results.
- Students evaluate and select new information resources and technological innovations based on the appropriateness to specific tasks.

6. *Technology problem-solving and decision-making tools*

- Students use technology resources for solving problems and making informed decisions.
- Students employ technology in the development of strategies for solving problems in the real world.

APPENDIX B

National Educational Technology Standards for Teachers (NETS•T)

All classroom teachers should be prepared to meet the following standards and performance indicators.

I. Technology Operations and Concepts

Teachers demonstrate a sound understanding of technology operations and concepts. Teachers:

A. demonstrate introductory knowledge, skills, and understanding of concepts related to technology (as described in the ISTE National Educational Technology Standards for Students).

B. demonstrate continual growth in technology knowledge and skills to stay abreast of current and emerging technologies.

II. Planning and Designing Learning Environments and Experiences

Teachers plan and design effective learning environments and experiences supported by technology. Teachers:

A. design developmentally appropriate learning opportunities that apply technology-enhanced instructional strategies to support the diverse needs of learners.

B. apply current research on teaching and learning with technology when planning learning environments and experiences.

C. identify and locate technology resources and evaluate them for accuracy and suitability.

D. plan for the management of technology resources within the context of learning activities.

E. plan strategies to manage student learning in a technology-enhanced environment.

III. Teaching, Learning, and the Curriculum

Teachers implement curriculum plans that include methods and strategies for applying technology to maximize student learning. Teachers:

A. facilitate technology-enhanced experiences that address content standards and student technology standards.

B. use technology to support learner-centered strategies that address the diverse needs of students.

C. apply technology to develop students' higher-order skills and creativity.

D. manage student learning activities in a technology-enhanced environment.

IV. Assessment and Evaluation

Teachers apply technology to facilitate a variety of effective assessment and evaluation strategies. Teachers:

A. apply technology in assessing student learning of subject matter using a variety of assessment techniques.

B. use technology resources to collect and analyze data, interpret results, and communicate findings to improve instructional practice and maximize student learning.

C. apply multiple methods of evaluation to determine students' appropriate use of technology resources for learning, communication, and productivity.

V. Productivity and Professional Practice

Teachers use technology to enhance their productivity and professional practice. Teachers:

A. use technology resources to engage in ongoing professional development and lifelong learning.

B. continually evaluate and reflect on professional practice to make informed decisions regarding the use of technology in support of student learning.

C. apply technology to increase productivity.

D. use technology to communicate and collaborate with peers, parents, and the larger community in order to nurture student learning.

VI. Social, Ethical, Legal, and Human Issues

Teachers understand the social, ethical, legal, and human issues surrounding the use of technology in PK–12 schools and apply that understanding in practice. Teachers:

A. model and teach legal and ethical practice related to technology use.

B. apply technology resources to enable and empower learners with diverse backgrounds, characteristics, and abilities.

C. identify and use technology resources that affirm diversity.

D. promote safe and healthy use of technology resources.

E. facilitate equitable access to technology resources for all students.